STUDY GUIDE

FUNDAMENTALS OF BIOSTATISTICS

FOURTH EDITION

BERNARD ROSNER
HARVARD UNIVERSITY

Duxbury Press
An Imprint of Wadsworth Publishing Company
 An International Thomson Publishing Company

Belmont • Albany • Bonn • Boston • Cincinnati • Detroit • London • Madrid • Melbourne
Mexico City • New York • Paris • San Francisco • Singapore • Tokyo • Toronto • Washington

Wadsworth Publishing Company
10 Davis Drive
Belmont, California 94002
USA

International Thomson Editores
Campos Eliseos 385, Piso 7
Col. Polanco
11560 México D.F. México

International Thomson Publishing Europe
Berkshire House 168–173
High Holborn
London, WC1V 7AA
England

International Thomson Publishing GmbH
Königswinterer Strasse 418
53227 Bonn
Germany

Thomas Nelson Australia
102 Dodds Street
South Melbourne 3205
Victoria, Australia

International Thomson Publishing Asia
221 Henderson Road
#05–10 Henderson Building
Singapore 0315

Nelson Canada
1120 Birchmount Road
Scarborough, Ontario
Canada M1K 5G4

International Thomson Publishing Japan
Hirakawacho Kyowa Building, 3F
2-2-1 Hirakawacho
Chiyoda-ku, Tokyo 102
Japan

PREFACE

This Study Guide is designed to be used with the text *Fundamentals of Biostatistics, Fourth Edition*, by Bernard Rosner. Its purpose is to provide a review of concepts about which you may have questions or could use a second explanation, as well as a substantial number of additional problems to practice what you have learned. This Study Guide contains over 600 problems that do not appear in the main text, along with their complete solutions.

The first twelve chapters correspond to Chapters 2 through 13 of *Fundamentals of Biostatistics*. These chapters contain a review of key concepts, using different examples from those given in the text, additional problems, and solutions to most problems.

The last chapter, "Miscellaneous Problems," is designed to give you more practice deciding which statistical methods to use in particular situations. It contains problems based on material from Chapters 6 through 13 (the statistical methods chapters) of the main text, arranged in random order so you will not automatically know from which chapters the problems are derived. I created this chapter based on requests from my students and hope it will help you achieve an important goal of this book—learning to identify appropriate statistical techniques.

The method of handling computations is similar to that used in the text. All intermediate results are carried to full precision (10+ significant digits) even though they are presented with fewer significant digits (usually 2–3) in the Study Guide. Thus, intermediate results may seem to be inconsistent with final results in some instances, although this is not the case. This method allows for greater accuracy of final results.

CONTENTS

CHAPTER 4   **Discrete Probability Distributions 29**

CHAPTER 5   **Continuous Probability Distributions 44**

CHAPTER 6   **Estimation 61**

CHAPTER 10   Hypothesis Testing: Categorical Data   135

CHAPTER 11   Regression and Correlation Analysis   173

NOTE:

There is no Chapter 1 in this Study Guide because the chapters correspond to those in *Fundamentals of Biostatistics,* Fourth Edition, in which Chapter 1 was a general overview and contained no problems or exercises.

CHAPTER TWO

DESCRIPTIVE STATISTICS

SECTION 2.1   Measures of Location

2.1.1   Arithmetic Mean

$$\bar{x} = \sum_{i=1}^{n} x_i/n = (x_1 + x_2 + \ldots + x_n)/n$$

Consider the data in Table 2.1. They represent serum-cholesterol levels from a group of hospital workers who were regularly eating a standard U.S. diet and who agreed to change their diet to a vegetarian diet for a 6-week period. Cholesterol levels were measured before and after adopting the diet. The mean serum-cholesterol level before adopting the diet is computed as follows:

$$\sum_{i=1}^{24} x_i = 4507, \quad \bar{x} = 4507/24 = 187.8 \text{ mg/dL}$$

Advantages

(1) It is representative of all the points.

(2) If the underlying distribution is Gaussian, then it is the most efficient estimator of the middle of the distribution.

(3) Many statistical tests are based on the arithmetic mean.

Disadvantages

(1) It is very sensitive to outliers; e.g., if one of the cholesterol levels were 800 rather than 200, then the mean would be increased by 25 mg/dL.

(2) It is inappropriate if the underlying distribution is far from being Gaussian; for example, serum triglycerides have a distribution that looks highly skewed.

TABLE 2.1 Serum-cholesterol levels before and after adopting a vegetarian diet (mg/dL)

Subject	Before	After	Before–After
1	195	146	49
2	145	155	−10
3	205	178	27
4	159	146	13
5	244	208	36
6	166	147	19
7	250	202	48
8	236	215	21
9	192	184	8
10	224	208	16
11	238	206	32
12	197	169	28
13	169	182	−13
14	158	127	31
15	151	149	2
16	197	178	19
17	180	161	19
18	222	187	35
19	168	176	−8
20	168	145	23
21	167	154	13
22	161	153	8
23	178	137	41
24	137	125	12
Mean	187.8	168.3	19.5
sd	33.2	26.8	16.8
n	24	24	24

2.1.2   **Alternatives to the Arithmetic Mean—Median**

One interesting property from the table is that the diet appears to work best in people with high levels versus people with low levels. How can we test if this is true? Divide the group in half, and look at cholesterol change in each half. To do this we must compute the median $\equiv$ 50% point in the distribution. Specifically,

$$\text{Median} = \left(\frac{n+1}{2}\right)\text{th largest point if } n \text{ is odd}$$

$$= \text{average of } \left[\frac{n}{2}\text{th} + \left(\frac{n}{2}+1\right)\text{th}\right] \text{ largest points if } n \text{ is even}$$

For example,

if $n = 7$, then the median = 4th largest point

if $n = 24$, then the median = average of (12th + 13th) largest point

2.1.3   **Stem-and-Leaf Plots**

How can we easily compute the median? We would have to order the data to obtain the 12th and 13th largest points. An easier way is to compute a *stem-and-leaf plot.* Divide each data value into a leaf (the least-significant digit or digits) and a stem (the most-significant digit or digits) and collect all data points with the same stem on a single row:

Cumulative total	Stems	Leaves
24	25	0
23	24	4
22	23	68
20	22	42
	21	
18	20	5
17.	19	5277
13	18	0
12	17	8
11	16	698871
5	15	981
2	14	5
1	13	7

$$\text{Median} = 179 = \frac{178 + 180}{2}$$

The stem-and-leaf plots of the change in cholesterol for the subgroups of people below and above the median are given as follows:

≤ 179		≥ 180	
4	1	4	98
3	1	3	625
2	3	2	718
1	3932	1	699
0	28	0	8
–0	8	–0	
–1	03	–1	
–2		–2	
–3		–3	

The change scores in the subgroups look quite different; the subgroup above the median is showing more change.

2.1.4 **Percentiles**

We can also use stem-and-leaf plots to obtain percentiles of the distribution.

p^{th} Percentile. If $np/100$ is an integer, then average the

$$\frac{np}{100}\text{th} + \left(\frac{np}{100} + 1\right)\text{th largest points}$$

Otherwise, $= \{[np] + 1\}$th largest point, where $[np] =$ largest integer $\leq np$. For example, to compute the 10th percentile of the baseline cholesterol distribution, also known as the lower decile, we have $n = 24$, $p = 10$, $np = 2.4$, $[np] = 2$, lower decile = 3rd largest point = 151 mg/dL. To compute the 90th percentile (or upper decile), $n = 24$, $p = 90$, $np = 21.6$, $[np] = 21$. Upper decile = 22nd largest point = 238 mg/dL.

Commonly used percentiles

10, 20, ... , 90% (deciles)
20, 40, ... , 80% (quintiles)
25, 50, 75% (quartiles)
33.3, 66.7% (tertiles)

Median

Advantages

(1) Always guarantees that 50% of the data values are on either side of the median.

(2) Insensitive to outliers. If one of the cholesterol values increased from 200 to 800, the median would remain at 179 but the mean would increase from 187 to 187 + 25 = 212.

Disadvantages

(1) It is not as efficient an estimator of the middle as the mean if the distribution really is Gaussian in that it is mostly sensitive to the middle of the distribution.

(2) Most statistical procedures are based on the mean.

For this data set, the distribution is only slightly skewed, and the mean may be adequate.

2.1.5 **Geometric Mean**

One way to get around the disadvantage (2) of the arithmetic mean is to transform the data onto a different scale to make the distribution more symmetric and compute the arithmetic mean on the new scale. The most popular such scale is the ln scale:

$$\ln(x_1), \ldots , \ln(x_n)$$

$$\overline{\ln x} = \frac{\ln(x_1) + \ldots + \ln(x_n)}{n}$$

The problem with this is that the average is in the ln scale rather than the original scale. Thus, we take the antilog of $\overline{\ln x}$ to obtain

$$GM = e^{\overline{\ln x}} = \text{geometric mean}$$

The ERG (electroretinogram) amplitude (μv) is a measure of electrical activity of the retina and is used to monitor retinal function in patients with retinitis pigmentosa, an often-blinding ocular condition. The following data were obtained from 10 patients to monitor the course of the condition over a 1-year period.

	Year 1 ERG amplitude (μv)	Year 2 ERG amplitude (μv)	Absolute change (μv)
1	1.9	1.4	0.5
2	3.9	3.9	0.0
3	64.4	46.2	18.2
4	25.9	19.1	6.8
5	4.0	2.5	1.5
6	0.9	1.6	−0.7
7	2.0	1.8	0.2
8	4.0	3.7	0.3
9	33.8	12.1	21.7
10	6.3	3.5	2.8

The distribution of values at each year is very skewed, with change scores dominated by people with high year-1 ERG amplitudes. The distribution in the log scale is much more symmetric. Let's compute the GM for year 1 and year 2.

Year 1

$$\overline{\ln x} = \frac{\ln(1.9) + \ldots + \ln(6.3)}{10} = \frac{0.64 + \ldots + 1.84}{10} = 1.8144$$

$$GM_1 = e^{1.8144} = 6.137 \mu v$$

Year 2

$$\overline{\ln x} = \frac{\ln(1.4) + \ldots + \ln(3.5)}{10} = \frac{0.34 + \ldots + 1.25}{10} = 1.5508$$

$$GM_2 = e^{1.5508} = 4.715 \mu v$$

We can quantify the % change by

$$\frac{GM_2}{GM_1} = \frac{4.715}{6.137} = 0.768 \approx 23.2\% \text{ decline}$$

Thus, the ERG has declined, on average, by 23.2% over 1 year.

Geometric Mean

Advantages

(1) Useful for certain types of skewed distributions

(2) Standard statistical procedures can be used on the transformed scale

Disadvantages

(1) Not appropriate for symmetric data

(2) More sensitive to outliers than the median but less so than the mean

SECTION 2.2   Measures of Spread

2.2.1   Range

The range = the interval from the smallest value to the largest value. This gives a quick feeling for the overall spread—but is misleading because it is solely influenced by the most extreme values; e.g., cholesterol data–initial readings; range = (137,250).

2.2.2   Quasi-Range

A quasi-range is similar to the range but is derived after excluding a specified percentage of the sample at each end; e.g., 10%, 90% in cholesterol data and specify the quasi-range as the interval from the 10% point to the 90% point. For example, cholesterol data

10% point = 3rd largest from bottom = 151

90% point = 3rd largest from top = 238

quasi-range = (151,238)

2.2.3   Standard Deviation, Variance

If the distribution is normal or near normal, then the standard deviation is more frequently used as a measure of spread.

$$s^2 = \text{sample variance} = \sum_{i=1}^{n} (x_i - \bar{x})^2 / (n - 1)$$

$$s = \text{sample standard deviation} = \sqrt{s^2}$$

Why s Rather than s^2?   We want an estimator of spread in the same units as $\bar{x}$; i.e., if units change by a factor of c, and the transformed data is referred to as y, then

$$\bar{y} = c\bar{x} \qquad s_y = cs_x \qquad \text{but } s_y^2 = c^2 s_x^2$$

Note that s can be related to $\bar{x}$ but s^2 cannot.

How can we use $\bar{x}$ and s to get a feeling for the spread of the distribution? If the distribution is normal, then

$\bar{x} \pm s$ comprises about 2/3 of the distribution

$\bar{x} \pm 2s$ (more precisely, 1.96s) comprises about 95% of the distribution

$\bar{x} \pm 2.5s$ (more precisely, 2.576s) comprises about 99% of the distribution

If the distribution is not normal or near normal, then the distribution is not well characterized by $\bar{x},s$. Use the percentiles in this case.

How Do We Compute s? An alternative computational form for s^2 is

$$\frac{\sum_{i=1}^{n} x_i^2 - \left(\sum_{i=1}^{n} x_i\right)^2/n}{n-1}$$

For example, for the cholesterol data, the variance and standard deviation of the before measurements are computed as follows:

$$\sum_{i=1}^{24} x_i = 4507, \quad \sum_{i=1}^{24} x_i^2 = 871{,}667$$

$$s^2 = \frac{871{,}667 - 4507^2/24}{23} = \frac{871{,}667 - 846{,}377.0}{23} = \frac{25{,}289.96}{23} = 1099.56$$

$$s = 33.2$$

Let's see how normal the distribution looks.

$$\bar{x} \pm 1.96s = 187.8 \pm 1.96(33.2) = (122.8, 252.8)$$

includes all points; it should include 95% (or 23 out of 24 points) under a normal distribution.

$$\bar{x} \pm 1s = 187.8 \pm 33.2 = (154.6, 221.0)$$

includes $15/24 = 62.5\%$ of points; it should be 2/3 under a normal distribution. The normal distribution appears to provide a reasonable approximation.

2.2.4 Coefficient of Variation (CV)

$$CV = 100\% \times \frac{s}{\bar{x}}$$

For the cholesterol data, $CV = 100\% \times \dfrac{33.2}{187.8} = 17.7\%$

The CV is used if the variability is thought to be related to the mean.

SECTION 2.3 Some Other Methods for Describing Data

2.3.1 Frequency Distribution

This is a listing of each value and how frequently it occurs (or in addition, the % of scores associated with each value). This can be done either based on the original values, or in grouped form; e.g., if we group the cholesterol change scores by 10-mg increments, then we would have

	Frequency	%
≥ 40.0,	3	13
≥ 30.0, < 40.0	4	17
≥ 20.0, < 30.0	4	17
≥ 10.0, < 20.0	7	29
≥ 0.0, < 10.0	3	13
≥ –10.0, < 0.0	2	8
≥ –20.0, < 10.0	1	4
	24	

This can be done either in numeric or graphic form. If in graphic form, it is often represented as a *bar graph*.

2.3.2   **Box Plot**

The rectangle displays the upper and lower quartiles, the median, arithmetic mean, and outlying values (if any). It is a concise way to look at the symmetry and range of a distribution.

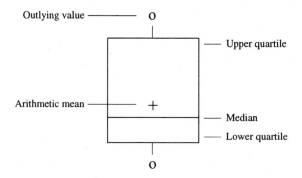

PROBLEMS

Suppose the origin for a data set is changed by adding a constant to each observation.

2.1 What is the effect on the median?

2.2 What is the effect on the mode?

2.3 What is the effect on the geometric mean?

2.4 What is the effect on the range?

Renal Disease
For a study of kidney disease, the following measurements were made on a sample of women working in several factories in Switzerland. They represent concentrations of bacteria in a standard-size urine specimen.

High concentrations of these bacteria may indicate possible kidney failure. The data are presented in Table 2.2.

2.5 Compute the arithmetic mean for this sample.

2.6 Compute the geometric mean for this sample.

2.7 Which do you think is a more appropriate measure of location?

Cardiovascular Disease
The mortality rates from heart disease (per 100,000 population) for each of the 50 states and the District of Columbia in 1973 are given in descending order in Table 2.3 [1].

TABLE 2.2 Concentration of bacteria in the urine in a sample of female factory workers in Switzerland

Concentration	Frequency
10^0	521
10^1	230
10^2	115
10^3	74
10^4	69
10^5	62
10^6	43
10^7	30
10^8	21
10^9	10
10^{10}	2

TABLE 2.3 Mortality rates from heart disease (per 100,000 population) for the 50 states and the District of Columbia in 1973

1	West Virginia	445.4	18	Wisconsin	369.8	35	DC	327.1
2	Pennsylvania	442.7	19	Vermont	369.2	36	South Carolina	322.4
3	Maine	427.3	20	Nebraska	368.9	37	Montana	319.1
4	Missouri	422.9	21	Tennessee	361.4	38	Maryland	315.9
5	Illinois	420.8	22	New Hampshire	358.2	39	Georgia	311.8
6	Florida	417.4	23	Indiana	356.4	40	Virginia	311.2
7	Rhode Island	414.4	24	North Dakota	353.3	41	California	310.6
8	Kentucky	407.6	25	Delaware	351.6	42	Wyoming	306.8
9	New York	406.7	26	Mississippi	351.6	43	Texas	300.6
10	Iowa	396.9	27	Louisiana	349.4	44	Idaho	297.4
11	Arkansas	396.8	28	Connecticut	340.3	45	Colorado	274.6
12	New Jersey	395.2	29	Oregon	338.7	46	Arizona	265.4
13	Massachusetts	394.0	30	Washington	334.2	47	Nevada	236.9
14	Kansas	391.7	31	Minnesota	332.7	48	Utah	214.2
15	Oklahoma	391.0	32	Michigan	330.2	49	New Mexico	194.0
16	Ohio	377.7	33	Alabama	329.1	50	Hawaii	169.0
17	South Dakota	376.2	34	North Carolina	328.4	51	Alaska	83.9

Consider this data set as a sample of size 51 ($x_1, x_2, \ldots, x_{51}$). If

$$\sum_{i=1}^{51} x_i = 17{,}409, \quad \left(\sum_{i=1}^{51} x_i\right)^2 = 303{,}073{,}281, \quad \text{and} \quad \sum_{i=1}^{51} x_i^2 = 6{,}191{,}677, \text{ then do the following:}$$

2.8 Compute the arithmetic mean of this sample.

2.9 Compute the median of this sample.

2.10 Compute the standard deviation of this sample.

2.11 The national mortality rate for heart disease in 1973 was 360.8 per 100,000. Why does this figure *not* correspond to your answer for Problem 2.8?

2.12 Does the differential in raw rates between Florida (417.4) and Georgia (311.8) actually imply that the risk of dying from heart disease is greater in Florida than in Georgia? Why or why not?

Nutrition

Table 2.4 shows the distribution of dietary vitamin-A intake as reported by 14 students who filled out a dietary questionnaire in class. The total intake is a combination of intake from individual food items and from vitamin pills. The units are in IU/100 (International Units/100).

2.13 Compute the mean and median from these data.

2.14 Compute the standard deviation and coefficient of variation from these data.

2.15 Suppose the data are expressed in IU rather than

TABLE 2.4 Distribution of dietary vitamin-A intake as reported by 14 students

Student number	Intake (IU/100)	Student number	Intake (IU/100)
1	31.1	8	48.1
2	21.5	9	24.4
3	74.7	10	13.4
4	95.5	11	37.1
5	19.4	12	21.3
6	64.8	13	78.5
7	108.7	14	17.7

IU/100. What are the mean, standard deviation, and coefficient of variation in the new units?

2.16 Construct a stem-and-leaf plot of the data on some convenient scale.

2.17 Do you think the mean or median is a more appropriate measure of location for this data set?

SOLUTIONS

2.1 Each data value is changed from x_i to $x_i + a$, for some constant a. The median also increases by a.

2.2 The mode increases by a.

2.3 The geometric mean is changed by an undetermined amount, because the geometric mean is given by antilog $[\Sigma \ln(x_i + a)/n]$ and there is no simple relationship between $\ln(x_i + a)$ and $\ln(x_i)$.

2.4 The range is not changed, since it is the distance between the largest and smallest values, and distances between points will not be changed by shifting the origin.

2.5 The arithmetic mean is given by

$$\frac{10^0(521) + \ldots + 10^{10}(2)}{521 + \ldots + 2}$$

$$= \frac{3.24 \times 10^{10}}{1177}$$

$$= 2.757 \times 10^7$$

2.6 To compute the geometric mean, we first compute the mean ln to the base 10 as follows:

$$\frac{521 \ln_{10}(10^0) + \ldots + 2 \ln_{10}(10^{10})}{521 + \ldots + 2}$$

$$= \frac{521 \times 0 + \ldots + 2 \times 10}{1177}$$

$$= \frac{2014}{1177} = 1.711$$

The geometric mean is then given by $10^{1.711} = 51.4$.

2.7 The geometric mean is more appropriate because the distribution is in powers of 10 and is very skewed. In the log scale, the distribution becomes less skewed, and the mean provides a more central measure of location. Notice that only 33 of the 1177 data points are greater than the arithmetic mean, while 426 of the 1177 points are greater than the geometric mean.

2.8 We have that $\bar{x} = 17,409/51 = 341.4$ per 100,000.

2.9 Since $n = 51$ is odd, the median is given by the $[(51+1)/2)]$th or 26th largest value = mortality rate for Mississippi = 351.6 per 100,000.

2.10 We have that

$$s^2 = \frac{\Sigma x_i^2 - (\Sigma x_i)^2/51}{50}$$

CHAPTER THREE

PROBABILITY

SECTION 3.1   Frequency Definition of Probability

What is probability? Suppose we want to set up a test for color blindness. We use a plate divided into four quadrants. One of four color quadrants has a particular color. The other three quadrants are of the same color, but different from the first quadrant. Thus, the probability that a color-blind person will pick the correct plate at random = 1/4. What does "at random" mean? Suppose the test is performed by many color-blind people and the following results are obtained:

Number of color-blind people	% correct
20	6/20 correct = .30
100	24/100 correct = .24
1000	255/1000 correct = .255

As the number of color-blind people taking the test is increased, the proportion of correct trials will approach a number p (in this case, .25), which we call the *probability*. This is the *frequency definition of probability*.

All probabilities must be between 0 and 1. Probabilities are defined over events. Two events are mutually exclusive if they cannot occur at the same time. Probabilities of mutually exclusive events must add; e.g., suppose we repeat the test 4 times for a single color-blind person. Let the event E_1 = exactly 1 out of 4 correct, E_2 = exactly 2 out of 4 correct:

$$Pr(E_1) + Pr(E_2) = Pr(E_1 \text{ or } E_2) = Pr(E_1 \cup E_2)$$

E_1 and E_2 are *mutually exclusive* events because they cannot occur at the same time.

SECTION 3.2   Multiplication Law of Probability

When can we multiply probabilities? Let the events A, B be defined by

    A = 1st selection is correct

    B = 2nd selection is correct

Then $Pr(A \cap B)$ = probability both selections are correct = $Pr(A) \times Pr(B) = 1/16$. Two probabilities can be multiplied if the events are *independent*.

Consider another example: suppose we have a group of 6-month-old children with two normal ears at their routine 6-month checkup. Suppose there is a 10% chance that a child will have fluid in the middle ear at an exam 1 month later in a specific ear, while the probability that both ears are affected (called "bilateral middle-ear effusion") is .07. Are the ears independent? No.

$$Pr \text{ (bilateral middle-ear effusion)} = .07 > .1 \times .1 = .01$$

This is an example of *dependent events*. The middle-ear status of both ears of the same child are dependent events, because there is often a common reason why both ears get infected at the same time (e.g., exposure of the child to other affected children in a day-care center).

SECTION 3.3   Addition Law of Probability

Let A = right ear affected, B = left ear affected.

What is $Pr(A \cup B) = Pr$(either ear affected)?

$$Pr(A \cup B) = Pr(A) + Pr(B) - Pr(A \cap B)$$

For the *ear* example, Pr (either ear affected) $= .1 + .1 - .07 = .13$:

13% have at least one ear affected

7% have bilateral middle-ear effusion (both ears affected)

6% have unilateral middle-ear effusion (only one ear affected)

For the *color plate* example, A = (1st selection correct), B = (2nd selection correct)

$$
\begin{aligned}
Pr(A \cup B) &= Pr(\text{at least 1 of 2 selections are correct}) \\
&= Pr(A) + Pr(B) - Pr(A \cap B) \\
&= Pr(A) + Pr(B) - Pr(A) \times Pr(B) \\
&= \frac{1}{4} + \frac{1}{4} - \left(\frac{1}{4}\right)^2 = \frac{7}{16}
\end{aligned}
$$

SECTION 3.4   Conditional Probability

The conditional probability of B given A is defined as $Pr(B \cap A)/Pr(A)$ and is denoted by $Pr(B|A)$. For the *ear* example, let A = right ear affected, B = left ear affected, and $\overline{A}$ = the event that the right ear is *not* affected.

$$Pr(B|A) = \frac{Pr(A \cap B)}{Pr(A)} = \frac{.07}{.10} = 70\% = \text{conditional probability of } B \text{ given } A$$

$$Pr(B|\overline{A}) = \frac{Pr(B \cap \overline{A})}{Pr(\overline{A})} = \frac{Pr(B) - Pr(A \cap B)}{.90} = \frac{.10 - .07}{.90} = \frac{.03}{.90} = \frac{1}{30} \approx .03$$

In words, $Pr(B|A)$ = probability that the left ear is affected given that the right ear is affected = 70%; $Pr(B|\overline{A})$ = probability that the left ear is affected given that the right ear is *not* affected $\doteq$ 3%.

3.4.1 **Relative Risk**

The Relative Risk of B given A is defined as $Pr(B|A)/Pr(B|\overline{A})$. For the ear example, if A = right ear affected and B = left ear affected, then

$$\text{Relative risk} = RR = \frac{Pr(B|A)}{Pr(B|\overline{A})} = \frac{7/10}{1/30} = 21$$

The left ear is 21 times as likely to be affected if the right ear is affected than if the right ear is unaffected.

There was an outbreak of Legionnaire's disease in Austin, Minnesota in 1957. Subsequent investigation focused on employment at a meat-packing plant as a possible cause. The illness rate per 1000 subjects among all adults in the town is given in the following table:

Employment status	%	Total	Number ill	Illness rate per 1000	RR
Employed at meat-packing plant	.19	4,718	46	9.7	6.1
Not employed at meat-packing plant	.81	19,897	32	1.6	
	1.0	24,615	78	3.2	

The relative risk (RR) = 9.7/1.6 = 6.1.

If two events are independent, then $Pr(B|A) = Pr(B|\overline{A}) = Pr(B)$ and RR = 1. For the color plates example, let A = 1st selection correct, B = 2nd selection correct; $Pr(B|\overline{A}) = Pr(B) = Pr(B|A) = 1/4$ and RR = 1.

SECTION 3.5 **Total Probability Rule**

The total probability rule specifies the relationship between conditional and unconditional probabilities:

$$Pr(B) = Pr(B|A)Pr(A) + Pr(B|\overline{A})Pr(\overline{A})$$

For example, in the case of Legionnaire's disease: A = work at meat-packing plant, B = Legionnaire's disease. Suppose Pr(A) = .19.

$$Pr(B) = \frac{3.2}{1000} = .19 \times \frac{9.7}{1000} + .81 \times \frac{1.6}{1000}$$

SECTION 3.6 **Sensitivity, Specificity, Predictive Values of Screening Tests**

The angiogram is the standard test used to diagnose the occurrence of stroke. However, some patients experience side effects from this test, and some investigators have attempted to use a noninvasive test as an alternative. 64 patients with transient monocular blindness, or TMB (where a person temporarily loses vision in one eye), were given both tests. The sample was selected to have about equal numbers of angiogram-positive and -negative patients.

Angiogram	Noninvasive test	n
−	−	21
−	+	8
+	−	3
+	+	32
		64

How can we compare the two tests? If we assume the angiogram is the gold standard, then

$$\text{Sensitivity is defined as } Pr(\text{test} + | \text{true}+) = 32/35 = .914$$
$$\text{Specificity is defined as } Pr(\text{test} - | \text{true}-) = 21/29 = .724$$

We would like to convert sensitivity and specificity into predictive values:

$$\text{Predictive value positive } (PV+) \text{ as defined as } Pr(\text{true}+ | \text{test}+)$$

It can be shown that

$$PV+ = \frac{\text{sensitivity} \times \text{prevalence}}{\text{sensitivity} \times \text{prevalence} + (1 - \text{specificity}) \times (1 - \text{prevalence})}$$

Assume the prevalence of strokes is 20% among TMB patients:

$$PV+ = \frac{.914 \times .20}{.914(.20) + .276(.80)} = \frac{.1829}{.1829 + .2207} = \frac{.1829}{.4035} = .453$$

$$\text{Predictive value negative } (PV-) \text{ as defined as } Pr(\text{true}- | \text{test}-)$$

It can be shown that

$$PV- = \frac{\text{specificity} \times (1- \text{prevalence})}{\text{specificity} \times (1 - \text{prevalence}) + (1 - \text{sensitivity})\text{prevalence}}$$

$$= \frac{.724(.80)}{.724(.80) + .086(.20)} = \frac{.5793}{.5793 + .0171} = \frac{.5793}{.5965} = .971$$

SECTION 3.7 Bayes' Theorem

The determination of predictive value positive and negative is a particular application of a more general principle (*Bayes' theorem*). If A = symptom(s) and B = disease, then

$$Pr(B|A) = \frac{Pr(A|B)Pr(B)}{Pr(A|B)Pr(B) + Pr(A|\overline{B})Pr(\overline{B})}$$

More generally, if there are k disease states such that each person has one and only one disease state (which could include being normal),

$$Pr(B_i|A) = \frac{Pr(A|B_i)Pr(B_i)}{\displaystyle\sum_{j=1}^{k} Pr(A|B_j)Pr(B_j)}, i = 1, \ldots, k$$

B_i = ith disease state, A = symptom.

PROBLEMS

Let A = {serum cholesterol = 250–299}, B = {serum cholesterol ≥ 300}, C = {serum cholesterol ≤ 280}.

3.1 Are the events A and B mutually exclusive?

3.2 Are the events A and C mutually exclusive?

3.3 Suppose $Pr(A)$ = .2, $Pr(B)$ = .1. What is Pr(serum cholesterol ≥ 250)?

3.4 What does $A \cup C$ mean?

3.5 What does $A \cap C$ mean?

3.6 What does $B \cup C$ mean?

3.7 What does $B \cap C$ mean?

3.8 Are the events B and C mutually exclusive?

3.9 What does the event $\overline{B}$ mean? What is its probability?

Suppose that the gender of successive offspring in the same family are independent events and that the probability of a male or female offspring is .5.

3.10 What is the probability of two successive female offspring?

3.11 What is the probability that exactly one of two successive children will be female?

3.12 Suppose that three successive offspring are male. What is the probability that a fourth child will be male?

Cardiovascular Disease

A survey was performed among people 65 years of age and older who underwent open-heart surgery. It was found that 30% of patients died within 90 days of the operation, whereas an additional 25% of those who survived 90 days died within 5 years after the operation.

3.13 What is the probability that a patient undergoing open-heart surgery will die within 5 years?

3.14 What is the mortality incidence (per patient month)

in patients receiving this operation in the first 90 days after the operation? (Assume that 90 days = 3 months.)

3.15 Answer the same question as in Problem 3.14 for the period from 90 days to 5 years after the operation.

3.16 Can you tell if the operation prolongs life from the data presented? If not, then what additional data do you need?

A study relating smoking history to several measures of cardiopulmonary disability was recently reported [1]. The data in Table 3.1 were presented relating the number of people with different disabilities according to cigarette-smoking status.

3.17 What is the prevalence of angina among light current smokers (< 15g/day)?

3.18 What is the relative risk of ex-smokers, light current smokers, and heavy current smokers, respectively, for shortness of breath as compared with nonsmokers?

3.19 Answer Problem 3.18 for angina.

3.20 Answer Problem 3.18 for possible infarction.

Pulmonary Disease

Pulmonary embolism is a relatively common condition that necessitates hospitalization and also often occurs in patients hospitalized for other reasons. An oxygen tension (arterial P_{O_2}) < 90 mm Hg is one of the important criteria used in diagnosing this condition. Suppose that the sensitivity of this test is 95%, the specificity is 75%, and the estimated prevalence is 20% (i.e., a doctor estimates that a patient has a 20% chance of pulmonary embolism before performing the test).

3.21 What is the predictive value positive of this test? What does it mean in words?

3.22 What is the predictive value negative of this test? What does it mean in words?

TABLE 3.1 Number of people with selected cardiopulmonary disabilities versus cigarette-smoking status

Disability	Cigarette-smoking status			
	None (n = 656)	Ex (n = 826)	Current < 15 g/day (n = 955)	Current ≥ 15 g/day (n = 654)
Shortness of breath	7	15	18	13
Angina	15	19	19	16
Possible infarction	3	7	8	6

3.23 Answer Problem 3.21 if the estimated prevalence is 80%.

3.24 Answer Problem 3.22 if the estimated prevalence is 80%.

Environmental Health, Pediatrics

3.25 Suppose that a company plans to build a lead smelter in a community and that the city council wishes to assess the health effects of the smelter. In particular, there is concern from previous literature that children living very close to the smelter will experience unusually high rates of lead poisoning in the first 3 years of life. Suppose that the projected rates of lead poisoning over this time period are 50 per 100,000 for those children living within 2 km of the smelter, 20 per 100,000 for children living > 2 km but ≤ 5 km from the smelter, and 5 per 100,000 for children living > 5 km from the smelter. If 80% of the children live more than 5 km from the smelter, 15% live > 2 km but ≤ 5 km from the smelter, and the remainder live ≤ 2 km from the smelter, then what is the overall probability that a child from this community will get lead poisoning?

Diabetes

The prevalence of diabetes in adults at least 20 years old has been studied in Tecumseh, Michigan [2]. The age–sex-specific prevalence (per 1000) is given in Table 3.2.

TABLE 3.2 Age-sex-specific prevalence of diabetes in Tecumseh, MI (per 1000)

Age group (years)	Sex	
	Male	Female
20–39	5	7
40–54	23	31
55+	57	89

Source: Reprinted with permission from the *American Journal of Epidemiology, 116*(6), 971–980.

3.26 Suppose we plan a new study in a town that consists of 48% males and 52% females. Of the males, 40% are ages 20–39, 32% are 40–54, and 28% are 55+. Of the females, 44% are ages 20–39, 37% are 40–54, and 19% are 55+. Assuming that the Tecumseh prevalence rates hold, what is the expected prevalence of diabetes in the new study?

3.27 What proportion of diabetics in the new study would be expected in each of the six age–sex groups?

Cancer

Table 3.3 shows the annual incidence rates for colon cancer, lung cancer, and stomach cancer in males ages 50 years and older from the Connecticut Tumor Registry, 1963–1965 [3].

TABLE 3.3 Average annual incidence per 100,000 males for colon, lung, and stomach cancer from the Connecticut Tumor Registry, 1963–1965.

Type of cancer	Ages		
	50–54	55–59	60–64
Colon	35.7	60.3	98.9
Lung	76.1	137.5	231.7
Stomach	20.8	39.1	46.0

Source: Reprinted from *Cancer Incidence in Five Continents II,* 1970, with permission of Springer-Verlag, Berlin.

3.28 What is the probability that a 57-year-old, disease-free male will develop lung cancer over the next year?

3.29 What is the probability that a 55-year-old, disease-free male will develop colon cancer over the next 5 years?

3.30 Suppose there is a cohort of 1000 50-year-old men who have never had cancer. How many colon cancers would be expected to develop in this cohort over a 15-year period?

3.31 Answer Problem 3.30 for lung cancer.

3.32 Answer Problem 3.30 for stomach cancer.

Cardiovascular Disease

An experiment was set up by a group from the University of Utah to use Bayes' rule to help make clinical diagnoses [4]. In particular, a detailed medical history questionnaire and electrocardiogram were administered to each patient referred to a cardiovascular laboratory and suspected of having congenital heart disease. From the experience of this laboratory and from estimates based on other published data, two sets of probabilities were generated:

(1) The unconditional probability of each of several disease states (refer to prevalence column in Table 3.4).

(2) The conditional probability of specific symptoms given specific disease states (refer to the rest of Table 3.4).

Thus, the probability that a person has chest pain given that he or she is normal is .05. Similarly, the proportion of persons with isolated pulmonary hypertension is .020. A

TABLE 3.4 Prevalence of symptoms and diagnoses for patients suspected of having congenital heart disease

Diagnosis	Prevalence	Symptoms						
		X_1	X_2	X_3	X_4	X_5	X_6	X_7
Y_1	.155	.49	.50	.01	.10	.05	.05	.01
Y_2	.126	.50	.50	.02	.50	.02	.40	.70
Y_3	.084	.55	.05	.25	.90	.05	.10	.95
Y_4	.020	.45	.45	.01	.95	.10	.10	.95
Y_5	.098	.10	.00	.20	.70	.01	.05	.40
Y_6	.391	.70	.15	.01	.30	.01	.15	.30
Y_7	.126	.60	.10	.30	.70	.10	.20	.70

Note: Y_1 = normal
   Y_2 = atrial septal defect without pulmonary stenosis or pulmonary hypertension[a]
   Y_3 = ventricular septal defect with valvular pulmonary stenosis
   Y_4 = isolated pulmonary hypertension[a]
   Y_5 = transposed great vessels
   Y_6 = ventricular septal defect without pulmonary hypertension[a]
   Y_7 = ventricular septal defect with pulmonary hypertension[a]
   X_1 = age 1–20 years old
   X_2 = age > 20 years old
   X_3 = mild cyanosis
   X_4 = easy fatigue
   X_5 = chest pain
   X_6 = repeated respiratory infections
   X_7 = EKG axis more than 110 degrees

[a]Pulmonary hypertension is defined as pulmonary artery pressure ≥ systemic arterial pressure.

Source: Reprinted with permission of *The American Medical Association* from *The Journal of the American Medical Association, 177*(3), 177–183, 1961. Copyright 1961, American Medical Association.

subset of the data is given in Table 3.4.

Assume that these diagnoses are the only ones possible and that a patient can have one and only one diagnosis.

3.33 What is the probability of having a symptom of chest pain given that you have isolated pulmonary hypertension?

3.34 What is the probability of being more than 20 years old in this clinic?

3.35 Suppose we assume that the probability of any set of symptoms are independent given a specific diagnosis (e.g., the probability of being > 20 years old and having both chest pain and mild cyanosis given that one is normal is .50 × .05 × .01 = .00025). What is the probability of being diagnosed as normal given that you have the following symptoms: (a) age 1–20 years, (b) repeated respiratory infections, and (c) easy fatigue?

3.36 What is the *most likely* diagnosis given that you have all the following symptoms: (a) mild cyanosis, (b) age > 20 years, (c) EKG axis more than 110 degrees? What is the second most likely diagnosis?

Suppose the symptom of an EKG axis of more than 110 degrees is used as a screening criterion for diagnosing atrial septal defect without pulmonary stenosis or pulmonary hypertension.

3.37 What is the sensitivity of this test?

3.38 What is the specificity of this test?

3.39 Suppose we want to use another symptom in addition to EKG axis more than 110 degrees to diagnose atrial septal defect without pulmonary stenosis or pulmonary hypertension. If we use the symptoms of age 1–20 years old and EKG axis more than 110 degrees, then what is the predictive value positive?

3.40 Suppose we want to use two symptoms to diagnose atrial septal defect without pulmonary stenosis or pulmonary hypertension (not necessarily including EKG axis more than 110 degrees). Which two symptoms can be used to maximize the predictive value positive? (Use a computer to answer this question.)

3.41 Answer Problem 3.40 using three symptoms rather than two. (Use a computer to answer this question.)

Cardiovascular Disease

The relationship between physical fitness and cardiovascular-disease mortality was recently studied in a group of railroad working men, ages 22–79 [5]. Data were presented relating baseline exercise-test heart rate and coronary heart-disease mortality (Table 3.5).

TABLE 3.5 Relationship between baseline exercise-test heart rate and coronary heart-disease mortality

Exercise-test heart rate (beats/min)	Coronary heart-disease mortality (20 years) (per 100)
≤ 105	9.1
106–115	8.7
116–127	11.6
> 127	13.2

Suppose that 20, 30, 30, and 20% of the population, respectively, have exercise-test heart rates of ≤ 105, 106–115, 116–127, > 127. Suppose a test is positive if exercise-test heart rate is > 127 beats/min and negative otherwise.

3.42 What is the probability of a positive test among men who have died over the 20-year period? Is there a name for this quantity?

3.43 What is the probability of a positive test among men who survived the 20-year period? Is there a name for this quantity?

3.44 What is the probability of death among men with a negative test? Is there a name for this quantity?

Cardiovascular Disease

Exercise testing has sometimes been used to diagnose patients with coronary-artery disease. One test criterion that has been used to identify those with disease is the abnormal ejection-fraction criterion; that is, an absolute rise of less than .05 with exercise. The validity of this noninvasive test was assessed in 196 patients versus coronary angiography, the gold standard, a procedure that can unequivocally diagnose the disease but the administration of which carries some risk for the patient. A sensitivity of 79% and a specificity of 68% were found for the exercise test in this group.

3.45 What does the sensitivity mean in words in this setting?

3.46 What does the specificity mean in words in this setting?

Suppose a new patient undergoes exercise testing and a physician feels before seeing the exercise-test results that the patient has a 20% chance of having coronary-artery disease.

3.47 If the exercise test is positive, then what is the probability that such a patient has disease?

3.48 Answer Problem 3.47 if the exercise test is negative.

Nutrition

The food-frequency questionnaire (FFQ) is a commonly used method for assessing dietary intake, whereby individuals are asked to record the number of times per week they usually eat each of about 100 food items over the previous year. It has the advantage of being easy to administer to large groups of people, but has the disadvantage of being subject to recall error. The gold-standard instrument for assessing dietary intake is the diet record (DR), where people are asked to record each food item eaten on a daily basis over a 1-week period. To investigate the accuracy of the FFQ, both the FFQ and the DR were administered to 173 participants in the United States. The reporting of alcohol consumption with each instrument is given in Table 3.6, where alcohol is coded as alcohol = some drinking, versus no alcohol = no drinking.

TABLE 3.6 Actual drinking habits as determined from the diet record cross-classified by self-reported drinking status from the food-frequency questionnaire

Diet record	Food-frequency questionnaire		
	Alcohol	No alcohol	Total
Alcohol	138	18	156
No alcohol	1	16	17
Total	139	34	173

3.49 What is the sensitivity of the FFQ?

3.50 What is the specificity of the FFQ?

Let us treat the 173 participants in this study as representative of people who would use the FFQ.

3.51 What is the predictive value positive of the FFQ?

3.52 What is the predictive value negative of the FFQ?

3.53 Suppose the questionnaire is to be administered in a different country where the proportion of drinkers is 80%. What would be the predictive value positive if administered in this setting?

Genetics

Two healthy parents have a child with a severe autosomal recessive condition that cannot be identified by prenatal diagnosis. They realize that the risk of this condition for subsequent offspring is 1/4, but wish to embark on a second pregnancy. During the early stages of the pregnancy, an ultrasound test determines that there are twins.

3.54 Suppose that there are monozygotic, or MZ (identical) twins. What is the probability that both twins are affected? one twin affected? neither twin affected? Are the outcomes for the two MZ twins independent or dependent events?

3.55 Suppose that there are dizygotic, or DZ (fraternal) twins. What is the probability that both twins are affected? one twin affected? neither twin affected? Are the outcomes for the two DZ twins independent or dependent events?

3.56 Suppose there is a 1/3 probability of MZ twins and a 2/3 probability of DZ twins. What is the overall probability that both twins are affected? one twin affected? neither affected?

3.57 Suppose we learn that both twins are affected but don't know whether they are MZ or DZ twins. What is the probability of MZ twins given this additional information?

Cerebrovascular Disease

Atrial fibrillation (AF) is a common cardiac condition in the elderly (e.g., former President George Bush has this condition) characterized by an abnormal heart rhythm that greatly increases the risk of stroke. The following estimates of the prevalence rate of AF and the incidence rate of stroke for people with and without AF by age from the Framingham Heart Study are given in Table 3.7 [6].

TABLE 3.7 Relationship between atrial fibrillation and stroke

Age group	Prevalence of AF (%)	Incidence rate of stroke per 1000 person-years	
		No AF	AF
60–69	1.8	4.5	21.2
70–79	4.7	9.0	48.9
80–89	10.2	14.3	71.4

3.58 What does an incidence rate of 48.9 strokes per 1000 person-years among 70–79-year-olds with AF mean in Table 3.7?

3.59 What is the relative risk of stroke for people with AF compared with people without AF in each age group? Does the relative risk seem to be the same for different age groups?

Suppose we screen 500 subjects from the general population of age 60–89, of whom 200 are 60–69, 200 are 70–79, and 100 are 80–89.

3.60 What is the incidence rate of stroke in the screened population over a 1-year period?

3.61 Suppose the study of 500 subjects is a "pilot study" for a larger study. How many subjects of age 60–89 need to be screened if we wish to observe an average of 50 strokes in the larger study over 1 year?

SOLUTIONS

3.1 Yes

3.2 No

3.3 .3

3.4 $A \cup C = \{$serum cholesterol $\leq 299\}$

3.5 $A \cap C = \{250 \leq$ serum cholesterol $\leq 280\}$

3.6 $B \cup C = \{$serum cholesterol ≤ 280 or $\geq 300\}$

3.7 $B \cap C$ is the empty set; that is, it can never occur.

3.8 Yes

3.9 $\overline{B} = \{$serum cholesterol $< 300\}$. $Pr(\overline{B}) = .9$

3.10 Let $A_1 = $ (1st offspring is a male), $A_2 = $ (2nd offspring is a male). $Pr(\overline{A}_1 \cap \overline{A}_2) = .5 \times .5 = .25$

3.11 $Pr(\overline{A}_1 \cap A_2) + Pr(A_1 \cap \overline{A}_2) = .5 \times .5 + .5 \times .5 = .50$

3.12 The probability $= .5$, because the sex of successive offspring are independent events.

3.13 Probability $= .30 + (1 - .30) \times .25 = .475$

3.14 10% per patient-month.

3.15 The mortality incidence per month = $0.25/57$ months = 0.44% per patient-month.

3.16 No. For comparison, mortality data on a control group of patients with the same clinical condition as the patients who underwent open-heart surgery, but who did not have the operation are needed.

3.17 The prevalence of angina among light smokers = $19/955 = .020$.

3.18 Relative risk of shortness of breath for
Ex-smokers vs. nonsmokers = $(15/826)/(7/656)$
$= .0182/.0107 = 1.7$
Light smokers vs. nonsmokers = $(18/955)/(7/656)$
$= .0188/.0107 = 1.8$
Heavy smokers vs. nonsmokers = $(13/654)/(7/656)$
$= .0199/.0107 = 1.9$

3.19 Relative risk of angina for
Ex-smokers vs. nonsmokers = $(19/826)/(15/656)$
$= .0230/.0229 = 1.0$
Light smokers vs. nonsmokers = $(19/955)/(15/656)$
$= .0199/.0229 = 0.9$
Heavy smokers vs. nonsmokers = $(16/654)/(15/656)$
$= .0245/.0229 = 1.1$

3.20 Relative risk of possible infarction for
Ex-smokers vs. nonsmokers = $(7/826)/(3/656)$
$= .0085/.0046 = 1.9$
Light smokers vs. nonsmokers = $(8/955)/(3/656)$
$= .0084/.0046 = 1.8$
Heavy smokers vs. nonsmokers = $(6/654)/(3/656)$
$= .0092/.0046 = 2.0$

3.21 We have that

$$PV+ = \frac{(x)(\text{sensitivity})}{(x)(\text{sensitivity}) + (1-x)(1-\text{specificity})}$$

where x = prevalence

$$= \frac{.20 \times .95}{.20 \times .95 + .80 \times .25}$$

$$= \frac{.19}{.39} = .487$$

It means that if this patient has a depressed arterial oxygen tension, then there is approximately a 50% chance that she will have a pulmonary embolism.

3.22 We have that

$$PV- = \frac{(1-x)(\text{specificity})}{(1-x)(\text{specificity}) + (x)(1-\text{sensitivity})}$$

$$= \frac{.80 \times .75}{.80 \times .75 + .20 \times .05}$$

$$= \frac{.60}{.61} = .984$$

It means that if this patient does not have a depressed arterial oxygen tension, then there is a 98.4% chance that she will *not* have a pulmonary embolism.

3.23 We have

$$PV+ = \frac{.80 \times .95}{.80 \times .95 + .20 \times .25}$$

$$= \frac{.76}{.81} = .938$$

3.24 We have

$$PV- = \frac{.20 \times .75}{.20 \times .75 + .80 \times .05}$$

$$= \frac{.15}{.19} = .789$$

3.25 Let A = {child gets lead poisoning}
B_1 = {child lives ≤ 2 km from the smelter}
B_2 = {child lives >2 but ≤ 5 km from the smelter}
B_3 = {child lives >5 km from the smelter}

We can write

$$Pr(A) = Pr(A \cap B_1) + Pr(A \cap B_2) + Pr(A \cap B_3)$$
$$= Pr(A|B_1)Pr(B_1) + Pr(A|B_2)Pr(B_2)$$
$$+ Pr(A|B_3)Pr(B_3)$$

We are given that $Pr(A|B_1) = 50/10^5$, $Pr(A|B_2) = 20/10^5$, $Pr(A|B_3) = 5/10^5$, $Pr(B_1) = .05$, $Pr(B_2) = .15$ and $Pr(B_3) = .80$. Therefore, it follows that

$$Pr(A) = (50/10^5)(.05) + (20/10^5)(.15) + (5/10^5)(.80)$$
$$= 9.5/10^5 = .000095$$

3.26 Let A = {diabetes}, B_1 = {20–39-year-old male}, B_2 = {40–54-year-old male}, B_3 = {55+-year-old male}, B_4 = {20–39-year-old female}, B_5 = {40–54-year-old female}, B_6 = {55+-year-old female}. We are given $Pr(A|B_1) = .005$, $Pr(A|B_2) = .023$, $Pr(A|B_3) = .057$, $Pr(A|B_4) = .007$, $Pr(A|B_5) = .031$, $Pr(A|B_6) = .089$. First, we compute the probabilities of the events $B_1, \ldots, B_6$.

$$Pr(B_1) = Pr(20\text{–}39|\text{male}) \times Pr(\text{male})$$
$$= .40 \times .48 = .192$$
$$Pr(B_2) = Pr(40\text{–}54|\text{male}) \times Pr(\text{male})$$
$$= .32 \times .48 = .154$$
$$Pr(B_3) = Pr(55+|\text{male}) \times Pr(\text{male})$$
$$= .28 \times .48 = .134$$

$Pr(B_4) = Pr(20\text{--}39 \,|\, \text{female}) \times Pr(\text{female})$

$= .44 \times .52 = .229$

$Pr(B_5) = Pr(40\text{--}54 \,|\, \text{female}) \times Pr(\text{female})$

$= .37 \times .52 = .192$

$Pr(B_6) = Pr(55+ \,|\, \text{female}) \times Pr(\text{female})$

$= .19 \times .52 = .099$

We now use the total probability rule as follows, whereby we have $Pr(A) = \sum_{i=1}^{6} Pr(A \,|\, B_i) \times Pr(B_i) = .005 \times .192 + \ldots + .089 \times .099 = .029$. Thus, the expected prevalence in the new study is 2.9%.

3.27   20–39 M: .034; 40–54 M: .124; 55+ M: .269; 20–39 F: .056; 40–54 F: .209; 55+ F: .308

3.28 $137.5/10^5 = .001375$

3.29

$Pr(not \text{ developing colon cancer over 5 years})$

$= Pr(\text{not developing colon cancer over the 1st year})$

$\times\, Pr(\text{not developing colon cancer over the 2nd year} \,|\, \text{disease free after 1 year})$

$\times \ldots \times Pr(\text{not developing colon cancer over the 5th year} \,|\, \text{disease free after 4 years})$

$= (1 - 60.3/10^5)^5 = .99699$

Thus,

$Pr(\text{developing colon cancer over 5 years})$

$= 1 - .99699 = .00301 = 301 \text{ per } 10^5$

3.30   By the same rationale as in Problem 3.29,

$Pr(not \text{ developing colon cancer over 15 years})$

$= Pr(\text{not developing colon cancer from age 50 to 54})$

$\times\, Pr(\text{not developing colon cancer from age 55 to 59} \,|\, \text{disease free after 5 years})$

$\times\, Pr(\text{not developing colon cancer from age 60 to 64} \,|\, \text{disease free after 10 years})$

$= (1 - 35.7/10^5)^5 \times (1 - 60.3/10^5)^5 \times (1 - 98.9/10^5)^5$

$= .998216 \times .996989 \times .995065 = .99030$

Therefore, $Pr(\text{developing colon cancer over 15 years}) = 1 - .99030 = .00970$. Thus, the expected number of colon cancers over 15 years among 1000 50-year-old men is $1000 \times .00970 = 9.7$.

3.31   Using methods similar to those used in Problem 3.30, we see that

$Pr(not \text{ developing lung cancer over 15 years})$

$= (1 - 76.1/10^5)^5 \times (1 - 137.5/10^5)^5$

$\times (1 - 231.7/10^5)^5$

$= .996201 \times .993144 \times .988469 = .97796$

Thus, $Pr(\text{developing lung cancer over 15 years}) = 1 - .97796 = .02204$ and the expected number of lung cancers over 15 years among 1000 50-year-old men is $1000 \times .02204 = 22.0$.

3.32   Using similar methods as in Problem 3.30, we see that

$Pr(not \text{ developing stomach cancer over 15 years})$

$= (1 - 20.8/10^5)^5 \times (1 - 39.1/10^5)^5 \times (1 - 46.0/10^5)^5$

$= .998960 \times .998047 \times .997702 = .99472$

Thus, $Pr(\text{developing stomach cancer over 15 years}) = 1 - .99472 = .00528$ and the expected number of stomach cancers over 15 years among 1000 50-year-old men is $1000 \times .00528 = 5.3$.

3.33   The probabilities in the prevalence column are $Pr(Y_j)$. The probabilities in the columns under the X_i are $Pr(X_i \,|\, Y_j)$. Thus, $Pr(\text{chest pain} \,|\, \text{isolated pulmonary hypertension}) = Pr(X_5 \,|\, Y_4) = .10$

3.34   We have

$Pr(\text{age} > 20 \text{ years}) = Pr(X_2)$

$= Pr(X_2 \cap Y_1) + Pr(X_2 \cap Y_2) + \ldots + Pr(X_2 \cap Y_7)$

$= Pr(X_2 \,|\, Y_1)Pr(Y_1) + \ldots + Pr(X_2 \,|\, Y_7)Pr(Y_7)$

$= .50(.155) + \ldots + .10(.126) = .225$

3.35   Let the event $(X_1 \cap X_6 \cap X_4) = E$. We wish to compute $Pr(Y_1 \,|\, E)$. By Bayes' theorem,

$$Pr(Y_1 \,|\, E) = \frac{Pr(E \,|\, Y_1)Pr(Y_1)}{\sum_{j=1}^{7} Pr(E \,|\, Y_j)Pr(Y_j)}$$

where, by independence,

$$Pr(E \,|\, Y_j) = Pr(X_1 \,|\, Y_j)Pr(X_6 \,|\, Y_j)Pr(X_4 \,|\, Y_j)$$

Thus,

$$Pr(Y_1 \,|\, E) = \frac{.49 \times .05 \times .10 \times .155}{.49 \times .05 \times .10 \times .155 + \ldots + .6 \times .2 \times .7 \times .126}$$

$$= \frac{.00037975}{.041236} = .009$$

3.36   Let $D = (X_2 \cap X_3 \cap X_7)$. We need to compute $Pr(Y_j \,|\, D)$ for all j and rank these probabilities. From Problem 3.35, all these probabilities share the common denominator:

$$\sum_{j=1}^{7} Pr(D \,|\, Y_j)Pr(Y_j)$$

So we need only calculate and rank the 7 numerators given by

$$Pr(X_2 \,|\, Y_j)Pr(X_3 \,|\, Y_j)Pr(X_7 \,|\, Y_j)Pr(Y_j), \quad j = 1, \ldots, 7$$

We have

j	**Numerator** $Pr(X_2 \mid Y_j)Pr(X_3 \mid Y_j)Pr(X_7 \mid Y_j)Pr(Y_j)$
1	7.750×10^{-6}
2	8.820×10^{-4}
3	9.975×10^{-4}
4	8.550×10^{-5}
5	0
6	1.760×10^{-4}
7	2.646×10^{-3}

Therefore, the most likely diagnosis is Y_7 and the second most likely diagnosis is Y_3.

3.37 X_7 is a screening criterion for Y_2. Sensitivity $= Pr(X_7 \mid Y_2) = .7$.

3.38 We have that

Specificity $= Pr(\overline{X}_7 \mid \overline{Y}_2)$

$$= \frac{Pr(\overline{X}_7 \cap \overline{Y}_2)}{Pr(\overline{Y}_2)}$$

$$= \frac{1 - Pr(X_7 \cup Y_2)}{1 - Pr(Y_2)}$$

$$= \frac{1 - Pr(X_7) - Pr(Y_2) + Pr(X_7 \cap Y_2)}{1 - Pr(Y_2)}$$

$$= \frac{1 - Pr(X_7) - Pr(Y_2) + Pr(X_7 \mid Y_2)Pr(Y_2)}{1 - Pr(Y_2)}$$

We have

$$Pr(X_7) = Pr(X_7 \mid Y_1)Pr(Y_1) + \ldots + Pr(X_7 \mid Y_7)Pr(Y_7)$$
$$= .43325$$

and

$$Pr(\overline{X}_7 \mid \overline{Y}_2) = \frac{1 - .43325 - .126 + .7 \times .126}{.874}$$

$$= .605 = \text{specificity}$$

3.39 We will use symptoms X_1 and X_7. To compute predictive value positive for this combination of symptoms, we have

$$Pr(Y_2 \mid X_1 \cap X_7) = \frac{Pr(X_1 \cap X_7 \mid Y_2)Pr(Y_2)}{\sum\limits_{i=1}^{7} Pr(X_1 \cap X_7 \mid Y_i)Pr(Y_i)}$$

$$= \frac{Pr(X_1 \mid Y_2)Pr(X_7 \mid Y_2)Pr(Y_2)}{\sum\limits_{i=1}^{7} Pr(X_1 \mid Y_i)Pr(X_7 \mid Y_i)Pr(Y_i)}$$

This information is displayed as follows for each Y_i:

i	$Pr(Y_i)$	$Pr(X_1 \mid Y_i)$	$Pr(X_7 \mid Y_i)$	$Pr(X_1 \mid Y_i)Pr(X_7 \mid Y_i)Pr(Y_i)$
1	.155	.49	.01	.00076
2	.126	.50	.70	.04410
3	.084	.55	.95	.04389
4	.020	.45	.95	.00855
5	.098	.10	.40	.00392
6	.391	.70	.30	.08211
7	.126	.60	.70	.05292
Total				.23625

Therefore,

$$Pr(Y_2 \mid X_1 \cap X_7) = \frac{.04410}{.23625} = .187 = PV+$$

3.40 For any two symptoms X_a and X_b, we can compute $PV+$ as follows:

$$Pr(Y_2 \mid X_a \cap X_b) = \frac{Pr(X_a \cap X_b \mid Y_2)Pr(Y_2)}{\sum\limits_{i=1}^{7} Pr(X_a \cap X_b \mid Y_i)Pr(Y_i)}$$

$$= \frac{Pr(X_a \mid Y_2)Pr(X_b \mid Y_2)Pr(Y_2)}{\sum\limits_{i=1}^{7} Pr(X_a \mid Y_i)Pr(X_b \mid Y_i)Pr(Y_i)}$$

We have evaluated this formula using a BASIC computer program for each combination of symptoms X_a and X_b, with the exception of symptoms 1 (age 1–20 years) and 2 (age > 20 years), which can never be present together. The results are given as follows:

a	b	PV+	a	b	PV+
1	3	.0307	3	4	.0204
1	4	.1363	3	5	.0097
1	5	.0675	3	6	.0817
1	6	.2764	3	7	.0307
1	7	.1867	4	5	.0685
2	3	.1671	4	6	.3401
2	4	.4039	4	7	.1706
2	5	.1557	5	6	.1949
2	6	.6041[a]	5	7	.0974
2	7	.5261	6	7	.4280

[a]Combination of two symptoms with the highest predictive value positive.

The combination of two symptoms that yields the highest predictive value positive is X_2 (age > 20 years old) and X_6 (repeated respiratory infections), ($PV+ = .6041$). The computer program used to obtain these predictive values was written in BASIC and is given as follows:

```
1 REM SAVED AS 'PV.BAS'
5 REM PROGRAM TO COMPUTE PREDICTIVE VALUE POSITIVE FOR ANY PAIRS OF SYMPTOMS
6 REM FOR ATRIAL SEPTAL DEFECT WITHOUT PULMONARY STENOSIS OR PULMONARY HYPERTENSION
10 DIM Y(7), X(7,7), TERM(7)
30 DATA .155,.126,.084,.020,.098,.391,.126
31 DATA .49,.50,.55,.45,.10,.70,.60
32 DATA .50,.50,.05,.45,.00,.15,.10
33 DATA .01,.02,.25,.01,.20,.01,.30
34 DATA .10,.50,.90,.95,.70,.30,.70
35 DATA .05,.02,.05,.10,.01,.01,.10
36 DATA .05,.40,.10,.10,.05,.15,.20
37 DATA .01,.70,.95,.95,.40,.30,.70
50 FOR I=1 TO 7: READ Y: Y(I)=Y: NEXT I
60 FOR J=1 TO 7: FOR I=1 TO 7: READ X: X(I,J)=X: NEXT I,J
65 REM Y(I) = UNCONDITIONAL PROBABILITY OF ITH DIAGNOSIS
70 REM X(I,J) = CONDITIONAL PROBABILITY OF JTH SYMPTOM FOR ITH DIAGNOSIS
80 FOR J=1 TO 6: FOR K=J+1 TO 7
85 SUM=0!
90 FOR I = 1 TO 7
92 TERM(I) = X(I,J)*X(I,K)*Y(I)
95 SUM=SUM+TERM(I): NEXT I
110 PV=TERM(2)/SUM
120 LPRINT USING "# # #.####";J,K,PV
130 NEXT K,J
```

3.41 A similar strategy can be used to compute the $PV+$ for any three symptoms X_a, X_b, X_c, (excluding triplets where symptoms 1 and 2 are together). The general formula is given by

$$Pr(Y_2 \mid X_a \cap X_b \cap X_c)$$
$$= \frac{Pr(X_a \mid Y_2)Pr(X_b \mid Y_2)Pr(X_c \mid Y_2)Pr(Y_2)}{\sum_{i=1}^{7} Pr(X_a \mid Y_i)Pr(X_b \mid Y_i)Pr(X_c \mid Y_i)Pr(Y_i)}$$

The results are given as follows:

a	b	c	PV+	a	b	c	PV+	a	b	c	PV+
1	3	4	.0215	2	3	4	.1382	3	4	5	.0066
1	3	5	.0085	2	3	5	.0495	3	4	6	.0588
1	3	6	.0747	2	3	6	.3358	3	4	7	.0203
1	3	7	.0300	2	3	7	.1840	3	5	6	.0223
1	4	5	.0622	2	4	5	.2020	3	5	7	.0093
1	4	6	.3056	2	4	6	.6766	3	6	7	.0822
1	4	7	.1643	2	4	7	.4868	4	5	6	.1634
1	5	6	.1713	2	5	6	.4387	4	5	7	.0674
1	5	7	.0871	2	5	7	.2908	4	6	7	.3866
1	6	7	.3833	2	6	7	.7559[a]	5	6	7	.2166

[a]Combination of three symptoms with the highest predictive value positive.

The combination of symptoms yielding the highest $PV+$ is symptom 2 (age >20 years), symptom 6 (repeated respiratory infections), and symptom 7 (EKG axis more than 110 degrees), ($PV+ = .7559$). The program used to compute the predictive value positive for combinations of three symptoms is given as follows:

```
1 REM SAVED AS 'PV.BAS'
5 REM PROGRAM TO COMPUTE PREDICTIVE VALUE POSITIVE FOR ANY TRIPLETS OF SYMPTOMS
6 REM FOR ATRIAL SEPTAL DEFECT WITHOUT PULMONARY STENOSIS OR PULMONARY HYPERTENSION
10 DIM Y(7), X(7,7), TERM(7)
30 DATA .155,.126,.084,.020,.098,.391,.126
31 DATA .49,.50,.55,.45,.10,.70,.60
32 DATA .50,.50,.05,.45,.00,.15,.10
33 DATA .01,.02,.25,.01,.20,.01,.30
34 DATA .10,.50,.90,.95,.70,.30,.70
35 DATA .05,.02,.05,.10,.01,.01,.10
36 DATA .05,.40,.10,.10,.05,.15,.20
37 DATA .01,.70,.95,.95,.40,.30,.70
50 FOR I=1 TO 7: READ Y: Y(I)=Y: NEXT I
60 FOR J=1 TO 7: FOR I=1 TO 7: READ X: X(I,J)=X: NEXT I,J
65 REM Y(I) = UNCONDITIONAL PROBABILITY OF ITH DIAGNOSIS
70 REM X(I,J) = CONDITIONAL PROBABILITY OF JTH SYMPTOM FOR ITH DIAGNOSIS
80 FOR J=1 TO 5: FOR K=J+1 TO 6: FOR L=K+1 TO 7
85 SUM=0!
90 FOR I = 1 TO 7
92 TERM(I) = X(I,J)*X(I,K)*X(I,L)*Y(I)
95 SUM=SUM+TERM(I): NEXT I
110 PV=TERM(2)/SUM
120 LPRINT USING "# # # #.####";J,K,L,PV
130 NEXT L,K,J
```

3.42 We wish to compute $Pr(\text{test}+ \mid \text{died})$. We use Bayes' theorem to solve this problem as follows:

$$Pr(\text{test}+ \mid \text{dead}) = \frac{Pr(\text{dead} \mid \text{test}+)Pr(\text{test}+)}{Pr(\text{dead})}$$

We have from Table 3.5 that $Pr(\text{dead} \mid \text{test}+) = 13.2/100 = .132$. Furthermore, we are given that $Pr(\text{test}+) = Pr(\text{exercise-test heart rate} > 127) = .20$. To compute $Pr(\text{dead})$, we use the relation

$Pr(\text{dead}) =$

$\quad Pr(\text{dead} \mid \text{heart rate} \le 105)Pr(\text{heart rate} \le 105)$

$\quad + \ldots + Pr(\text{dead} \mid \text{heart rate} > 127) \, Pr(\text{heart rate} > 127)$

$\quad = .091(.20) + .087(.30) + .116(.30) + .132(.20) = .1055$

Therefore, we have that

$$Pr(\text{test}+ \mid \text{dead}) = \frac{.132(.20)}{.1055} = .250$$

Thus, there is a 25% probability that people who die of coronary disease over a 20-year follow-up period will have a positive test at baseline. This is the *sensitivity* of the test.

3.43 We wish to compute $Pr(\text{test}+ \mid \text{alive})$. From Bayes' theorem, we have

$$Pr(\text{test}+ \mid \text{alive}) = \frac{Pr(\text{alive} \mid \text{test}+)Pr(\text{test}+)}{Pr(\text{alive})}$$

From Problem 3.42, we note that $Pr(\text{alive} \mid \text{test}+) = 1 - Pr(\text{dead} \mid \text{test}+) = 1 - .132 = .868$. Furthermore, $Pr(\text{test}+) = .20$, $Pr(\text{alive}) = 1 - Pr(\text{dead}) = 1 - .1055 = .8945$. Thus, we have

$$Pr(\text{test}+ \mid \text{alive}) = \frac{.868(.20)}{.8945} = .194$$

Thus, there is a 19.4% probability that people who do not die of coronary disease over a 20-year follow-up period will have a positive test at baseline. This probability = $1 - $ specificity.

3.44 We want $Pr(\text{death} \mid \text{negative test}) = Pr(\text{death} \mid$ heart rate $\leq 127)$. This is given by

$$\frac{Pr(\text{death} \cap \text{heart rate} \leq 127)}{Pr(\text{heart rate} \leq 127)}$$

$$= \frac{\begin{array}{c}Pr(\text{death} \cap \leq 105) \\ + \, Pr(\text{death} \cap 106\text{--}115) + Pr(\text{death} \cap 116\text{--}127)\end{array}}{.20 + .30 + .30}$$

$$= \frac{\begin{array}{c}[Pr(\text{death} \mid \leq 105)Pr(\leq 105) \\ + \, Pr(\text{death} \mid 106\text{--}115)Pr(106\text{--}115) \\ + \, Pr(\text{death} \mid 116\text{--}127)Pr(116\text{--}127)]\end{array}}{.80}$$

$$= \frac{.0791}{.80} = .099$$

Thus, there is a 9.9% probability of death over 20 years among men with a negative test. This is $1 - predictive$ *value negative.* Men who have a negative test, but who subsequently die of coronary heart disease are referred to as *false negatives.* The false-negative rate = 9.9%.

3.45 The sensitivity = the probability that the exercise test will be positive given that a patient has a positive angiogram.

3.46 The specificity = the probability that the exercise test will be negative given that a patient has a negative angiogram.

3.47 We use Bayes' theorem here. Let A = exercise-test positive, B = patient has disease. We wish to compute

$$Pr(B \mid A) = \frac{Pr(A \mid B)Pr(B)}{Pr(A \mid B)\, Pr(B) + Pr(A \mid \bar{B})Pr(\bar{B})}$$

where $Pr(A \mid B) = .79$, $Pr(A \mid \bar{B}) = 1 - Pr(\bar{A} \mid \bar{B}) = 1 - .68 = .32$, $Pr(B) = .20$. Therefore, we have

$$Pr(B \mid A) = \frac{.79 \times .20}{.79 \times .20 + .32 \times .80}$$

$$= \frac{.158}{.414} = .382$$

Therefore there is a 38.2% chance that the patient has disease given that she has a positive exercise test.

3.48 We wish to compute $Pr(B \mid \bar{A})$. We have from Bayes' theorem that

$$Pr(B \mid \bar{A}) = \frac{Pr(\bar{A} \mid B)Pr(B)}{Pr(\bar{A} \mid B)Pr(B) + Pr(\bar{A} \mid \bar{B})Pr(\bar{B})}$$

$$= \frac{(1 - .79)(.20)}{(1 - .79)(.20) + .68(.80)}$$

$$= \frac{.042}{.042 + .544} = \frac{.042}{.586} = .072$$

Thus there is only a 7.2% chance that the patient has disease if her exercise test is negative.

3.49 The sensitivity = $138/156 = .885$.

3.50 The specificity = $16/17 = .941$.

3.51 The predictive value positive = $138/139 = .993$.

3.52 The predictive value negative = $16/34 = .471$.

3.53 We use the formula

$$PV+ =$$
$$\frac{\text{prevalence} \times \text{sensitivity}}{\text{prevalence} \times \text{sensitivity} + (1 - \text{prevalence})(1 - \text{specificity})}$$

$$= \frac{.8(.885)}{.8(.885) + .2(1 - .941)}$$

$$= \frac{.708}{.719} = .984$$

3.54 Since both twins must have the same outcome, the probability that both twins are affected = $1/4$, the probability that one twin is affected = 0, the probability that neither twin is affected = $3/4$. The outcomes are dependent events because

$$\frac{1}{4} = Pr(A_1 \cap A_2) = Pr(A_1)$$

$$= Pr(A_2) \neq Pr(A_1) \times Pr(A_2) = \frac{1}{16}$$

3.55 The outcomes for the two DZ twins are independent events. We have Pr(both affected) $= (1/4)^2 = 1/16$, Pr(one affected) $= 2(1/4)(3/4) = 6/16 = 3/8$, Pr(neither affected) $= (3/4)^2 = 9/16$.

3.56 We use the total probability rule. Let B_2, B_1, B_0 be the events that 2, 1, and zero twins are affected. We have

$$Pr(B_2) = Pr(B_2 \mid MZ)Pr(MZ) + Pr(B_2 \mid DZ)Pr(DZ)$$
$$= (1/4)(1/3) + (1/16)(2/3) = 1/12 + 2/48$$
$$= 6/48 = 1/8$$

Similarly,

$$Pr(B_1) = 0(1/3) + (3/8)(2/3) = 1/4$$
$$Pr(B_0) = (3/4)(1/3) + (9/16)(2/3) - 1/4 + 3/8 = 5/8$$

3.57 We use Bayes' theorem. We wish to compute $Pr(MZ \mid B_2)$. We have

$$Pr(MZ \mid B_2) = \frac{Pr(B_2 \mid MZ)Pr(MZ)}{Pr(B_2 \mid MZ)Pr(MZ) + Pr(B_2 \mid DZ)Pr(DZ)}$$

From Problems 3.54 and 3.55, we have $Pr(B_2 \mid MZ) = 1/4$, $Pr(B_2 \mid DZ) = 1/16$. Also from Problem 3.56, $Pr(MZ) = 1/3$, $Pr(DZ) = 2/3$. Thus

$$Pr(MZ \mid B_2) = \frac{(1/4)(1/3)}{(1/4)(1/3) + (1/16)(2/3)}$$
$$= \frac{1/12}{1/12 + 1/24} = 2/3$$

3.58 An incidence rate of 48.9 strokes per 1000 person-years among 70–79-year-olds with AF means that if a group of 70–79-year-old people with AF are followed for 1 year, then on average 48.9 out of every 1000 such people will develop stroke over the next year.

3.59 The relative risk (RR) = incidence rate of stroke among people with AF/incidence rate of stroke among people without AF. Thus,

$$RR_{60-69} = 21.2/4.5 = 4.7$$
$$RR_{70-79} = 48.9/9.0 = 5.4$$
$$RR_{80-89} = 71.4/14.3 = 5.0$$

Thus, the relative risk is approximately 5 for each age group. This means that for persons in a given age group, those with AF are about 5 times as likely to develop a new case of stroke as those without AF.

3.60 We use the total probability rule. First, we find the incidence rate of stroke within a given age group using the formula

$$\text{Incidence rate of stroke} =$$
prevalence of AF $\times$ incidence rate of stroke for people with AF $+$ (1 − prevalence of AF) $\times$ incidence rate of stroke for people without AF

Thus we have

Incidence rate$_{60-69}$ $= .018(21.2/1000) + .982(4.5/1000)$
$= 4.80$ per 1000

Incidence rate$_{70-79}$ $= .047(48.9/1000) + .953(9.0/1000)$
$= 10.88$ per 1000

Incidence rate$_{80-89}$ $= .102(71.4/1000) + .898(14.3/1000)$
$= 20.12$ per 1000

The incidence rate of stroke roughly doubles in successive decades of life. We again use the total probability rule to obtain the overall incidence as follows:

$$\text{Incidence rate} = (200 \times \text{incidence rate}_{60-69}$$
$$+ 200 \times \text{incidence rate}_{70-79}$$
$$+ 100 \times \text{incidence rate}_{80-89})/500$$
$$= \frac{200(4.80/1000) + 200(10.88/1000) + 100(20.12/1000)}{500}$$
$$= \frac{5.148}{500} = 0.0103 \approx 10.3 \text{ per } 1000$$

3.61 If n subjects are enrolled, then the expected number of strokes over 1 year $= nI$, where I is the 1-year incidence rate. Thus, we can solve for n and obtain $n = 50/.0103 = 4856.6$. Thus we need to enroll 4857 subjects to ensure that the expected number of strokes over 1 year $= 50$. An alternative design would be to enroll fewer subjects but follow the subjects over a longer period of time.

REFERENCES

[1] Tenkanen, L., Teppo, L., & Hakulinen, T. (1987). Smoking and cardiac symptoms as predictors of lung cancer. *Journal of Chronic Disease, 40*(12), 1121–1128.

[2] Butler, W. J., Ostrander, L. D., Jr., Carman, W. J., & Lamphiear, D. E. (1982). Diabetes mellitus in Tecumseh, Michigan: Prevalence, incidence and associated conditions. *American Journal of Epidemiology, 116*(6), 971–980.

[3] Doll, R., Muir, C., & Waterhouse, J. (Eds.). (1970). *Cancer incidence in five continents II.* Berlin: Springer-Verlag.

[4] Warner, H., Toronto, A., Veasey, L. G., & Stephenson, R. (1961). A mathematical approach to medical diagnosis. *JAMA, 177*(3), 177–183.

[5] Slattery, M. L., & Jacobs, D. R., Jr. (1988). Physical fitness and cardiovascular disease mortality: The U.S. railroad study. *American Journal of Epidemiology, 127*(3). 571–580.

[6] Wolf, P. A., Abbott, R. D., & Kannel, W. B. (1987). Atrial fibrillation: A major contributor to stroke in the elderly. The Framingham Study. *Archives of Internal Medicine, 147*(9), 1561–1564.

CHAPTER FOUR

DISCRETE PROBABILITY DISTRIBUTIONS

REVIEW OF KEY CONCEPTS

SECTION 4.1 **Random Variable**

A **random variable** X is a numerically valued quantity that takes on specific values with different probabilities. The relationship between the values and their associated probabilities is called a **probability mass function.**

Example: Color plates Suppose we administer the color-blindness test 4 times to an individual subject. We want to determine the probability of a specified number of correct selections (k) out of 4 trials if the subject is color blind. The number of correct selections X is a random variable that takes on the values 0, 1, ... , 4 with probabilities $p_0, p_1, ... , p_4$. We can use the *binomial distribution* to obtain the probability mass function (or probability distribution) of the number of correct selections in this case. ∎

SECTION 4.2 **Combinations, Permutations, and Factorial**

Combinations

A *combination* is defined as the number of ways of selecting k objects out of n, where the order of selection does not matter. A combination is denoted by $\binom{n}{k}$ or $_nC_k$ and refers to the number of combinations of n things taken k at a time.
It can be shown that

$$_nC_k = \binom{n}{k} = \frac{n(n-1) \times ... \times (n-k+1)}{k(k-1) \times ... \times 1}$$

Permutations

A *permutation* is defined as the number of ways of selecting k objects out of n, where the order of selection matters. It is denoted by $_nP_k$ and refers to the number of permutations of n things taken k at a time. It can be shown that

$$_nP_k = n \times (n-1) \times ... \times (n-k+1)$$

Factorial

There is a special symbol that is used to denote $_kP_k = k(k - 1) \times \ldots \times 2 \times 1$ called $k! = k$ *factorial*; e.g., $4! = 4 \times 3 \times 2 \times 1 = 24$; $6! = 6 \times 5 \times 4 \times 3 \times 2 \times 1 = 720$. By convention, $0! \equiv 1$. A combination can also be written in terms of factorials as follows:

$$_nC_k = \frac{n!}{k!(n - k)!}$$

SECTION 4.3   Binomial Probability Distribution

The binomial probability distribution is given by

$$Pr(X = k) = \binom{n}{k} p^k q^{n-k} \qquad k = 0, 1, \ldots, n$$

We can use this distribution to model the probability distribution of the number of successes among n trials where each trial has probability of success $= p$, $q = 1 - p$ and the trials are independent.

We can write the binomial distribution in terms of factorials as follows:

$$Pr(X = k) = \binom{n}{k} p^k q^{n-k} = \frac{n(n - 1) \times \ldots \times (n - k + 1)}{k!} p^k q^{n-k}$$

$$= \frac{n!}{k!(n-k)!} p^k q^{n-k}$$

For example, what is the probability of obtaining 2 successes in 4 trials in the color-plate example? In this case, $n = 4$, $p = 1/4$, and

$$Pr(X = 2) = \binom{4}{2}\left(\frac{1}{4}\right)^2\left(\frac{3}{4}\right)^2 = 6\left(\frac{1}{4}\right)^2\left(\frac{3}{4}\right)^2 = .211$$

SECTION 4.4   Methods for Using the Binomial Distribution

(1) Calculate individual probabilities (brute-force method).
(2) Use binomial tables (as given in Table 1 in the Appendix of the text for selected values of n and k).
(3) Use recursion rule for binomial probabilities (see section 4.5, below).
(4) Use Poisson approximation to binomial (explanation on p. 33). Used if $n \geq 100$, $p < .01$.
(5) Use normal approximation to binomial (explanation on p. 47). Used if $npq \geq 5$.

SECTION 4.5   Recursion Rule for Binomial Probabilities

It is tedious to compute binomial probabilities for each k. Instead, we can use a recursion rule to relate successive binomial probabilities for the same n and p.

$$Pr(X = k + 1) = \left(\frac{n - k}{k + 1}\right) \times \left(\frac{p}{q}\right) \times Pr(X = k)$$

For example, if we consider the color-blindness test, suppose we wish to generate all the binomial probabilities for $n = 4, p = .25$; i.e.,

$$Pr(X = 0), \ldots, Pr(X = 4)$$

$$Pr(X = 0) = \binom{4}{0}(.25)^0(.75)^4 = \left(\frac{3}{4}\right)^4 = \frac{81}{256} = .316$$

$$Pr(X = 1) = \frac{4}{1} \times \frac{1}{3} \times \frac{81}{256} = \frac{4}{3}Pr(0) = .422$$

$$Pr(X = 2) = \frac{3}{2} \times \frac{1}{3} \times Pr(1) = \frac{1}{2}Pr(1) = .211$$

$$Pr(X = 3) = \frac{2}{3} \times \frac{1}{3} \times Pr(2) = \frac{2}{9}Pr(2) = .047$$

$$Pr(X = 4) = \frac{1}{4} \times \frac{1}{3} \times Pr(3) = \frac{1}{12}Pr(3) = .004$$

Note that in this case, the probabilities could be obtained from the binomial tables (see Table 1 in the Appendix of the text).

n	k	p = 0.25
4	0	.3164
	1	.4219
	2	.2109
	3	.0469
	4	.0039

SECTION 4.6   Expected Value of the Binomial Distribution

In much the same way that we use the sample mean as a measure of location for a sample, we use the "population mean" or expected value (denoted by $E(X)$ or μ) as a measure of location for a random variable.

For a general random variable,

$$E(X) = \sum_{i=1}^{m} x_i Pr(X = x_i) = \mu$$

that is, the expected value is a weighted average of the m possible values, weighted by their respective probabilities.

For the binomial distribution, the general formula reduces to

$$E(X) = np$$

For example, suppose we give the color-blindness test to 100 color-blind people.

$$n = 100, p = 1/4, \quad E(X) = 25$$

We would expect that 25 out of 100 color-blind people will pass the test.

SECTION 4.7   Variance of the Binomial Distribution

We can also describe the spread of a random variable by its variance that is denoted by $Var(X)$ and is defined by

$$\sigma^2 = Var(X) = \sum_{i=1}^{m}(x_i - \mu)^2 Pr(X = x_i)$$

as opposed to the sample variance s^2. It gives us a sense of the spread of the possible values relative to the mean (or expected value). The standard deviation of a random variable is defined by

$$= \sigma = \sqrt{\sigma^2} = \sqrt{Var(X)}$$

For the binomial distribution,

$$Var(X) = npq$$

$$\text{If } n = 100, p = .01, Var(X) = 0.99 \approx 1$$

$$n = 100, p = .99, Var(X) = 0.99 \approx 1$$

$$n = 100, p = \frac{1}{4}, Var(X) = 18.75$$

$$n = 100, p = \frac{1}{2}, Var(X) = 25$$

For a given n, the binomial distribution is most variable for $p = 1/2$ and least variable as p approaches 0 or 1.

SECTION 4.8   Poisson Distribution

Suppose a hospital has observed 1 case of a rare cancer every 2 years over a long time period. It suddenly gets 3 cases in one year. How unlikely is this, or more specifically, what is the probability of obtaining at least 3 cases in 1 year?

The hospital knows that it serves a large number of people but does not know the exact number. The Poisson distribution can be used to model the probability distribution of the number of cases over a time period of T years.

Assumptions

(1) $Pr(1 \text{ new case in time } \Delta t) \approx \lambda\Delta t$, where λ = number of cases per unit time, and Δt is a small increment of time.

(2) Stationarity; i.e., incidence of new cases per unit time is the same over the entire time period T.

(3) Independence; probabilities of events over different time intervals are independent random variables.

Under these assumptions,

$$Pr(k \text{ cases in time } T) = \frac{(\lambda T)^k e^{-\lambda T}}{k!} \quad k = 0, 1, \ldots,$$

In our case, $T = 1, \lambda = 1/2$, we want $Pr(X \geq 3)$. We have

$$Pr(0) = e^{-1/2} = .607$$

$$Pr(1) = .5e^{-1/2} = .303$$

$$Pr(2) = \frac{(.5)^2}{2} e^{-1/2} = .076$$

$$Pr(3+) = 1 - (.607 + .303 + .076) = .014$$

It is unlikely that 3 or more cases will occur in 1 year.

SECTION 4.9 Recursion Rule for Poisson Probabilities

We can use the Poisson tables (Table 2, Appendix, text) to obtain probabilities under the Poisson distribution. The distribution is quantified by a single parameter $\mu = \lambda T$. For values of μ not in the table, we can use the following recursion rule:

$$P(X = k + 1) = \frac{\lambda T}{k + 1} P(X = k)$$

For example, let us use the recursion rule to compute the probability of observing 3+ events in 1 year if the incidence rate is 1 case every 2 years. We have $\lambda = 0.5$, $T = 1$, $\lambda T = 0.5$.

$$Pr(0) = e^{-.5} = .607$$

$$Pr(1) = \frac{1}{2} \times .607 = .303$$

$$Pr(2) = \frac{.5}{2} \times .303 = .076$$

$$Pr(3+) = 1 - (.607 + .303 + .076) = .014$$

SECTION 4.10 Expected Value and Variance of the Poisson Distribution

$$E(X) = \lambda T = Var(X) = \lambda T = \mu$$

A good way to identify when a distribution is likely to be Poisson is to compare the sample mean and the sample variance. If these quantities are approximately the same, then a Poisson distribution will often fit well.

SECTION 4.11 Poisson Approximation to the Binomial Distribution

If $n \geq 100$, $p \leq .01$, we can approximate a binomial distribution with parameters n and p by a Poisson distribution with parameter $\mu = np$. The reason for using this approximation is that it is easier to use the Poisson than the binomial distribution, especially when n is large. Specifically, for a given n, p, and k, we approximate the binomial probability

$$Pr(X = k) = \binom{n}{k} p^k q^{nk} \quad \text{by} \quad Pr(X = k) = \frac{e^{np}(np)^k}{k!}$$

PROBLEMS

4.1 Evaluate the number of ways of selecting 4 objects out of 10 if the order of selection matters. What term is used to denote this quantity?

4.2 Evaluate the number of ways of selecting 4 objects out of 10 if the order of selection does *not* matter. What term is used to denote this quantity?

Suppose that the probability that a person will develop hypertension over a lifetime is 20%.

4.3 What is the probability distribution of the number of hypertensives over a lifetime among 20 students graduating from the same high school class?

4.4 What is the probability that exactly 4 people out of 50 aged 60–64 will die after receiving the flu vaccine if the probability that 1 person will die is .028?

4.5 What is the probability that at least 4 people will die after receiving the vaccine?

4.6 What is the expected number of deaths following the flu vaccine?

4.7 What is the standard deviation of the number of deaths following the flu vaccine?

Health-Services Administration
The in-hospital mortality rate for 16 clinical conditions at 981 hospitals was recently reported [1]. It was reported that in-hospital mortality was 10.5% for coronary-bypass surgery and 5.0% for total hip replacement. Suppose an institution changes from an academic institution to a private for-profit institution. They find that after the change, of the first 20 patients receiving coronary-bypass surgery, 5 die, while of 20 patients receiving total hip replacement, 4 die.

4.8 What is the probability that of 20 patients receiving coronary-bypass surgery, exactly 5 will die in-hospital, if this hospital is representative of the total pool of 981 hospitals?

4.9 What is the probability of at least 5 deaths among the coronary-bypass patients?

4.10 What is the probability of no more than 5 deaths among the coronary-bypass patients?

4.11 What is the probability that exactly 4 will die among the hip-replacement patients?

4.12 What is the probability that at least 4 will die among the hip-replacement patients?

4.13 What is the probability of 4 or fewer deaths among the hip-replacement patients?

4.14 Can you draw any conclusions based on the results

in Problems 4.8–4.13 regarding any effects of the change in hospital administration on in-hospital mortality rates?

Cardiovascular Disease
The rate of myocardial infarction (MI) in 50–59-year-old, disease-free women is approximately 2 per 1000 per year or 10 per 1000 over 5 years. Suppose that 3 MI's are reported over 5 years among 1000 postmenopausal women initially disease free who have been taking postmenopausal hormones.

4.15 Use the binomial distribution to see if this experience represents an unusually small number of events based on the overall rate.

4.16 Answer Problem 4.15 using the Poisson approximation to the binomial distribution.

4.17 Compare your answers in Problems 4.15 and 4.16.

Cardiovascular Disease
4.18 A new hypothesis in the etiology of heart disease is that aspirin intake of 325 mg per day reduces subsequent cardiovascular mortality in men with a prior heart attack. Suppose that in a pilot study of 50 men who received 1 tablet per day (325 mg), only 2 die over a 3-year period from cardiovascular disease. How likely is it that not more than 2 men will die if the underlying 3-year mortality rate is 10% in such men?

Pediatrics
A hospital administrator wants to construct a special-care nursery for low-birthweight infants (≤2500 g) and wants to have some idea as to the number of beds she should allocate to the nursery. She is willing to assume that the recovery period of each baby is exactly 4 days and thus is interested in the expected number of premature births over the period.

4.19 If the number of premature births in any 4-day period is binomially distributed with parameters $n = 25$ and $p = .1$, then find the probability of 0, 1, 2, ... , 7 premature births over this period.

4.20 The administrator wishes to allocate x beds where the probability of having more than x premature births over a 4-day period is less than 5%. What is the smallest value of x that satisfies this criterion?

4.21 Answer Problem 4.20 for 1%.

Cancer
The incidence rate of malignant melanoma is suspected to be increasing over time. To document this increase, a questionnaire was mailed to 100,000 U.S. nurses in 1976

and 1978, asking about any current or previous tumors. Thirty new cases of malignant melanoma were found to have developed over the 2-year period among women with no previous cancers in 1976.

4.22 If the annual incidence rate from previous cancer-registry data is 10 per 100,000, then what is the expected number of new cases over 2 years?

4.23 Are the preceding results consistent or inconsistent with the cancer-registry data? Specifically, what is the probability of observing at least 30 new cases over a 2-year period if the cancer-registry incidence rate is correct?

Accident Epidemiology

Suppose the annual number of traffic fatalities at a given intersection follows a Poisson distribution with parameter $\mu = 10$.

4.24 What is the probability of observing exactly 10 traffic fatalities in 1992?

4.25 What is the probability of observing exactly 25 traffic fatalities over the 2-year period from January 1, 1990, to December 31, 1991?

4.26 Suppose that the traffic intersection is redesigned with better lighting, and 12 traffic fatalities are observed over the next 2 years. Is this rate a meaningful improvement over the previous rate of traffic fatalities?

Pulmonary Disease, Environmental Health

Suppose the number of people seen for violent asthma attacks in the emergency ward of a hospital over a 1-day period is usually Poisson distributed with parameter $\lambda = 1.5$.

4.27 What is the probability of observing 5 or more cases over a 2-day period?

On a particular 2-day period, the air-pollution levels increase dramatically and the distribution of attacks over a 1-day period is now estimated to be Poisson distributed with parameter $\lambda = 3$.

4.28 Answer Problem 4.27 under these assumptions.

4.29 If 10 days out of every year are high-pollution days, then what is the expected number of asthma cases seen in the emergency ward over a 1-year period?

Demography

The data set in Table 4.1 is an example of current life-table data for persons living in the United States in 1986 [2]. P_x represents the probability of living for the next year given that one is currently x years old. The entries in the table for age x (referred to as l) are obtained from the formula

$$l_0 = 100,000 \qquad l_x = l_0 \times P_0 \times P_1 \times \ldots \times P_{x-1},$$
$$x = 1, 2, \ldots, 100$$

Assume that the *current* death rates hold not only for the year 1986 but for all the subsequent years as well.

4.30 What is the probability of living to age 65 given that one is 21 in 1986? Does this vary by sex and/or race (white vs. black)?

4.31 What is the probability of dying exactly between the ages of 56 and 57 given that a black female is 21 in 1986?

4.32 Suppose 100 white males of age 21 in 1986 live in a particular town and that 5 of them die before reaching the age of 30. Is this event unusual? Specifically, how likely are 5 or more men to die before reaching the age of 30?

4.33 Answer Problem 4.32 for a black male.

4.34 Suppose we are not willing to assume that the P_x's remain constant in years subsequent to 1986. Can Problems 4.30–4.33 still be answered? If not, what additional information is needed?

Pulmonary Disease

Each year approximately 4% of current smokers attempt to quit smoking, and 50% of those who try to quit are successful in the sense that they abstain from smoking for at least 1 year from the date they quit.

4.35 What is the probability that a current smoker will quit for at least 1 year?

4.36 What is the probability that among 100 current smokers, at least 5 will quit smoking for at least 1 year?

An educational program was conducted among smokers who attempt to quit to maximize the likelihood that such individuals would continue to abstain for the long term.

4.37 Suppose that of 20 people who enter the program when they first stop smoking, 15 still abstain from smoking 1 year later. Can the program be considered successful?

Cancer, Epidemiology

A frequent design for biomedical investigations is the case-control study. A group of **cases** is identified on the basis of having a particular disease (e.g., lung-cancer patients in a cancer registry), and a group of **controls** is chosen (e.g., patients in the same registry with cancer of the esophagus) such that every case is *matched* with 1 or more controls. That is, the case and control(s) are matched in every sense except that the controls do not have the disease trait. We can then look at whether or not some other factor (such as smoking) is associated with the disease trait. Obtaining "exact" matches is usually

TABLE 4.1 Number of survivors at single years of age, out of 100,000 born alive, by race and sex: United States, 1986

Age	All races			White			All other					
							Total			Black		
	Both sexes	Male	Female	Both sexes	Male	Female	Both sexes	Male	Female	Both sexes	Male	Female
0	100,000	100,000	100,000	100,000	100,000	100,000	100,000	100,000	100,000	100,000	100,000	100,000
1	98,964	98,845	99,090	99,106	98,998	99,220	98,426	98,262	98,597	98,190	97,996	98,391
2	98,892	98,764	99,028	99,040	98,923	99,164	98,332	98,158	98,513	98,085	97,882	98,296
3	98,838	98,704	98,980	98,991	98,869	99,121	98,256	98,075	98,444	98,000	97,790	98,218
4	98,796	98,658	98,942	98,953	98,828	99,087	98,194	98,007	98,388	97,932	97,716	98,155
5	98,762	98,620	98,912	98,923	98,794	99,060	98,143	97,951	98,342	97,876	97,655	98,104
6	98,733	98,587	98,887	98,897	98,765	99,038	98,100	97,904	98,304	97,830	97,605	98,062
7	98,707	98,557	98,866	98,874	98,738	99,019	98,064	97,863	98,272	97,792	97,563	98,028
8	98,684	98,530	98,847	98,853	98,712	99,002	98,033	97,828	98,246	97,760	97,526	97,999
9	98,663	98,505	98,830	98,834	98,689	98,987	98,006	97,797	98,223	97,732	97,494	97,975
10	98,645	98,484	98,815	98,817	98,669	98,973	97,982	97,769	98,203	97,706	97,464	97,954
11	98,628	98,464	98,801	98,802	98,651	98,960	97,959	97,742	98,184	97,681	97,435	97,934
12	98,610	98,443	98,786	98,785	98,632	98,946	97,935	97,713	98,166	97,655	97,404	97,914
13	98,587	98,415	98,768	98,763	98,606	98,929	97,907	97,677	98,146	97,625	97,365	97,893
14	98,554	98,372	98,745	98,730	98,564	98,906	97,871	97,629	98,123	97,587	97,314	97,869
15	98,507	98,309	98,716	98,683	98,501	98,876	97,825	97,565	98,095	97,538	97,247	97,840
16	98,445	98,223	98,678	98,620	98,414	98,838	97,767	97,484	98,061	97,477	97,161	97,805
17	98,369	98,116	98,633	98,542	98,306	98,792	97,696	97,384	98,021	97,403	97,056	97,763
18	98,280	97,991	98,583	98,452	98,180	98,741	97,612	97,264	97,975	97,315	96,930	97,715
19	98,183	97,852	98,530	98,355	98,041	98,688	97,515	97,123	97,923	97,214	96,781	97,662
20	98,081	97,702	98,477	98,254	97,894	98,635	97,405	96,959	97,867	97,098	96,607	97,603
21	97,974	97,542	98,424	98,150	97,740	98,583	97,281	96,771	97,805	96,967	96,407	97,538
22	97,861	97,372	98,370	98,043	97,578	98,532	97,143	96,560	97,737	96,821	96,181	97,467
23	97,745	97,196	98,316	97,935	97,412	98,482	96,993	96,331	97,664	96,661	95,933	97,390
24	97,628	97,018	98,261	97,827	97,247	98,432	96,834	96,089	97,586	96,490	95,670	97,306
25	97,512	96,843	98,204	97,720	97,086	98,381	96,669	95,839	97,502	96,311	95,395	97,216
26	97,397	96,672	98,147	97,616	96,931	98,330	96,499	95,582	97,413	96,124	95,110	97,119
27	97,283	96,504	98,088	97,514	96,780	98,278	96,322	95,317	97,318	95,927	94,814	97,015
28	97,168	96,336	98,027	97,412	96,632	98,225	96,137	95,043	97,217	95,719	94,503	96,902
29	97,050	96,164	97,964	97,309	96,481	98,171	95,942	94,756	97,107	95,497	94,174	96,778
30	96,927	95,986	97,897	97,202	96,325	98,115	95,735	94,455	96,988	95,258	93,823	96,642
31	96,798	95,800	97,826	97,090	96,162	98,056	95,515	94,139	96,858	95,001	93,450	96,492
32	96,663	95,606	97,751	96,973	95,993	97,994	95,283	93,807	96,718	94,727	93,054	96,329
33	96,522	95,405	97,672	96,852	95,818	97,928	95,038	93,459	96,568	94,436	92,636	96,154
34	96,377	95,199	97,588	96,727	95,639	97,859	94,781	93,093	96,411	94,131	92,196	95,970
35	96,227	94,988	97,500	96,598	95,457	97,786	94,512	92,708	96,247	93,812	91,735	95,779
36	96,072	94,771	97,407	96,465	95,271	97,708	94,230	92,303	96,076	93,480	91,252	95,581
37	95,910	94,546	97,308	96,326	95,079	97,625	93,934	91,877	95,896	93,132	90,745	95,373
38	95,739	94,311	97,202	96,180	94,878	97,534	93,622	91,429	95,705	92,765	90,212	95,152
39	95,557	94,065	97,086	96,024	94,667	97,434	93,291	90,958	95,498	92,373	89,649	94,911
40	95,363	93,805	96,958	95,856	94,443	97,323	92,938	90,462	95,271	91,952	89,052	94,646
41	95,153	93,528	96,816	95,674	94,203	97,199	92,560	89,939	95,022	91,500	88,419	94,353
42	94,927	93,232	96,659	95,476	93,945	97,061	92,156	89,386	94,751	91,015	87,749	94,033
43	94,683	92,916	96,487	95,261	93,668	96,909	91,727	88,806	94,458	90,501	87,046	93,687
44	94,421	92,579	96,300	95,029	93,371	96,743	91,276	88,201	94,145	89,962	86,315	93,318
45	94,139	92,218	96,097	94,778	93,051	96,563	90,803	87,572	93,812	89,400	85,559	92,928
46	93,836	91,831	95,878	94,507	92,705	96,367	90,307	86,919	93,458	88,816	84,781	92,518
47	93,508	91,413	95,639	94,212	92,329	96,153	89,784	86,237	93,081	88,206	83,974	92,082
48	93,151	90,960	95,377	93,888	91,919	95,917	89,226	85,514	92,673	87,559	83,125	91,614
49	92,758	90,464	95,087	93,531	91,469	95,656	88,620	84,733	92,226	86,862	82,217	91,104
50	92,325	89,919	94,766	93,137	90,973	95,365	87,957	83,881	91,733	86,104	81,237	90,545
51	91,848	89,320	94,409	92,701	90,427	95,042	87,230	82,951	91,189	85,280	80,178	89,931
52	91,324	88,665	94,016	92,221	89,827	94,684	86,440	81,945	90,594	84,389	79,042	89,261
53	90,752	87,949	93,586	91,694	89,167	94,291	85,591	80,869	89,951	83,437	77,836	88,540
54	90,130	87,169	93,122	91,117	88,441	93,864	84,693	79,735	89,268	82,433	76,575	87,774
55	89,458	86,322	92,623	90,488	87,644	93,401	83,749	78,549	88,548	81,382	75,266	86,968
56	88,733	85,405	92,088	89,803	86,773	92,901	82,763	77,316	87,794	80,287	73,915	86,124
57	87,951	84,413	91,513	89,058	85,823	92,360	81,729	76,030	87,000	79,142	72,514	85,236
58	87,103	83,339	90,888	88,247	84,788	91,770	80,630	74,674	86,148	77,930	71,046	84,284
59	86,180	82,173	90,203	87,361	83,661	91,121	79,444	73,227	85,215	76,626	69,487	83,244
60	85,173	80,908	89,449	86,393	82,435	90,406	78,156	71,675	84,185	75,215	67,822	82,097
61	84,077	79,539	88,619	85,338	81,105	89,619	76,757	70,011	83,047	73,687	66,042	80,834
62	82,891	78,065	87,712	84,193	79,669	88,758	75,254	68,243	81,808	72,051	64,157	79,461
63	81,618	76,492	86,731	82,961	78,131	87,823	73,665	66,388	80,485	70,328	62,189	77,999
64	80,264	74,827	85,681	81,645	76,497	86,817	72,016	64,473	79,104	68,548	60,167	76,479
65	78,833	73,076	84,565	80,246	74,770	85,740	70,325	62,516	77,684	66,732	58,113	74,922
66	77,327	71,244	83,381	78,766	72,955	84,590	68,600	60,526	76,232	64,890	56,039	73,337
67	75,740	69,325	82,122	77,199	71,046	83,360	66,834	58,499	74,737	63,015	53,941	71,715
68	74,059	67,305	80,776	75,532	69,027	82,040	65,011	56,423	73,180	61,092	51,805	70,035
69	72,267	65,161	79,330	73,750	66,877	80,618	63,109	54,279	71,534	59,098	49,612	68,270
70	70,353	62,881	77,772	71,841	64,581	79,082	61,111	52,055	69,778	57,018	47,350	66,401
71	68,312	60,462	76,096	69,801	62,139	77,427	59,013	49,749	67,903	54,848	45,018	64,419
72	66,149	57,916	74,299	67,634	59,561	75,649	56,823	47,373	65,916	52,598	42,630	62,332
73	63,872	55,256	72,383	65,347	56,860	73,748	54,557	44,945	63,830	50,284	40,203	60,154
74	61,491	52,503	70,350	62,949	54,057	71,724	52,236	42,489	61,666	47,926	37,763	57,904
75	59,016	49,675	68,200	60,450	51,169	69,577	49,877	40,023	59,437	45,540	35,329	55,597
76	56,452	46,785	65,931	57,853	48,210	67,303	47,486	37,558	57,145	43,133	32,914	53,236
77	53,800	43,842	63,538	55,162	45,191	64,896	45,061	35,097	54,783	40,706	30,522	50,815
78	51,061	40,855	61,012	52,376	42,122	62,350	42,594	32,639	52,339	38,254	28,154	48,324
79	48,235	37,833	58,347	49,498	39,013	59,657	40,078	30,181	49,799	35,773	25,811	45,749
80	45,324	34,789	55,535	46,530	35,879	56,812	37,506	27,723	47,151	33,260	23,495	43,081
81	42,333	31,739	52,570	43,478	32,737	53,809	34,876	25,270	44,387	30,716	21,215	40,314
82	39,269	28,705	49,450	40,350	29,611	50,643	32,192	22,831	41,506	28,147	18,982	37,447
83	36,144	25,712	46,172	37,158	26,528	47,313	29,461	20,419	38,509	25,564	16,812	34,485
84	32,972	22,791	42,736	33,916	23,520	43,817	26,695	18,053	35,405	22,982	14,728	31,435
85	29,771	19,977	39,143	30,642	20,625	40,155	23,912	15,755	32,206	20,419	12,755	28,312

TABLE 4.2 Selection of controls for a case–control study

Cases			Controls			
Case number	Age	Sex	Control number	Age group	Sex	Frequency
1	36	M	01–15	21–30	M	15
2	50	F	16–21	21–30	F	6
3	24	F	22–27	31–40	M	6
4	22	M	28–45	31–40	F	18
5	35	M	46–48	41–50	M	3
			49–54	41–50	F	6
			55–66	51–60	M	12
			67–69	51–60	F	3
			70–78	61–70	M	9
			79–84	61–70	F	6

impossible, and several characteristics are selected to use for potential matches, such as age and sex. Suppose the group of cases and controls is as given in Table 4.2.

Suppose that the match is performed so that each control has the same age group and sex as its corresponding case. An example of a 1-to-1 match would be

Case number	1	2	3	4	5
Control number	24	51	18	14	22

An example of a 2-to-1 match would be

Case number	1	2	3	4	5
Control number	24, 26	51, 49	18, 21	14, 06	22, 27

Assume that the order within each group of matched controls in a many-to-one matching does not matter.

4.38 How many ways can 1-to-1 matches be assigned for age and sex?

4.39 Suppose the designers of the study get lazy and match only for age. How many ways can this matching be done?

4.40 If the groups were matched only on age, then what is the probability that the groups will be matched for sex as well if each match is equally likely?

4.41 Answer Problem 4.38 for 2-to-1 matches.

4.42 Answer Problem 4.39 for 2-to-1 matches.

4.43 Answer Problem 4.40 for 2-to-1 matches.

4.44 Answer Problem 4.38 for 3-to-1 matches.

4.45 Answer Problem 4.39 for 3-to-1 matches.

4.46 Answer Problem 4.40 for 3-to-1 matches.

Infectious Disease

An outbreak of acute gastroenteritis occurred at a nursing home in Baltimore, Maryland, in December 1980 [3]. A total of 46 out of 98 residents of the nursing home became ill. People living in the nursing home shared rooms: 13 rooms contained 2 occupants, 4 rooms contained 3 occupants, and 15 rooms contained 4 occupants. One question that arises is whether or not a geographical clustering of disease occurred for persons living in the same room.

4.47 If the binomial distribution holds, what is the probability distribution of the number of affected people in rooms with 2 occupants? That is, what is the probability of finding 0 affected people? 1 affected person? 2 affected people?

4.48 Answer Problem 4.47 for the probability distribution of the number of affected people in rooms with 3 occupants.

4.49 Answer Problem 4.47 for the probability distribution of the number of affected people in rooms with 4 occupants. A summary of the number of affected people and the total number of people in a room is given in Table 4.3.

4.50 One useful summary measure of geographical clustering is the number of rooms with 2 or more affected occupants. If the binomial distribution holds, what is the

TABLE 4.3 Number of affected people and total number of people in a room for an outbreak of acute gastroenteritis in a nursing home in Baltimore, Maryland

People in room	Total number of rooms	Number of rooms with				
		0 affected people	1 affected person	2 affected people	3 affected people	4 affected people
2	13	5	4	4	0	0
3	4	1	2	0	1	0
4	15	2	4	3	5	1

Source: Reprinted with permission from the *American Journal of Epidemiology, 116*(6), 940–948, 1982.

expected number of rooms with 2 or more affected occupants over the entire nursing home?

4.51 Compare the observed number of rooms with 2 or more affected occupants with the expected number of rooms. Does this comparison give any evidence that clustering of disease occurs within rooms?

Cancer

The incidence rate of malignant melanoma in women ages 35–59 is approximately 11 new cases per 100,000 women per year. A study is planned to follow 10,000 women with excessive exposure to sunlight.

4.52 What is the expected number of cases among 10,000 women over 4 years? (Assume no excess risk due to sunlight exposure.)

4.53 Suppose that 9 new cases are observed in this period. How unusual a finding is this?

Cancer

A study was performed in Woburn, MA, looking at the rate of leukemia among children ($\leq$ age 19) in the community in comparison to statewide leukemia rates. Suppose there are 12,000 children in the community that have lived there for a 10-year period and 12 leukemias have occurred in 10 years.

4.54 If the statewide incidence rate of leukemia in children is 5 events per 100,000 children per year (i.e., per 100,000 person-years) then how many leukemias would be expected in Woburn over the 10-year period if the statewide rates were applicable?

4.55 What is the probability of obtaining 12 events over a 10-year period if statewide incidence rates were applicable?

4.56 What is the probability of obtaining at least 12 events over the 10-year period if the statewide incidence rates were applicable?

4.57 How do you interpret the results in Problem 4.56?

SOLUTIONS

4.1 $10 \times 9 \times 8 \times 7 = 5040$. This is referred to as $_{10}P_4$ or the number of permutations of 10 things taken 4 at a time.

4.2
$$\frac{10 \times 9 \times 8 \times 7}{4 \times 3 \times 2 \times 1} = 210$$

This is referred to as $_{10}C_4$ or the number of combinations of 10 things taken 4 at a time.

4.3 X = number of hypertensives over a lifetime is binomially distributed with parameters $n = 20$, $p = .2$; that is,

$$Pr(X = x) = {_{20}}C_x(.2)^x(.8)^{20-x} \quad x = 0, 1, \ldots, 20$$

or, from Table 1 (Appendix, text):

TABLE 4.4

x	Pr(X = x)	x	Pr(X = x)
0	.0115	7	.0546
1	.0576	8	.0222
2	.1369	9	.0074
3	.2054	10	.0020
4	.2182	11	.0005
5	.1746	12	.0001
6	.1091	13–20	.0000

4.4 We wish to compute

$$Pr(X = 4) = {}_{50}C_4\,.028^4.972^{46} = .038$$

4.5 $Pr(X \geq 4) = 1 - Pr(X \leq 3)$. We use the recursion rule to evaluate $Pr(X \leq 3)$ as follows:

$$Pr(X = 0) = (.972)^{50} = .2417$$

$$Pr(X = 1) = \frac{50}{1} \times \frac{.028}{.972} \times .2417 = .3482$$

$$Pr(X = 2) = \frac{49}{2} \times \frac{.028}{.972} \times .3482 = .2457$$

$$Pr(X = 3) = \frac{48}{3} \times \frac{.028}{.972} \times .2457 = .1133$$

Therefore,

$$Pr(X \geq 4) = 1 - (.2417 + .3482 + .2457 + .1133)$$
$$= 1 - .9488 = .051$$

4.6 $E(X) = 50 \times .028 = 1.4 = \mu$

4.7 We have $Var(X) = npq = 50 \times .028 \times .972 = 1.361 = \sigma^2$. Thus, $\sigma = \sqrt{1.361} = 1.17$.

4.8 We have a binomial distribution with $n = 20$, $p = .105$. Let X = number of in-hospital deaths after coronary-bypass surgery. We have that

$$Pr(X = 5) = {}_{20}C_5(.105)^5(1 - .105)^{15}$$

$$= \frac{20 \times 19 \times 18 \times 17 \times 16}{5 \times 4 \times 3 \times 2 \times 1}(.105)^5(.895)^{15}$$

$$= 15{,}504(1.2763 \times 10^{-5})\,(.1894) = .037$$

4.9 We must compute $Pr(X \geq 5) = 1 - Pr(X \leq 4)$. We will use the recursion rule as follows:

$$Pr(X = 0) = .895^{20} = .1088$$
$$Pr(X = 1) = (20/1)(.105/.895)(.1088) = .2552$$
$$Pr(X = 2) = (19/2)(.105/.895)(.2552) = .2844$$
$$Pr(X = 3) = (18/3)\,(.105/.895)(.2844) = .2002$$
$$Pr(X = 4) = (17/4)(.105/.895)(.2002) = .0998$$

Thus, $Pr(X \geq 5) = 1 - (.1088 + \ldots + .0998) = .052$

4.10 Compute $Pr(X \leq 5) = .9484 + .0375 = .986$

4.11 We have a binomial distribution with $n = 20$, $p = .05$. Let Y = number of in-hospital deaths after hip replacement. To compute $Pr(Y = 4) = {}_{20}C_4(.05)^4(.95)^{16}$, refer to the binomial tables (Table 1 in the Appendix of the text) under $n = 20$, $p = .05$, $k = 4$ and obtain $.0133 \cong .013$.

4.12 From the binomial tables under $n = 20$, $p = .05$, $Pr(Y \geq 4) = 1 - Pr(Y \leq 3) = 1 - (.3585 + .3774 + .1887 + .0596) = 1 - .9842 = .016$.

4.13 From the answer to Problems 4.11 and 4.12, $Pr(Y \leq 4) = Pr(Y \leq 3) + Pr(Y = 4) = .9842 + .0133 = .9975 \cong .998$.

4.14 From Problem 4.9, $Pr(X \geq 5) = .052$, and from Problem 4.12, $Pr(Y \geq 4) = .016$. Therefore, there is a 5.2% probability of observing at least 5 coronary-bypass in-hospital deaths if the old underlying rate of .105 holds and there is a 1.6% probability of observing at least 4 in-hospital deaths after hip replacement if the old underlying rate of .05 holds. Since .05 is the traditional cutoff for unusual probabilities, we would conclude that the in-hospital mortality from hip replacement has increased, while the results for coronary-bypass surgery suggest a similar trend.

4.15 The observed number of events = 3, which is less than the expected number of events = $1000 \times (10/1000) = 10.0$. Thus, we wish to evaluate

$$Pr(X \leq 3) = \sum_{k=0}^{3} {}_{1000}C_k(.01)^k(.99)^{1000-k}$$

We use the recursion rule as follows:

$$Pr(X = 0) = (.99)^{1000} = 4.317 \times 10^{-5}$$

$$Pr(X = 1) = \frac{1000}{1} \times \frac{.01}{.99} \times 4.317 \times 10^{-5} = .000436$$

$$Pr(X = 2) = \frac{999}{2} \times \frac{.01}{.99} \times .000436 = .002200$$

$$Pr(X = 3) = \frac{998}{3} \times \frac{.01}{.99} \times .002200 = .007393$$

Therefore, $Pr(X \leq 3) = 4.317 \times 10^{-5} + \ldots + .007393 = .010$. Since this probability is low, this represents an unusually small number of events in this subgroup.

4.16 Since $n = 1000 \geq 100$ and $p = .01$, we can approximate the binomial distribution by a Poisson distribution with parameter $\mu = np = 1000 \times .01 = 10$. We refer to the exact Poisson tables (Table 2 in the Appendix of the text) under $\mu = 10$ and note that $Pr(X = 0) = .0000$, $Pr(X = 1) = .0005$, $Pr(X = 2) = .0023$, $Pr(X = 3) = .0076$. Therefore, $Pr(X \leq 3) = .0000 + .0005 + .0023 + .0076 = .010$.

4.17 The two answers agree to three decimal places.

4.18 The probability that exactly 2 out of 50 men will die over a 3-year period is given by

$$\binom{50}{2}(.1)^2(.9)^{48}$$

The probability that not more than two men will die out of 50 = $Pr(X \leq 2)$ is given by

$$\sum_{k=0}^{2}\binom{50}{k}(.1)^{k}(.9)^{50-k}$$

This expression can be evaluated from the recursion rule for binomial probabilities.

$$Pr(0) = (.9)^{50} = .00515$$

$$Pr(1) = \frac{50}{1} \times \frac{.1}{.9} \times Pr(0) = .02863$$

$$Pr(2) = \frac{49}{2} \times \frac{.1}{.9} \times Pr(1) = .07794$$

Thus, $Pr(X \le 2) = .00515 + .02863 + .07794 = .112$

4.19 We use the recursion rule as follows:

$$Pr(0) = (.9)^{25} = .0718$$

$$Pr(1) = \frac{25}{1} \times \frac{.1}{.9} \times .0718 = .1994$$

$$Pr(2) = \frac{24}{2} \times \frac{.1}{.9} \times .1994 = .2659$$

$$Pr(3) = \frac{23}{3} \times \frac{.1}{.9} \times .2659 = .2265$$

$$Pr(4) = \frac{22}{4} \times \frac{.1}{.9} \times .2265 = .1384$$

$$Pr(5) = \frac{21}{5} \times \frac{.1}{.9} \times .1384 = .0646$$

$$Pr(6) = \frac{20}{6} \times \frac{.1}{.9} \times .0646 = .0239$$

$$Pr(7) = \frac{19}{7} \times \frac{.1}{.9} \times .0239 = .0072$$

4.20 We see from the results of Problem 4.19 that

$$Pr(X > 0) = 1 - Pr(X \le 0)$$
$$= 1 - .0718 = .928$$

$$Pr(X > 1) = Pr(X \ge 2)$$
$$= 1 - (.0718 + .1994) = .729$$

$$Pr(X > 2) = Pr(X \ge 3)$$
$$= 1 - (.0718 + .1994 + .2659) = .463$$

$$Pr(X > 3) = Pr(X \ge 4)$$
$$= 1 - (.0718 + \ldots + .2265) = .236$$

$$Pr(X > 4) = Pr(X \ge 5)$$
$$= 1 - (.0718 + \ldots + .1384) = .098$$

$$Pr(X > 5) = Pr(X \ge 6)$$
$$= 1 - (.0718 + \ldots + .0646) = .033$$

Therefore X should be 5, since this is the smallest number of beds such that the probability of having more than this number of premature births in a 4-day period is less than .05.

4.21 If we continue with the method used in the solution of Problem 4.20, then we have $Pr(X > 6) = Pr(X \ge 7) = .0095$. Therefore, in order to lower the probability to less than .01, we need 6 beds.

4.22 The expected number of new cases $= 100{,}000 \times (10/100{,}000) \times 2 = 20$.

4.23 The number of new cases of malignant melanoma is binomially distributed with parameters $n = 100{,}000$ and $p = (10/100{,}000) \times 2 = 20/100{,}000$. We will approximate this distribution by a Poisson distribution with parameter $\mu = np = 20$. We wish to compute $Pr(X \ge 30 \mid \mu = 20)$. We refer to Table 2 (see Appendix in text) under $\mu = 20$ to obtain $Pr(X \ge 30) = .0083 + .0054 + \ldots + .0001 = .022$. These results disagree with the cancer-registry data and suggest that the rate of malignant melanoma has increased.

4.24 From Table 2 (see Appendix in the text), $Pr(X = 10 \mid \mu = 10) = .1251$.

4.25 We have $\mu = 2 \times 10 = 20$. Referring to Table 2 (Appendix, text), we note that $Pr(X = 25 \mid \mu = 20) = .0446$.

4.26 We wish to compute $Pr(X \le 12 \mid \mu = 20)$. From Table 2 (Appendix, text) we obtain $Pr(X \le 12) = .0000 + \ldots + .0176 = .039$. Since this probability is $< .05$, this represents a meaningful improvement over the previous rate of traffic fatalities.

4.27 We have that $\lambda = 1.5$, $T = 2$, and thus, $\mu = \lambda T = 3.0$. We wish to compute $Pr(X \ge 5) = 1 - Pr(X \le 4)$. We refer to Table 2 (Appendix, text) under $\mu = 3$ and obtain $Pr(X \ge 5) = 1 - (.0498 + \ldots + .1680) = .185$.

4.28 We have that $\lambda = 3$, $T = 2$, and thus, $\mu = \lambda T = 6.0$. Referring to Table 2 (Appendix, text), we find that $Pr(X \ge 5) = 1 - Pr(X \le 4) = 1 - (.0025 + \ldots + .1339) = .715$.

4.29 Since the expected number of cases on a high-pollution day is 3 and on a low-pollution day is 1.5, it follows that the expected number of cases in one year $= 10(3) + 355(1.5) = 562.5$.

4.30 The probability of living to age 65 given that one is age 21 is given by

$$Pr = \frac{l_{22}}{l_{21}} \times \frac{l_{23}}{l_{22}} \times \ldots \times \frac{l_{64}}{l_{63}} \times \frac{l_{65}}{l_{64}} = \frac{l_{65}}{l_{21}}$$

We will compute these probabilities separately for white males, white females, black males, and black females. The results are given in Table 4.5.

The probabilities vary substantially by sex and race, with the highest probability for white females (.870) and the lowest probability for black males (.603).

TABLE 4.5 Probability of living to age 65 given that one is alive at age 21, by sex and race

	l_{65}	l_{21}	l_{65}/l_{21} = probability
White male	74,770	97,740	.765
White female	85,740	98,583	.870
Black male	58,113	96,407	.603
Black female	74,922	97,538	.768

4.31 The probability of dying exactly between the ages of 56 and 57 given that one is age 21 in 1986 = the probability that one lives to age 56 and then dies before age 57:

$$\frac{l_{56}}{l_{21}} \times \left(1 - \frac{l_{57}}{l_{56}}\right) = \frac{86,124}{97,538} \times \left(1 - \frac{85,236}{86,124}\right)$$

$$= .8830 \times (1 - .9897) = .8830 \times .0103 = .009$$

4.32 First compute the probability of a single white male dying before age 30 given that he is age 21 in 1986. This probability is given by

$$Pr = 1 - \frac{l_{30}}{l_{21}} = 1 - \frac{96,325}{97,740} = 1 - .9855 = .0145$$

We are now interested in the distribution of the number of deaths among 100 men, which follows a binomial distribution with parameters $n = 100$, $p = .0145$. Thus, the probability of k deaths is given by

$$Pr(k) = \binom{100}{k}(.0145)^k(.9855)^{100-k}$$

$$k = 0, 1, 2, \dots, 100$$

We specifically want $Pr(k \geq 5) = 1 - Pr(k \leq 4)$. Use the recursion rule for binomial probabilities:

$$Pr(0) = (.9855)^{100} = .2326$$

$$Pr(1) = \frac{100}{1}\left(\frac{.0145}{.9855}\right)(.2326) = .3417$$

$$Pr(2) = \frac{99}{2}\left(\frac{.0145}{.9855}\right)(.3417) = .2485$$

$$Pr(3) = \frac{98}{3}\left(\frac{.0145}{.9855}\right)(.2485) = .1192$$

$$Pr(4) = \frac{97}{4}\left(\frac{.0145}{.9855}\right)(.1192) = .0425$$

Thus,

$$Pr(k \leq 4) = .2326 + \dots + .0425 = .9846$$

$$Pr(k \geq 5) = 1 - .9846 = .0154$$

This is somewhat unusual.

4.33 The probability of a single black male dying before age 30 given that he is 21 in 1986 is

$$Pr = 1 - \frac{l_{30}}{l_{21}} = 1 - \frac{93,823}{96,407} = 1 - .9732 = .0268$$

If k = number of deaths among 100 black males, then $Pr(k \geq 5) = 1 - Pr(k \leq 4)$.

We have

$$Pr(0) = (.9732)^{100} = .0661$$

$$Pr(1) = \frac{100}{1}\left(\frac{.0268}{.9732}\right)(.0661) = .1820$$

$$Pr(2) = \frac{99}{2}\left(\frac{.0268}{.9732}\right)(.1820) = .2481$$

$$Pr(3) = \frac{98}{3}\left(\frac{.0268}{.9732}\right)(.2481) = .2232$$

$$Pr(4) = \frac{97}{4}\left(\frac{.0268}{.9732}\right)(.2232) = .1491$$

Thus,

$$Pr(k \leq 4) = .0661 + \dots + .1491 = .8685$$

$$Pr(k \geq 5) = 1 - .8685 = .1315$$

Therefore, this is a much more common event for black males than for white males.

4.34 The questions *cannot* be answered if we do not assume that the P_x's remain constant. Current life tables for each succeeding year after 1986 would be needed. For example, in computing the probability that a man will live to age 23 given that he is age 21 in 1986, the probability that he will reach age 22 in 1987 must be multiplied by the probability that he will reach age 23 in 1988 given that he was age 22 in 1987. We must assume that the latter probability is the same as the probability that he will reach age 23 in 1987 given that he was age 22 in 1986, which is not necessarily the case. This assumption is especially risky if events happening over a large number of years are being considered, since death rates have tended to go down. The probability that 65-year-old people will live 1 extra year given that they were 65 in the year 2051 is likely to be quite different from the corresponding probability if they were 65 in the year 1986. Life tables that follow the *same* group of people over time are called *cohort* life tables as opposed to the *current* life-table data given in this example. These cohort life tables are especially useful in providing baseline mortality data for epidemiologic studies of mortality in high-risk groups.

4.35 .02

4.36 We wish to compute $Pr(X \geq 5) = 1 - Pr(X \leq 4)$, where X is binomially distributed with parameters $n = 100$, $p = .02$. We use the recursion rule as follows:

$$Pr(X = 0) = (.98)^{100} = .1326$$

$$Pr(X = 1) = \frac{100}{1} \times \frac{.02}{.98} \times .1326 = .2707$$

$$Pr(X = 2) = \frac{99}{2} \times \frac{.02}{.98} \times .2707 = .2734$$

$$Pr(X = 3) = \frac{98}{3} \times \frac{.02}{.98} \times .2734 = .1823$$

$$Pr(X = 4) = \frac{97}{4} \times \frac{.02}{.98} \times .1823 = .0902$$

Therefore, $Pr(X \geq 5) = 1 - (.1326 + ... + .0902) = 1 - .949 = .051$.

4.37 We wish to compute $Pr(X \geq 15)$ where X is binomially distributed with parameters $n = 20$, $p = .50$. We refer to Table 1 (Appendix, text) and find that $Pr(X \geq 15) = .0148 + ... + .0000 = .021$. Since this probability is low ($<.05$), we can consider the program a success (i.e., the success rate is probably better than 50%).

4.38 The number of ways is $6 \times 6 \times 6 \times 15 \times 5 = 16,200$.

4.39 The number of ways is $24 \times 9 \times 21 \times 20 \times 23 = 2,086,560$.

4.40 We have

$$Pr(\text{age–sex match} \mid \text{age match}) =$$

$$\frac{\text{number of age–sex matches}}{\text{number of age matches}} = \frac{16,200}{2,086,560} = .0078$$

4.41 The number of ways $= {}_6C_2 \times {}_6C_2 \times {}_6C_2 \times {}_{15}C_2 \times {}_4C_2 = 15 \times 15 \times 15 \times 105 \times 6 = 2,126,250 \doteq 2.13 \times 10^6$.

4.42 The number of ways $= {}_{24}C_2 \times {}_9C_2 \times {}_{21}C_2 \times {}_{19}C_2 \times {}_{22}C_2 = 276 \times 36 \times 210 \times 171 \times 231 \doteq 8.24 \times 10^{10}$.

4.43 The probability $= 2.13 \times 10^6 / (8.24 \times 10^{10}) = 2.58 \times 10^{-5}$.

4.44 The number of ways $= {}_6C_3 \times {}_6C_3 \times {}_6C_3 \times {}_{15}C_3 \times {}_3C_3 = 20 \times 20 \times 20 \times 455 \times 1 = 3.64 \times 10^6$.

4.45 The number of ways $= {}_{24}C_3 \times {}_9C_3 \times {}_{21}C_3 \times {}_{18}C_3 \times {}_{21}C_3 = 2024 \times 84 \times 1330 \times 816 \times 1330 \doteq 2.45 \times 10^{14}$.

4.46 The probability $= 3.64 \times 10^6 / (2.45 \times 10^{14}) = 1.48 \times 10^{-8}$.

4.47 Overall, the probability of disease $= 46/98 = .4694$. We use a binomial distribution with parameters $n = 2$, $p = .4694$. We have

$$Pr(X = 0) = .5306^2 = .282$$

$$Pr(X = 1) = {}_2C_1(.4694)(.5306) = .498$$

$$Pr(X = 2) = .4694^2 = .220$$

4.48 We use a binomial distribution with parameters $n = 3$, $p = .4694$. We have

$$Pr(X = 0) = .5306^3 = .149$$

$$Pr(X = 1) = {}_3C_1(.4694)(.5306)^2 = .396$$

$$Pr(X = 2) = {}_3C_2(.4694)^2(.5306) = .351$$

$$Pr(X = 3) = .4694^3 = .103$$

4.49 We use a binomial distribution with parameters $n = 4$, $p = .4694$. We have

$$Pr(X = 0) = .5306^4 = .079$$

$$Pr(X = 1) = {}_4C_1(.4694)(.5306)^3 = .280$$

$$Pr(X = 2) = {}_4C_2(.4694)^2(.5306)^2 = .372$$

$$Pr(X = 3) = {}_4C_3(.4694)^3(.5306) = .219$$

$$Pr(X = 4) = (.4694)^4 = .049$$

4.50 The expected number is obtained as follows:

$$\mu = 13 \times Pr(X \geq 2 \mid n = 2) + 4 \times Pr(X \geq 2 \mid n = 3)$$
$$+ 15 \times Pr(X \geq 2 \mid n = 4)$$
$$= 13 \times .220 + 4 \times (.351 + .103)$$
$$+ 15 \times (.372 + .219 + .049)$$
$$= 2.86 + 1.82 + 9.60 = 14.3$$

4.51 From Table 4.3, there are $4 + 1 + 9 = 14$ rooms with 2 or more affected people. This is virtually identical to the expected number given in the solution to Problem 4.50 (14.3) and gives little evidence of clustering. We discuss the issue of testing for goodness-of-fit of probability models in more detail in Chapter 10 of the text.

4.52 The expected number of cases

$$= 4 \times (11/100,000) \times 10,000 = 4.4.$$

4.53 We have a binomial distribution X with $n = 10,000$ and $p = 44/100,000$. We will approximate this distribution by a Poisson distribution Y with parameter $\mu = np = 4.4$. To compute $Pr(Y \geq 9 \mid \mu = 4.4)$, we use the recursion rule as follows:

$$Pr(Y = 0) = e^{-4.4} = .0123$$

$$Pr(Y = 1) = (4.4/1)(.0123) = .0540$$

$$Pr(Y = 2) = (4.4/2)(.0540) = .1188$$

$$Pr(Y = 3) = (4.4/3)(.1188) = .1743$$

$$Pr(Y = 4) = (4.4/4)(.1743) = .1917$$

$Pr(Y = 5) = (4.4/5)(.1917) = .1687$

$Pr(Y = 6) = (4.4/6)(.1687) = .1237$

$Pr(Y = 7) = (4.4/7)(.1237) = .0778$

$Pr(Y = 8) = (4.4/8)(.0778) = .0428$

Therefore, $Pr(Y \geq 9) = 1 - Pr(Y \leq 8) = 1 - (.0123 + \ldots + .0428) = 1 - .9642 = .036$.

4.54 There were $12,000 \times 10 = 120,000$ person-years. Therefore, we would expect $(5/100,000 \times 120,000) = 6$ events over 10 years.

4.55 From the Poisson table (Table 2, Appendix, text)

$Pr(X = 12 \mid \mu = 6) = .0113$.

4.56 Let $X =$ number of events over 10 years. We will assume X is a Poisson random variable with $\mu = 6$ and wish to compute $Pr(X \geq 12)$. From the Poisson table (Table 2, Appendix, text) we have that $Pr(X \geq 12 \mid \mu = 6) = .0113 + .0052 + .0022 + .0009 + .0003 + .0001 = .020$.

4.57 Since the probability of obtaining 12 or more events is $< .05$, it follows that $\mu = 6$ is an unlikely value and that the true mean is likely to be higher than 6; i.e., there is an excess risk of leukemia in Woburn.

REFERENCES

[1] Shortell, S. M., & Hughes, E. F. X. (1988). The effects of regulation, competition, and ownership on mortality rates among hospital inpatients. *New England Journal of Medicine, 318*(17), 1100–1107.

[2] U.S. Department of Health and Human Services. (1986). *Vital statistics of the United States, 1986.* Washington, DC: U.S. Department of Health and Human Services.

[3] Kaplan, J. E., Schonberger, L. B., Varano, G., Jackman, N., Bied, J., & Gary, G. W. (1982). An outbreak of acute nonbacterial gastroenteritis in a nursing home: Demonstration of person-to-person transmission by temporal clustering of cases. *American Journal of Epidemiology, 116*(6), 940–948.

CHAPTER FIVE

CONTINUOUS PROBABILITY DISTRIBUTIONS

FEV (forced expiratory volume) is an important measure of lung function, representing the volume of air expelled in 1 second in a spirometry test. From the Tecumseh study, the distribution of FEV in children 10–14 years old has mean 2.28 liters, sd 0.56 liters. How can we define abnormal lung function in this age group? We could pick the lower 5% point or 10% point, or we could pick some absolute standard such as under 1.5 L and find the proportion of people who fall below this cutoff point. In any event, we need to make an assumption about the underlying distribution to calculate these probabilities. To accomplish this, we need to develop the concept of a *probability density function*.

SECTION 5.1 Probability Density Function

For distributions that take on a few discrete values, we could define probabilities of specific events; e.g., $Pr(0)$, $Pr(1)$, ... , $Pr(4)$ successes out of 4 color plates. For continuous distributions, there are an effectively infinite number of possibilities, which cannot be enumerated. How can we describe such distributions? By using a *probability density function*.

(1) $f(x) \approx \dfrac{Pr(x < X < x + \Delta x)}{\Delta x}$ = "density" of probability at x as Δx approaches 0.

(2) The total area under the density curve must equal 1.

(3) The area under the density curve from a to b
   = $Pr(a < x < b)$.

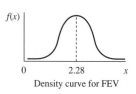

Density curve for FEV

Example: FEV in 10–14-year-old children Probabilities get less dense as one gets further from 2.28.   ■

The density curve does not have to be symmetric except in special cases such as for the normal distribution.

Example: Distribution of serum triglycerides.   ■

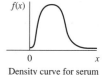

Density curve for serum triglycerides

SECTION 5.2 **Expected Value and Variance of a Continuous Random Variable**

In Chapter 4, we learned that for a discrete random variable

$$E(X) = \Sigma x_i Pr(X = x_i) = \mu$$

$$Var(X) = \Sigma(x_i - \mu)^2 Pr(X = x_i) = \sigma^2$$

A similar definition holds for continuous random variables:

$$E(X) = \int_{-\infty}^{\infty} xf(x)dx = \mu$$

$$Var(X) = \sigma^2 = \int_{-\infty}^{\infty} (x - \mu)^2 f(x)dx$$

In both cases, μ = average value for the random variable, σ^2 – average squared distance of the random variable from the mean.

SECTION 5.3 **Normal Probability Density Function**

The normal distribution is the most commonly used continuous distribution. Its density function is given by

$$f(x) = \frac{1}{\sqrt{2\pi}\,\sigma} \exp\left[-\frac{1}{2}\left(\frac{x - \mu}{\sigma}\right)^2\right] - \infty < x < \infty$$

and is denoted as a $N(\mu, \sigma^2)$ distribution. Many random variables can only take on positive values but are well approximated by a normal distribution anyway, because the probability of obtaining a value < 0 is negligible. The normal density is a function of two parameters: μ, σ^2. These are the mean and variance for the distribution.

SECTION 5.4 **Empirical and Symmetry Properties of the Normal Distribution**

(1) Distribution is symmetric about μ.

(2) The points of inflection of the normal density are at $\mu + \sigma$ and $\mu - \sigma$, respectively.

(3) The height of the normal density is inversely proportional to σ.

(4) Probability $\cong 2/3$ that a normally distributed random variable will fall within
$\mu \pm 1\sigma$

$\cong 95\%$ that a normally distributed random variable will fall within
$\mu \pm 2\sigma$ (actually $\mu \pm 1.96\sigma$)

$\cong 99\%$ that a normally distributed random variable will fall within
$\mu \pm 2.5\sigma$ (actually $\mu \pm 2.576\sigma$)

Example: Distribution of FEV in 10–14-year-old children is approximately normal with $\mu = 2.28$ L, $\sigma = 0.56$ L. ∎

Range of FEV values	Approximate probability
2.28 ± 0.56 = (1.72, 2.84)	2/3
2.28 ± 1.12 = (1.16, 3.40)	95%
2.28 ± 1.40 = (0.88, 3.68)	99%

SECTION 5.5   Calculation of Probabilities for a Standard Normal Distribution

Suppose we want to compute $Pr(X < 1.5$ L$)$ for FEV in 10–14-year-old children. How can we do this? We first calculate probabilities for a standard normal distribution; i.e., a normal distribution with mean = 0 and variance = 1. In Table 3, column A (in the Appendix of the main text), we can obtain $\Phi(x)$ = cumulative probability distribution function $= \int_{-\infty}^{x} f(x)dx = Pr(X \le x)$ for positive values of x for a standard normal distribution. For negative values of x, we use the symmetry relationship $\Phi(x) = 1 - \Phi(-x)$. In general, $Pr(a \le X \le b) = \Phi(b) - \Phi(a)$.

Example: If $X \sim N(0, 1)$, what is $Pr(X \le 1.55)$, $Pr(X \ge 1.55)$, $Pr(1.55 \le X \le 2.75)$, $Pr(-1.55 \le X \le 2.75)$? (The tilde, $\sim$, stands for "is distributed as.")

$$Pr(X \le 1.55) = \Phi(1.55) = .9394$$
$$Pr(X \ge 1.55) = 1 - \Phi(1.55) = .0606$$
$$Pr(1.55 \le X \le 2.75) = \Phi(2.75) - \Phi(1.55) = .9970 - .9394 = .0576$$
$$Pr(-1.55 \le X \le 2.75) = \Phi(2.75) - \Phi(-1.55) = \Phi(2.75) - [1 - \Phi(1.55)]$$
$$= \Phi(2.75) + \Phi(1.55) - 1$$
$$= .9970 + .9394 - 1 = .9364 \qquad \blacksquare$$

In general, we will not distinguish between $Pr(X < x)$ and $Pr(X \le x)$ for normal distributions, since the probabilities of specific values for continuous distributions (e.g., $Pr(X = x)$) is zero.

SECTION 5.6   Calculation of Probabilities for a General Normal Distribution

Suppose we want to know $Pr(a < X < b)$ where $X \sim N(\mu, \sigma^2)$

$$Pr(a < X < b) = \Phi\left(\frac{b - \mu}{\sigma}\right) - \Phi\left(\frac{a - \mu}{\sigma}\right)$$
$$Pr(X < a) = \Phi\left(\frac{a - \mu}{\sigma}\right)$$
$$Pr(X > b) = 1 - Pr(X < b) = 1 - \Phi\left(\frac{b - \mu}{\sigma}\right)$$

Thus, all computations for a general normal distribution can be performed based on percentiles of a standard normal distribution such as those given in Table 3 (Appendix, text).

Example: FEV in 10–14-year-old children

$$Pr(X < 1.5 \text{ L}) = \Phi\left(\frac{1.5 - 2.28}{.56}\right) = \Phi\left(\frac{-.78}{.56}\right) = \Phi(-1.393) = 1 - \Phi(1.393)$$

$$= 1 - .9182 = .082$$

Thus 8.2% of children will have FEV < 1.5 L. ∎

SECTION 5.7   Inverse Normal Distribution

(1) We define z_p to be the *p*th percentile of a standard normal distribution. We refer to the function z_p as the *inverse normal distribution function*. The probability is *p* that $Z \le z_p$ where Z follows a standard normal distribution (i.e., $\Phi(z_p) = p$).

(2) The *p*th percentile of a general normal distribution with mean $= \mu$ and variance $= \sigma^2$ is $\mu + z_p\sigma$.

Example: What is the 5th percentile of the distribution of the FEV in 10–14-year-old children? From Table 3 (Appendix of the text), the 5th percentile of a standard normal distribution $= -1.645$. Therefore, the 5th percentile of the FEV distribution $= 2.28 - 1.645(0.56) = 1.36$ L. Thus, the interval FEV < 1.36 L is another way to specify an abnormal range of FEV values. ∎

SECTION 5.8   Normal Approximation to the Binomial Distribution

In children presenting with bilateral effusion (fluid in the middle ear of both ears) about 2/3 of children still have bilateral effusion after 4 weeks if left untreated. Suppose that if treated with antibiotics 118/198 or 59.6% have bilateral effusion after 4 weeks. Is treatment actually beneficial, or can these results be due to chance? Let $X =$ number of treated children with bilateral effusion after 4 weeks. In general $Pr(X = k) = \binom{n}{k}p^k q^{n-k}$; in other words, X will follow a binomial distribution with parameters *n* and *p*. Suppose treatment was not beneficial at all. In this case, $n = 198$, $p = 2/3$. We want to know the probability that 118 or fewer children will have bilateral efusion if the treatment is not beneficial at all. This is given by

$$Pr(X \le 118) = \sum_{k=0}^{118}\binom{198}{k}\left(\frac{2}{3}\right)^k\left(\frac{1}{3}\right)^{198-k}$$

This is very tedious to compute even with the recursion rule, and instead we use the *normal approximation to the binomial distribution*.

We approximate the binomial random variable X by a normal random variable Y with the same mean (np) and variance (npq); i.e., $N(np, npq)$. This approximation should only be used if $npq \ge 5$. Under these conditions,

$$Pr(a \le X \le b) \approx \Phi\left(\frac{b + 1/2 - \mu}{\sigma}\right) - \Phi\left(\frac{a - 1/2 - \mu}{\sigma}\right)$$

$$= \Phi\left(\frac{b + 1/2 - np}{\sqrt{npq}}\right) - \Phi\left(\frac{a - 1/2 - np}{\sqrt{npq}}\right)$$

$$Pr(X \le a) \approx \Phi\left(\frac{a + 1/2 - \mu}{\sigma}\right) = \Phi\left(\frac{a + 1/2 - np}{\sqrt{npq}}\right)$$

$$Pr(X \ge b) = 1 - Pr(X \le b - 1) \cong 1 - \Phi\left(\frac{b - 1/2 - \mu}{\sigma}\right)$$

$$= 1 - \Phi\left(\frac{b - 1/2 - np}{\sqrt{npq}}\right)$$

Note: One can disregard the equal sign for Y but not for X, since Y is a continuous random variable, while X is a discrete random variable.

Example: Middle-ear effusion We wish to compute $Pr(X \le 118)$. We have $n = 198$, $p = 2/3$, $q = 1/3$, $a = 118$, $np = 132$, $npq = 44 \ge 5$. Therefore, we can use a normal approximation.

$$Pr(X \le 118) \approx Pr(Y \le 118.5) = \Phi\left(\frac{118.5 - 132}{\sqrt{44}}\right) = \Phi\left(\frac{-13.5}{6.63}\right) = \Phi(-2.04)$$

$$= 1 - \Phi(2.04) = 1 - .979 = .021 < .05$$

Since this probability is $< .05$, we conclude that treatment is beneficial. ∎

Example: Color plates Suppose we identify a group of 100 people with a particular type of color blindness. We give each person the color plate test once and find that 32 identify the correct quadrant of the color plate. Are they doing significantly better than random? If they are randomly selecting the quadrant, then X = number of correct selections will be binomially distributed with parameters $n = 100$, $p = 1/4$. Since $npq = 18.75 \ge 5$, we can use a normal approximation. We compute $Pr(X \ge 32)$ as follows:

$$Pr(X \ge 32) \approx Pr(Y \ge 31.5) = 1 - \Phi\left(\frac{31.5 - 25}{\sqrt{18.75}}\right) = 1 - \Phi\left(\frac{6.5}{4.33}\right)$$

$$= 1 - \Phi(1.50) = 1 - .933 = .067 > .05$$

This is not significantly different from random. ∎

SECTION 5.9   Normal Approximation to the Poisson Distribution

Suppose we are studying a particular form of influenza. Usually 150 cases a year occur in a state. We observe 200 cases in 1993. Is this an unusual occurrence? We assume a Poisson distribution for X = number of cases occurring in 1 year with parameters $\lambda = 150$, $T = 1$. We wish to compute $Pr(X \ge 200)$. This will be tedious to do. Instead, we use a normal approximation. We approximate X by $Y \sim N(\mu, \mu)$, where $\mu = \lambda T$.

$$Pr(X = k) \cong Pr\left(k - \frac{1}{2} \le Y \le k + \frac{1}{2}\right) \text{ if } k > 0$$

$$Pr(X = 0) \cong Pr\left(Y \le \frac{1}{2}\right)$$

$$Pr(a \le X \le b) \approx \Phi\left(\frac{b + 1/2 - \mu}{\sqrt{\mu}}\right) - \Phi\left(\frac{a - 1/2 - \mu}{\sqrt{\mu}}\right), \text{ where } 9 > 0$$

$$Pr(X \leq a) \approx \Phi\left(\frac{a + 1/2 - \mu}{\sqrt{\mu}}\right)$$

This approximation should only be used if $\mu \geq 10$.

Example: Influenza

$$Pr(X \geq 200) \approx Pr(Y \geq 199.5) = 1 - \Phi\left(\frac{199.5 - 150}{\sqrt{150}}\right) = 1 - \Phi\left(\frac{49.5}{12.25}\right)$$

$$= 1 - \Phi(4.04) < 10^{-4}$$

Thus, there is a significant increase in the number of cases in 1993.   ∎

PROBLEMS

5.1 What are the deciles of the standard normal distribution; that is, the 10, 20, 30, ... , 90 percentiles?

5.2 What are the quartiles of the standard normal distribution?

Hypertension
Blood pressure (BP) in childhood tends to increase with age, but differently for boys and girls. Suppose that for both boys and girls, mean systolic blood pressure is 95 mm Hg at 3 years of age and increases 1.5 mm Hg per year up to the age of 13. Furthermore, starting at age 13, the mean increases by 2 mm Hg per year for boys and 1 mm Hg per year for girls up to the age of 18. Finally, assume that blood pressure is normally distributed and that the standard deviation is 12 mm Hg for all age–sex groups.

5.3 What is the probability that an 11-year-old boy will have an SBP greater than 130 mm Hg?

5.4 What is the probability that a 15-year-old girl will have an SBP between 100 and 120 mm Hg?

5.5 What proportion of 17-year-old boys have SBP between 120 and 140 mm Hg?

5.6 What is the probability that of 200 15-year-old boys, at least 10 will have SBP of 130 mm Hg or greater?

5.7 What level of SBP is at the 80th percentile for 7-year-old boys?

5.8 What level of SBP is at the 70th percentile for 12-year-old girls?

5.9 Suppose that a task force of pediatricians decides that children over the 95th percentile, but not over the 99th percentile, for their age–sex group should be encouraged to take preventive nonpharmacologic measures to reduce their blood pressure, whereas those children over the 99th percentile should receive antihypertensive drug therapy. Construct a table giving the appropriate BP levels to identify these groups for boys and girls for each year of age from 3 to 18.

Cancer
The incidence of breast cancer in 40–59-year-old women is approximately 1 new case per 1000 women per year.

5.10 What is the incidence of breast cancer over 10 years in women initially 40 years old?

5.11 Suppose we are planning a study based on an enrollment of 10,000 women. What is the probability of obtaining at least 120 new breast-cancer cases over a 10-year follow-up period?

Accident Epidemiology
Suppose the annual death rate from motor-vehicle accidents in 1980 was 10 per 100,000 in urban areas of the United States.

5.12 In an urban state with a population of 2 million, what is the probability of observing not more than 150 traffic fatalities in a given year?

In rural areas of the United States the annual death rate from motor-vehicle accidents is on the order of 100 per 100,000 population.

5.13 In a rural state with a population of 100,000, what is the probability of observing not more than 80 traffic fatalities in a given year?

5.14 How large should x be so that the probability of observing not more than x traffic fatalities in a given year in a rural state with population 100,000 is 5%?

5.15 How large should x be so that the probability of observing x or more traffic fatalities in a given year, in an urban state with population 500,000 is 10%?

Obstetrics

Assume that birthweights are normally distributed with a mean of 3400 g and a standard deviation of 700 g.

5.16 Find the probability of a low-birthweight child, where low birthweight is defined as ≤ 2500 g.

5.17 Find the probability of a very low birthweight child, where very low birthweight is defined as ≤ 2000 g.

5.18 Assuming that successive deliveries by the same woman have the same probability of being low birthweight, what is the probability that a woman with exactly 3 deliveries will have 2 or more low-birthweight deliveries?

Renal Disease

The presence of bacteria in a urine sample (bacteriuria) is sometimes associated with symptoms of kidney disease. Assume that a determination of bacteriuria has been made over a large population at one point in time and that 5% of those sampled are positive for bacteriuria.

5.19 Suppose that 500 people from this population are sampled. What is the probability that 50 or more people would be positive for bacteriuria?

Ophthalmology

A study was conducted of patients with retinitis pigmentosa, an ocular condition where pigment appears over the retina, resulting in substantial loss of vision in many cases [1]. The study was based on 94 patients who were seen annually at a baseline visit and at three annual follow-up visits. In this study, 90 patients provided visual-field measurements at each of the four examinations and are the subjects of the following data analyses. Visual field was transformed to the ln scale to better approximate normality and yielded the data given in Table 5.1.

5.20 Assuming the change in visual field over 3 years is normally distributed when using the ln scale, what is the

TABLE 5.1 Visual-field measurements in retinitis-pigmentosa patients

Year of examination	Mean[a]	Standard deviation	n
Year 0 (baseline)	8.15	1.23	90
Year 3	8.01	1.33	90
Year 0–year 3	0.14	0.66	90

[a]ln (area of visual field) in degrees squared.

Source: Reprinted with permission of the *American Journal of Ophthalmology, 99,* 240–251, 1985.

proportion of patients who showed a decline in visual field over 3 years?

5.21 What percentage of patients would be expected to show a decline of at least 20% in visual field over 3 years? (*Note:* In the ln scale this is equivalent to a decline of at least ln (1/0.8) = 0.223).

5.22 Answer Problem 5.21 for a 50% decline over 3 years.

Cancer

A study of the Massachusetts Department of Health found 46 deaths due to cancer among women in the city of Bedford over the period 1974–1978, where 30 deaths had been expected from statewide rates [2].

5.23 Write an expression for the probability of observing exactly k deaths due to cancer over this period if the statewide rates are correct.

5.24 Can the occurrence of 46 deaths due to cancer be attributed to chance? Specifically, what is the probability of observing at least 46 deaths due to cancer if the statewide rates are correct?

Hypertension

Suppose we want to recruit subjects for a hypertension treatment study and we feel that 10% of the population to be sampled is hypertensive.

5.25 If 100 subjects are required for the study and perfect cooperation is assumed, then how many people need to be sampled to be 80% sure of ascertaining at least 100 hypertensives?

5.26 How many people need to be sampled to be 90% sure of ascertaining at least 100 hypertensives?

Nutrition

The distribution of serum levels of alpha tocopherol (serum vitamin E) is approximately normal with mean 860 μg/dL and standard deviation 340 μg/dL.

5.27 What percentage of people have serum alpha tocopherol levels between 400 and 1000 μg/dL?

5.28 Suppose a person is identified as having toxic levels of alpha tocopherol if his or her serum level is >2000 μg/dL. What percentage of people will be so identified?

5.29 A study is undertaken for evidence of toxicity of 2000 people who regularly take vitamin-E supplements. The investigators found that 4 have serum alpha tocopherol levels >2000 μg/dL. Is this an unusual number of people with toxic levels of serum alpha tocopherol?

Epidemiology

A major problem in performing longitudinal studies in medicine is that people initially entered into a study are lost to follow-up for various reasons.

5.30 Suppose we wish to evaluate our data after 2 years and anticipate that the probability a patient will be available for study after 2 years is 90%. How many patients should be entered into the study to be 80% sure of having at least 100 patients left at the end of this period?

5.31 How many patients should be entered to be 90% sure of having at least 150 patients after 4 years if the probability of remaining in the study after 4 years is 80%?

Pulmonary Disease

5.32 The usual annual death rate from asthma in England over the period 1862–1962 for people aged 5–34 was approximately 1 per 100,000. Suppose that in 1963 twenty deaths were observed in a group of 1 million people in this age group living in Britain. Is this number of deaths inconsistent with the preceding 100-year rate? In particular, what is the probability of observing 20 or more deaths in 1 year in a group of 1 million people? *Note:* This finding is both statistically and medically interesting, since it was found that the excess risk could be attributed to certain aerosols used by asthmatics in Britain during the period 1963–1967. The rate returned to normal during the period 1968–1972, when these types of aerosols were no longer used. (For further information see Speizer et al. [3].)

Hypertension

Blood-pressure measurements are known to be variable, and repeated measurements are essential to accurately characterize a person's BP status. Suppose a person is measured on n visits with k measurements per visit and the average of all nk DBP measurements ($\bar{x}$) is used to classify a person as to BP status. Specifically, if $\bar{x} \geq 95$ mm Hg, then the person is classified as hypertensive; if $\bar{x} < 90$ mm Hg, then the person is classified as normotensive; and if $\bar{x} \geq 90$ mm Hg, and < 95 mm Hg, the person is classified as borderline. It is also assumed that a person's "true" blood pressure is μ, representing an average over a large number of visits with a large number of measurements per visit, and that $\bar{x}$ is normally distributed with mean μ and variance $= 27.7/n + 7.9/(nk)$.

5.33 If a person's true diastolic blood pressure is 100 mm Hg, then what is the probability that the person will be classified accurately (as hypertensive) if a single measurement is taken at 1 visit?

5.34 Is the probability in Problem 5.33 a measure of sensitivity, specificity, or predictive value?

5.35 If a person's true blood pressure is 85 mm Hg, then what is the probability that the person will be accurately classified (as normotensive) if 3 measurements are taken at each of 2 visits?

5.36 Is the probability in Problem 5.35 a measure of sensitivity, specificity, or predictive value?

5.37 Suppose we decide to take 2 measurements per visit. How many visits are needed so that the sensitivity and specificity would each be at least 95%?

Hypertension

A study is planned to look at the effect of sodium restriction on lowering blood pressure. Nutritional counseling sessions are planned for the participants to encourage dietary sodium restriction. An important component in the study is validating the extent to which individuals comply with a sodium-restricted diet. This is usually accomplished by obtaining urine specimens and measuring sodium in the urine.

Assume that in free-living individuals (individuals with no sodium restriction) 24-hour urinary sodium excretion is normally distributed with mean 160.5 mEq/24 hrs and standard deviation = 57.4 mEq/24 hrs.

5.38 If 100 mEq/24 hr is the cutoff value chosen to measure compliance, then what percentage of noncompliant individuals—that is, individuals who do not restrict sodium—will have a 24-hour sodium level below this cutoff point?

Suppose that in an experimental study it is found that people who are on a sodium-restricted diet have 24-hour sodium excretion that is normally distributed with mean 57.5 mEq/24 hrs and standard deviation of 11.3 mEq/24 hrs.

5.39 What is the probability that a person who is on a sodium-restricted diet will have a 24-hour urinary sodium level above the cutoff point (100 mEq/24 hrs)?

5.40 Suppose the investigators in the study wish to change the cutoff value (from 100 mEq/24 hrs) to another value such that the misclassification probabilities in Problems 5.38 and 5.39 are the same. What should the new cutoff value be, and what are the misclassification probabilities corresponding to your answers to Problems 5.38 and 5.39?

Table 5.2 shows data reported relating alcohol consumption and blood-pressure level among 18–39-year-old males [4].

TABLE 5.2 The relationship of blood pressure level to drinking behavior

	Systolic blood pressure (SBP)			Diastolic blood pressure (DBP)			Elevated blood pressure[b]
	Mean	sd	n	Mean	sd	n	
Nondrinker	120.2	10.7	96	74.8	10.2	96	11.5%
Heavy drinker[a]	123.5	12.8	124	78.5	9.2	124	17.7%

[a] >2.0 drinks per day.
[b] Either SBP ≥ 140 mm Hg or DBP ≥ 90 mm Hg.

5.41 If we assume that the distributions of SBP are normal, then what proportion of nondrinkers have SBP ≥ 140 mm Hg?

5.42 What proportion of heavy drinkers have SBP ≥ 140 mm Hg?

5.43 What proportion of heavy drinkers have "isolated systolic hypertension"; i.e., SBP ≥ 140 mm Hg but DBP < 90 mm Hg?

5.44 What proportion of heavy drinkers have "isolated diastolic hypertension"; i.e., DBP ≥ 90 mm Hg but SBP < 140 mm Hg?

Hypertension
The Pediatric Task Force Report on Blood Pressure Control in Children [5] reports blood-pressure norms for children by age and sex group. The mean ± standard deviation for 17-year-old boys for diastolic blood pressure (DBP) is 63.7 ± 11.4 mm Hg based on a large sample.

5.45 Suppose the 90th percentile of the distribution is the cutoff for elevated BP. If we assume that the distribution of blood pressure for 17-year-old boys is normally distributed, then what is the cutoff in mm Hg?

5.46 Another approach for defining elevated BP is to use 90 mm Hg as the cutoff (the standard for elevated adult DBP). What percentage of 17-year-old boys would have elevated BP using this approach?

5.47 Suppose there are 200 17-year-old boys in the 11th grade of whom 25 have elevated BP using the criteria in Problem 5.45. Is this an unusually high number of boys with elevated BP? Why or why not?

Environmental Health

5.48 A study was conducted relating particulate air pollution and daily mortality in Steubenville, Ohio [6]. On average over the last 10 years, there have been 3 deaths per day. Suppose that on 90 high-pollution days (where the total suspended particulates are in the highest quartile among all days) the death rate is 3.2 deaths per day or 288 deaths observed over the 90 high-pollution days. Is there an unusual number of deaths on high-pollution days?

Mental Health
A study was performed among three groups of male prisoners. 36 subjects between the ages of 18 and 45 were selected from the general inmate population at a Connecticut prison [7]. Three groups were identified from the inmate population: group A, a group of men with chronic aggressive behavior, who were in prison for

TABLE 5.3 Physical variables in three groups of male prisoners

	(A) Aggressive	(B) Socially dominant	(C) Nonaggressive
Testosterone (μg/mL ± SD)	10.10 ± 2.29[a]	8.36 ± 2.36[b]	5.99 ± 1.20
Weight (lb)	173.0 ± 13.8	182.2 ± 14.4	168.0 ± 25.2
Height (in)	71.2 ± 2.4	71.3 ± 2.2	69.5 ± 3.5
Age (yr)	29.08 ± 6.34	27.25 ± 3.59	28.4 ± 6.57

[a] Significant difference from nonaggressive group ($P < 0.01$).
[b] Significant difference from nonaggressive group ($P < 0.001$).

violent crimes such as aggravated assault or murder (the aggressive group); group B, a group of socially dominant men who were in prison for a variety of nonviolent crimes such as theft, check passing, or drug-related felonies and had asserted themselves into prestigious positions in prison hierarchy (the socially dominant group); group C, a nonaggressive group who were in prison for nonviolent crimes and were not socially dominant in prison hierarchy (the nonaggressive group). Group C was a group of "average" inmates and served as a control group. 12 volunteers were obtained from each group who were willing to take blood tests. Blood samples were obtained on three consecutive days and a 3-day average value was computed for each subject. The mean ±1 sd for plasma testosterone and other variables are given in Table 5.3.

5.49 Suppose we assume that the distribution of mean plasma testosterone within each group is normally distributed. If values of mean plasma testosterone > 10 μg/mL are considered high, then what percentage of subjects from each group would have high values?

5.50 The actual distribution of plasma-testosterone values within each group are plotted in Figure 5.1. Compare the observed and expected number of high values according to your answer to Problem 5.49. Do you think a normal distribution is fitting the data well? Why or why not?

5.51 The authors also remark that "there was a small variation in the plasma-testosterone values on successive days for each individual subject (± 1.32 μg/mL standard deviation)." What is the coefficient of variation using the overall mean over all three groups ($n = 36$) as a typical mean level for a subject?

FIGURE 5.1
A plot of the mean values of plasma testosterone (μg/mL) for 36 subjects divided into the three groups: aggressive, social dominant, and nonaggressive

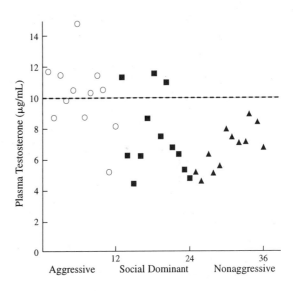

SOLUTIONS

5.1 We have

x	Φ(x)	x	Φ(x)
−1.28	.10	0.25	.60
−0.84	.20	0.52	.70
−0.52	.30	0.84	.80
−0.25	.40	1.28	.90
0.00	.50		

5.2 We have

x	−0.67	0.00	0.67
Φ(x)	.25	.50	.75

5.3 The mean systolic blood pressure (SBP) for an 11-year-old boy $= 95 + 1.5 \times (11 - 3) = 107$. If X represents SBP for an 11-year-old boy, then $X \sim N(107,144)$. Therefore,

$$Pr(X > 130) = 1 - \Phi\left(\frac{130 - 107}{12}\right) = 1 - \Phi\left(\frac{23}{12}\right)$$

$$= 1 - \Phi(1.917) = 1 - .9724 = .028$$

5.4 The mean SBP for a 15-year-old girl = 95 + 1.5 × (13 − 3) + 1 × (15 − 13) = 112, while the standard deviation is 12. Therefore, if Y represents SBP for a 15-year-old girl, then

$$Pr(100 < Y < 120)$$

$$= \Phi\left(\frac{120 - 112}{12}\right) - \Phi\left(\frac{100 - 112}{12}\right)$$

$$= \Phi(0.667) - \Phi(-1.00) = .7475 - [1 - \Phi(1.00)]$$

$$= .7475 - .1587 = .589$$

5.5 The mean SBP for a 17-year-old boy = 95 + 1.5 × (13 − 3) + 2 × (17 − 13) = 118, while the standard deviation is 12. If Z represents SBP for a 17-year-old boy, then

$$Pr(120 < Z < 140)$$

$$= \Phi\left(\frac{140 - 118}{12}\right) - \Phi\left(\frac{120 - 118}{12}\right)$$

$$= \Phi(1.833) - \Phi(0.167) = .9666 - .5662 = .400$$

5.6 We first calculate the probability that one 15-year-old boy will have an SBP > 130. Let X_1 represent SBP for a 15-year-old boy. We have that the mean SBP = 95 + 1.5 × (13 − 3) + 2 × (15 − 13) = 114. Therefore,

$$Pr(X_1 > 130) = 1 - \Phi\left(\frac{130 - 114}{12}\right) = 1 - \Phi\left(\frac{16}{12}\right)$$

$$= 1 - \Phi(1.333) = .0912$$

Now let X represent the number of boys among 200 15-year-old boys with SBP > 130. X is binomially distributed with parameters $n = 200$, $p = .0912$. We wish to compute $Pr(X \geq 10)$. We will approximate X by a normal distribution Y with mean = np = 200 × .0912 = 18.24 and variance = npq = 200 × .0912 × .9088 = 16.58. We have

$$Pr(X \geq 10) \cong Pr(Y > 9.5) = 1 - \Phi\left(\frac{9.5 - 18.24}{\sqrt{16.58}}\right)$$

$$= 1 - \Phi\left(\frac{-8.742}{4.072}\right)$$

$$= 1 - \Phi(-2.147) = \Phi(2.147) = .984$$

Therefore, there is a 98.4% probability that at least 10 of the 200 15-year-old boys will have an SBP of at least 130.

5.7 The mean SBP for 7-year-old boys is given by 95 + 1.5 × (7 − 3) = 101. From Table 3 (Appendix, text), the 80% point for a $N(0, 1)$ distribution = $z_{.80}$ = 0.84.

Therefore, since the distribution of SBP for 7-year-old boys is normal, with mean 101 and standard deviation 12, the 80% point = 101 + 0.84 × 12 = 111.1.

5.8 From Table 3 (Appendix, text) the 70% point for a $N(0, 1)$ distribution = $z_{.70}$ = 0.52. Systolic blood pressure for 12-year-old girls is normally distributed with mean = 95 + 1.5 × (12 − 3) = 108.5 and standard deviation = 12. Therefore, the 70% point for 12-year-old girls is given by 108.5 + 0.52 × 12 = 114.7.

5.9 Since $z_{.95} = 1.645$, $z_{.99} = 2.326$, it follows that in general, the 95th and 99th percentiles are given by $\mu + 1.645\sigma$ and $\mu + 2.326\sigma$ respectively, where μ = age–sex-specific mean SBP and $\sigma = 12$. For example, for 3-year-old boys, the 95th percentile = 95 + 1.645 × 12 = 114.7; the 99th percentile = 95 + 2.326 × 12 = 122.9. These percentiles are presented in the table on page 55 for each year of age for 3–18-year-old boys and girls.

Therefore, for 3-year-old boys, the nonpharmacologic measure group is defined by SBP ≥ 115 and ≤ 122, while the antihypertensive-drug-therapy group is defined by SBP ≥ 123. A similar strategy is followed for the other age–sex subgroups.

5.10 .010

5.11 The number of breast-cancer cases (X) is binomially distributed with parameters $n = 10,000$, $p = .010$. This distribution is approximated by a normal distribution (Y) with mean = np = 100, variance = npq = 99. We have

$$Pr(X \geq 120) \approx Pr(Y \geq 119.5)$$

$$= 1 - \Phi[(119.5 - 100)/\sqrt{99}]$$

$$= 1 - \Phi(1.96) = .025$$

5.12 Let X = number of traffic fatalities in the urban state in 1 year. We will assume that X is Poisson-distributed with parameter $\mu = 2 \times 10^6 \times (10/10^5)$ = 200. We wish to compute $Pr(X \leq 150)$. We will approximate X by a normal distribution Y with mean = 200, variance = 200. Compute

$$Pr(X \leq 150) \cong Pr(Y \leq 150.5)$$

$$= \Phi\left(\frac{150.5 - 200}{\sqrt{200}}\right)$$

$$= \Phi(-3.50) = 1 - \Phi(3.50) = .0002$$

5.13 Let X = number of traffic fatalities in the rural state in 1 year. We will assume that X is Poisson-distributed with parameter = $\mu = 10^5 \times (100/10^5)$ = 100. To compute $Pr(X \leq 80)$, approximate X by a normal distribution Y with mean = 100, variance = 100. Therefore,

Boys					Girls			
		Percentiles					Percentiles	
Age	μ	95th	99th		Age	μ	95th	99th
3	95.0	114.7	122.9		3	95.0	114.7	122.9
4	96.5	116.2	124.4		4	96.5	116.2	124.4
5	98.0	117.7	125.9		5	98.0	117.7	125.9
6	99.5	119.2	127.4		6	99.5	119.2	127.4
7	101.0	120.7	128.9		7	101.0	120.7	128.9
8	102.5	122.2	130.4		8	102.5	122.2	130.4
9	104.0	123.7	131.9		9	104.0	123.7	131.9
10	105.5	125.2	133.4		10	105.5	125.2	133.4
11	107.0	126.7	134.9		11	107.0	126.7	134.9
12	108.5	128.2	136.4		12	108.5	128.2	136.4
13	110.0	129.7	137.9		13	110.0	129.7	137.9
14	112.0	131.7	139.9		14	111.0	130.7	138.9
15	114.0	133.7	141.9		15	112.0	131.7	139.9
16	116.0	135.7	143.9		16	113.0	132.7	140.9
17	118.0	137.7	145.9		17	114.0	133.7	141.9
18	120.0	139.7	147.9		18	115.0	134.7	142.9

$$Pr(X \le 80) \cong Pr(Y \le 80.5)$$
$$= \Phi\left(\frac{80.5 - 100}{\sqrt{100}}\right)$$
$$= \Phi(-1.95) = 1 - \Phi(1.95) = .026$$

5.14 We wish to compute x such that $Pr(X \le x) = .05$. We approximate X by a $N(100,100)$ distribution (Y). Therefore, $Pr(X \le x)$ is approximated by

$$Pr(Y \le x + .5) = \Phi\left(\frac{x + .5 - 100}{10}\right) = .05 = \Phi(-1.645)$$

or

$$\frac{x + .5 - 100}{10} = -1.645$$

Solving for x, we obtain $x = 83.05$. Thus, x should be 83.

5.15 We wish to compute x such that $Pr(X \ge x) = .10$, where X is Poisson-distributed with parameter $\mu = 500{,}000 \times (10/10^5) = 50$. We approximate X by a $N(50,50)$ distribution (Y). Therefore,

$$Pr(X \ge x) \cong Pr(Y \ge x - .5)$$
$$= 1 - \Phi\left(\frac{x - .5 - 50}{\sqrt{50}}\right) = .10$$

or

$$\Phi\left(\frac{50 - x + .5}{\sqrt{50}}\right) = \Phi(-1.28)$$

or

$$\frac{50 - x + .5}{\sqrt{50}} = -1.28$$

Solving for x, we obtain $x = 59.6$. Thus, x should be 60.

5.16 If $X =$ birthweight, then $X \sim N(3400, 700^2)$. Thus,

$$Pr(X \le 2500) = \Phi\left(\frac{2500 - 3400}{700}\right) = \Phi(-1.286)$$
$$= 1 - \Phi(1.286)$$
$$= 1 - .9007 = .099$$

5.17
$$Pr(X \le 2000) = \Phi\left(\frac{2000 - 3400}{700}\right) = \Phi(-2.00)$$
$$= 1 - \Phi(2.00) = 1 - .9772 = .023$$

5.18 Let $X =$ number of low-birthweight deliveries. X is binomially distributed with parameters $n = 3$, $p = .099$. We wish to compute $Pr(X \ge 2)$.

$$Pr(X \ge 2) = \binom{3}{2}(.099)^2(.901) + \binom{3}{3}(.099)^3$$
$$= .0266 + .0010 = .028$$

5.19 Let X be the number of people positive for bacteriuria out of 500. X follows a binomial distribution with parameters $n = 500$ and $p = .05$. Compute $Pr(X \ge 50)$.

X will be approximated by a normal random variable Y with mean $= 500 \times .05 = 25$ and variance $= 500 \times .05 \times .95 = 23.75$. Compute

$$Pr(X \geq 50) \approx Pr(Y \geq 49.5)$$
$$= 1 - \Phi\left(\frac{49.5 - 25}{\sqrt{23.75}}\right)$$
$$= 1 - \Phi(5.03) < .001$$

5.20 Let $X = \ln$ (visual-field year 0) $- \ln$ (visual-field year 3). We assume that $X \sim N(0.14, 0.66^2)$. A patient loses visual field if $X > 0$. Thus, we wish to compute

$$Pr(X > 0) = 1 - \Phi\left(\frac{0 - 0.14}{0.66}\right)$$
$$= 1 - \Phi(-0.212) = \Phi(0.212) = .584$$

Thus, about 58% of patients would be expected to decline in visual field over 3 years.

5.21 We wish to evaluate $Pr(X > 0.223)$. We have

$$Pr(X > 0.223) = 1 - \Phi\left(\frac{0.223 - 0.14}{0.66}\right)$$
$$= 1 - \Phi(0.126) = .450$$

Thus, about 45% of patients would be expected to decline at least 20% in visual field over 3 years.

5.22 The equivalent decline in the ln scale is $\ln(1/0.5) = 0.693$. We have

$$Pr(X > 0.693) = 1 - \Phi\left(\frac{0.693 - 0.14}{0.66}\right)$$
$$= 1 - \Phi(0.838) = .201$$

Thus, about 20% of patients would be expected to decline at least 50% in visual field over 3 years.

5.23 Suppose there are n women living in the city of Bedford during 1974–1978 and that the probability of getting cancer for an individual woman over this period $= p$. If $X =$ the number of women who actually get cancer, then X is binomially distributed with parameters n and p. Since n is large and p is small, we can approximate this distribution by a Poisson distribution with parameter $\mu = np = 30$. Thus,

$$Pr(k \text{ deaths}) = \frac{e^{-30} 30^k}{k!}, \; k = 0, 1, 2, \dots$$

5.24 We wish to compute $Pr(X \geq 46)$. This is tedious to do with the Poisson distribution. Instead, we approximate the Poisson random variable X by a normal random variable Y with mean $=$ variance $= 30$. Thus,

$$Pr(X \geq 46) \cong Pr(Y \geq 45.5)$$
$$= 1 - \Phi\left(\frac{45.5 - 30}{\sqrt{30}}\right)$$
$$= 1 - \Phi(2.83) = 1 - .9977 = .002$$

Thus, it is unlikely that the statewide cancer rate applies to the city of Bedford.

5.25 Suppose n people are sampled for the study. The number of hypertensives X ascertained from this sample will be binomially distributed with parameters n and $p = .10$. For large n, from the normal approximation to the binomial, it follows that the distribution of X can be approximated by a normal distribution Y with mean $.1n$ and variance $n \times .1 \times .9 = .09n$. Thus, we want n to be large enough so that $Pr(X \geq 100) = .8$. We have

$$Pr(X \geq 100) \approx Pr(Y \geq 99.5) = 1 - Pr(Y \leq 99.5)$$
$$= 1 - \Phi\left(\frac{99.5 - .1n}{\sqrt{.09n}}\right) = .8$$

or

$$\Phi\left(\frac{99.5 - .1n}{\sqrt{.09n}}\right) = .2$$

However, from the normal tables, $\Phi(0.84) = .8$ or $\Phi(-0.84) = .2$. Thus,

$$\frac{99.5 - .1n}{\sqrt{.09n}} = -0.84$$

or

$$99.5 - .1n = -0.84\sqrt{.09n}$$

or

$$.1n - 0.252\sqrt{n} - 99.5 = 0$$

If z^2 is substituted for n, this equation can be rewritten as $.1z^2 - 0.252z - 99.5 = 0$. The solution of this quadratic equation is given by

$$z = \sqrt{n} = \frac{0.252 \pm \sqrt{0.252^2 + 4(99.5)(.1)}}{2(.1)}$$
$$= \frac{0.252 \pm \sqrt{39.864}}{.2} = \frac{0.252 \pm 6.314}{.2}$$
$$= -30.309 \quad \text{or} \quad 32.829$$

or $n = z^2 = (-30.309)^2$ or $(32.829)^2 = 918.62$ or 1077.73. Since n must be larger than $100/.1 = 1000$, we have $n = 1077.73$, or 1078. Thus, 1078 people need to be sampled to be 80% sure of recruiting 100 hypertensives.

5.26 The same approach is used as in Problem 5.25. Find n such that

$$Pr(Y \geq 99.5) = 1 - \Phi\left(\frac{99.5 - .1n}{\sqrt{.09n}}\right) = .9$$

or

$$\Phi\left(\frac{99.5 - .1n}{\sqrt{.09n}}\right) = .1$$

From the normal tables, $\Phi(1.28) = .9$ or $\Phi(-1.28) = .1$. Thus,

$$\frac{99.5 - .1n}{\sqrt{.09n}} = -1.28$$

or $\qquad 99.5 - .1n = -0.384\sqrt{n}$

or $\quad .1n - 0.384\sqrt{n} - 99.5 = 0$

Substitute z^2 for n and obtain the quadratic equation

$$.1z^2 - 0.384z - 99.5 = 0$$

The solution is given by

$$z = \frac{0.384 \pm \sqrt{0.384^2 + 4(.1)(99.5)}}{2(.1)}$$

$$= \frac{0.384 \pm \sqrt{39.947}}{.2} = \frac{0.384 \pm 6.320}{.2}$$

$$= -29.68 \text{ or } 33.52$$

Thus,

$$n = z^2 = (-29.68)^2 = 881.02 \text{ or } (33.52)^2 = 1123.72$$

Again use the root such that $n > 1000$, and thus $n = 1123.72$, or 1124 if rounded up to the nearest integer. Thus, 1124 people need to be sampled to be 90% sure of recruiting 100 hypertensives.

5.27 Let the random variable X represent serum alpha tocopherol. We have that $X \sim N(860, 340^2)$. We wish to compute

$$Pr(400 < X < 1000)$$

$$= \Phi\left(\frac{1000 - 860}{340}\right) - \Phi\left(\frac{400 - 860}{340}\right)$$

$$= \Phi(0.412) - \Phi(-1.353)$$

$$= \Phi(0.412) - [1 - \Phi(1.353)]$$

$$= .6597 - .0880 = .572$$

Thus, 57% of people have serum alpha-tocopherol levels between 400 and 1000.

5.28 We wish to compute

$$Pr(X > 2000) = 1 - \Phi\left(\frac{2000 - 860}{340}\right)$$

$$= 1 - \Phi(3.353) = .0004$$

Thus, $<1\%$ of people will have toxic levels of serum alpha tocopherol.

5.29 Let Y be the number of people with toxic levels of serum alpha tocopherol. From Problem 5.28, Y is binomially distributed with parameters $n = 2000$ and $p = .0004$. We wish to compute $Pr(Y \geq 4) = 1 - Pr(Y \leq 3)$. We use the recursion rule as follows:

$$Pr(Y = 0) = .9996^{2000} = .4494$$

$$Pr(Y = 1) = \frac{2000}{1} \times \frac{.0004}{.9996} \times .4494 = .3595$$

$$Pr(Y = 2) = \frac{1999}{2} \times \frac{.0004}{.9996} \times .3595 = .1437$$

$$Pr(Y = 3) = \frac{1998}{3} \times \frac{.0004}{.9996} \times .1437 = .0383$$

Thus, $Pr(Y \geq 4) = 1 - (.4494 + ... + .0383) = .009$. Therefore, there are an unusual number of people with toxic levels in this group.

5.30 Suppose we enter n patients into the study. The number of patients, X, remaining in the study after two years will be binomially distributed with parameters n and $p = .90$. We want to choose n such that $Pr(X \geq 100) = .80$. We can approximate X by a normal random variable Y with mean $.9n$ and variance $n \times .9 \times .1 = .09n$. We have

$$Pr(X \geq 100) \cong Pr(Y \geq 99.5)$$

$$= 1 - \Phi\left(\frac{99.5 - .9n}{\sqrt{.09n}}\right) = .80$$

or $\qquad 0.20 = \Phi\left(\frac{99.5 - .9n}{\sqrt{.09n}}\right) = \Phi(-0.84)$

Thus, we set $(99.5 - .9n)/\sqrt{.09n} = -0.84$, and obtain the equation

$$99.5 - .9n = -.252\sqrt{n}$$

or $\qquad .9n - .252\sqrt{n} - 99.5 = 0$

We substitute z^2 for n and rewrite the equation in the form

$$.9z^2 - .252z - 99.5 = 0$$

The solution to this quadratic equation is given by

$$z = \sqrt{n} = \frac{.252 \pm \sqrt{.252^2 + 4(.9)(99.5)}}{2(.9)}$$

$$= \frac{.252 \pm \sqrt{358.26}}{1.8} = \frac{.252 \pm 18.928}{1.8}$$

$$= -10.375 \text{ or } 10.655$$

Since $\sqrt{n}$ must be positive, we set $\sqrt{n} = 10.655$. Thus $n = 10.655^2 = 113.5$ or 114 people. Thus, 114 people must be entered into the study in order to be 80% sure of having 100 patients left in the study after 2 years.

5.31 Let $n =$ the number of patients entered into the study. The number of patients, X, remaining in the study after 4 years will be binomially distributed with parameters n and $p = .8$. We want to choose n such that $Pr(X \geq$

150) = .90. We can approximate X by a normal random variable Y with mean $.8n$ and variance $n \times .8 \times .2 = .16n$. We approximate $Pr(X \geq 150)$ by

$$Pr(Y \geq 149.5) = 1 - \Phi\left(\frac{149.5 - .8n}{\sqrt{.16n}}\right) = .90$$

Thus,

$$\Phi\left(\frac{149.5 - .8n}{\sqrt{.16n}}\right) = .10 = \Phi(-1.28)$$

and we set

$$\frac{149.5 - .8n}{\sqrt{.16n}} = -1.28$$

and obtain the equation

$$149.5 - .8n = -.512\sqrt{n}$$

or $\qquad .8n - .512\sqrt{n} - 149.5 = 0$

We substitute z^2 for n and rewrite the equation in the form

$$.8z^2 - .512z - 149.5 = 0$$

The solution to this quadratic equation is given by

$$z = \sqrt{n}$$

$$= \frac{.512 \pm \sqrt{.512^2 + 4(.8)(149.5)}}{2(.8)}$$

$$= \frac{.512 \pm \sqrt{478.662}}{1.6}$$

$$= \frac{.512 \pm 21.878}{1.6}$$

$$= -13.354 \quad \text{or} \quad 13.994$$

Since $\sqrt{n}$ must be positive, we set $\sqrt{n} = 13.994$ or $n = 195.8$ or 196 people. Thus, 196 people must be entered into the study in order to be 90% sure of having 150 patients left in the study after 4 years.

5.32 The number of deaths is a binomial random variable. However, since n is large (10^6), and p is small (10^{-5}), we can approximate this random variable by a Poisson distribution. Also, since $npq \cong 10$, we could use a normal approximation. We will use the normal approximation here since it is easier to use. We have $\mu = np = 10$, $\sigma^2 = npq \cong 10$. Thus, X = number of deaths $\sim N(10, 10)$. We have

$$Pr(\geq 20 \text{ deaths}) \cong 1 - \Phi\left(\frac{19.5 - 10}{\sqrt{10}}\right)$$

$$= 1 - \Phi(3.004)$$

$$= .001$$

Thus, it is very unlikely that the same death rate prevailed in 1963.

5.33 If $n = 1, k = 1$, then $\bar{x} \sim N(100, 27.7/1 + 7.9/1) = N(100, 35.6)$. We wish to compute

$$Pr(\bar{x} \geq 95) = 1 - \Phi\left(\frac{95 - 100}{\sqrt{35.6}}\right)$$

$$= 1 - \Phi(-0.838)$$

$$= \Phi(0.838) = .799$$

Thus, there is an 80% chance of detecting such a person as hypertensive.

5.34 The probability in Problem 5.33 = $Pr(\bar{x} \geq 95 \mid \mu = 100) = Pr(\text{test}+ \mid \text{true}+) = $ sensitivity.

5.35 If $n = 2, k = 3$, then $\bar{x} \sim N(85, 27.7/2 + 7.9/6) = N(85, 15.2)$. We wish to compute

$$Pr(\bar{x} < 90) = \Phi\left(\frac{90 - 85}{\sqrt{15.2}}\right)$$

$$= \Phi(1.284) = .900$$

Thus, there is a 90% chance of detecting such a person as normotensive.

5.36 The probability in Problem 5.35 = $Pr(\bar{x} < 90 \mid \mu = 85) = Pr(\text{test}- \mid \text{true}-) = $ specificity.

5.37 If we take 2 measurements per visit, then $\bar{x} \sim N[\mu, 27.7/n + 7.9/(2n)] = N(\mu, 31.65/n)$. We have that

$$\text{Sensitivity} = Pr[\bar{x} \geq 95 \mid \bar{x} \sim N(100, 31.65/n)]$$

$$= 1 - \Phi\left(\frac{95 - 100}{\sqrt{31.65/n}}\right) = .95$$

or $\qquad \Phi\left(\frac{-5\sqrt{n}}{\sqrt{31.65}}\right) = .05$

or $\qquad \Phi\left(\frac{5\sqrt{n}}{\sqrt{31.65}}\right) = .95$

Since $\Phi(1.645) = .95$, it follows that

$$\frac{5\sqrt{n}}{\sqrt{31.65}} = 1.645$$

or $\qquad \sqrt{n} = \frac{1.645 \times \sqrt{31.65}}{5} = 1.851$

or $\qquad n = 1.851^2 = 3.43$

Thus, we would need at least 4 visits to ensure that the sensitivity is 95%. Similarly, we have that

$$\text{Specificity} = Pr[\bar{x} < 90 \mid \bar{x} \sim N(85, 31.65/n)]$$

$$= \Phi\left(\frac{90 - 85}{\sqrt{31.65/n}}\right)$$

$$= \Phi\left(\frac{5\sqrt{n}}{\sqrt{31.65}}\right) = .95$$

This is exactly the same criterion as was used earlier for calculating sensitivity. Therefore, we need at least 4 visits so that the sensitivity and specificity are each at least 95%.

5.38 If a person does not restrict sodium, then his or her urinary sodium excretion will follow a normal distribution with mean = 160.5 and sd = 57.4. Therefore, the probability that his or her urinary sodium will fall below 100 is given by $\Phi[(100 - 160.5)/57.4] = \Phi(-1.054) = 1 - \Phi(1.054) = 1 - .854 = .146$. Thus, there is a 14.6% chance that a noncompliant person will have a urinary sodium < 100 mEq/24 hours.

5.39 If a person is on a sodium-restricted diet, then his or her urinary sodium excretion will follow a normal distribution with mean = 57.5 and sd = 11.3. In this case, the probability that his or her urinary sodium will be >100 = $1 - \Phi[(100 - 57.5)/11.3] = 1 - \Phi(3.76) = .0001$. Thus, the probability that a compliant person has a urinary sodium > 100 is negligible (.0001).

5.40 Suppose the new cutoff value = x. The misclassification probabilities in Problems 5.38 and 5.39 can be written as $\Phi[(x - 160.5)/57.4]$ and $1 - \Phi[(x - 57.5)/11.3] = \Phi[(57.5 - x)/11.3]$, respectively. We want these probabilities to be the same. Thus we set

$$\frac{x - 160.5}{57.4} = \frac{57.5 - x}{11.3}$$

Solving for x, we obtain $x = 74.4$. Thus, the cutoff value should be 74.4 mEq/24 hours. The misclassification probabilities corresponding to this cutoff are $\Phi[(74.4 - 160.5)/57.4] = \Phi(-1.499) = 1 - \Phi(1.499) = 1 - .9331 = .067 = \Phi[(57.5 - 74.4)/11.3]$. Thus, there is a 6.7% probability that people not on a sodium-restricted diet will be designated as compliers, and also a 6.7% probability that people on a sodium-restricted diet will be designated as noncompliers.

5.41 We have that

$$Pr(SBP \geq 140) = 1 - Pr(SBP < 140)$$
$$= 1 - \Phi\left(\frac{140 - 120.2}{10.7}\right)$$
$$= 1 - \Phi(1.85)$$
$$= 1 - .968 = .032$$

5.42

$$Pr(SBP \geq 140) = 1 - Pr(SBP < 140)$$
$$= 1 - \Phi\left(\frac{140 - 123.5}{12.8}\right)$$
$$= 1 - \Phi(1.289)$$
$$= 1 - .901 = .099$$

5.43 Let $A = \{SBP \geq 140$ mm Hg$\}$, $B = \{DBP \geq 90$ mm Hg$\}$. We wish to compute $Pr(A \cap \overline{B})$. We are given that $Pr(A \cup B) = .177$.

Furthermore, from the addition law of probability we have

$$Pr(A \cup B) = Pr(A) + Pr(B) - Pr(A \cap B)$$
$$= Pr(B) + Pr(A \cap \overline{B})$$

Therefore, by subtraction we have

$$Pr(A \cap \overline{B}) = Pr(A \cup B) - Pr(B)$$

To compute $Pr(B)$, we have

$$Pr(B) = Pr(DBP \geq 90) = 1 - Pr(DBP < 90)$$
$$= 1 - \Phi\left(\frac{90 - 78.5}{9.2}\right)$$
$$= 1 - \Phi(1.25)$$
$$= 1 - .8944$$
$$= .106$$

Thus, $Pr(A \cap \overline{B}) = .177 - .106 = .071$.

5.44 We wish to compute $Pr(\overline{A} \cap B)$. If we reverse the roles of A and B, we have

$$Pr(\overline{A} \cap B) = Pr(A \cup B) - Pr(A)$$

From Problem 5.42 we have $Pr(A) = .099$. Therefore,

$$Pr(\overline{A} \cap B) = .177 - .099 = .078$$

5.45 The 90th percentile of a standard normal distribution = 1.28. Thus, the 90th percentile for diastolic blood pressure (DBP) for 17-year-old boys = $63.7 + 11.4(1.28) = 78.3$ mm Hg.

5.46 Let X = DBP for 17-year-old boys. We wish to compute $Pr(X \geq 90)$. We have

$$Pr(X \geq 90) = 1 - \Phi\left(\frac{90 - 63.7}{11.4}\right)$$
$$= 1 - \Phi(2.307)$$
$$= 1 - .9895 = .0105$$

Thus, approximately 1% of 17-year-old boys have DBP ≥ 90 mm Hg.

5.47 We wish to compute $Pr(X \geq 25)$ where X is a random variable depicting the number of boys with elevated DBP. X is binomially distributed with parameters $n = 200$ and $p = .10$. We will approximate X by a normal random variable Y with mean = $np = 200(.1) = 20$ and variance = $npq = 200(.1)(.9) = 18$. Thus,

$$Pr(X \geq 25) \cong Pr(Y \geq 24.5)$$
$$= 1 - \Phi\left(\frac{24.5 - 20}{\sqrt{18}}\right)$$
$$= 1 - \Phi(1.061)$$
$$= 1 - .856 = .144$$

Thus, since this probability is $> .05$, there are not an unusually high number of boys with elevated DBP.

5.48 We observe 288 deaths over 90 high-pollution days, whereas we expect 270 deaths if pollution has no effect on mortality. If pollution has no effect on mortality, then the observed number of deaths over 90 days (X) should be Poisson-distributed with parameters, $\lambda = 3.0$ and $T = 90$. Thus, we wish to compute

$$Pr(X \geq 288 \mid \lambda = 3.0, T = 90) = \sum_{k=288}^{\infty} e^{-\mu} \mu^k/k!$$

$$= 1 - \sum_{k=0}^{287} e^{-\mu} \mu^k/k!, \text{ where } \mu = \lambda T = 3(90) = 270$$

This is very tedious to compute, and instead we use the normal approximation (Y) to the Poisson distribution, which has mean = variance = $\mu = 270$. We wish to compute $Pr(Y \geq 287.5) = 1 - Pr(Y \leq 287.5) = 1 - \Phi[(287.5 - 270)/\sqrt{270}] = 1 - \Phi(17.5/16.43) = 1 - \Phi(1.065) = 1 - .857 = .143$. Since this probability is $> .05$, there are *not* an unusual number of deaths on high-pollution days.

5.49 If the distribution of X = mean plasma testosterone within a group is normally distributed with mean = μ and variance = σ^2, then the probability of a high value is given by

$$Pr(X > 10) = 1 - Pr(X < 10)$$

$$= 1 - \Phi\left(\frac{10 - \mu}{\sigma}\right)$$

Thus, referring to Table 5.3, we have

Group A:

$$Pr(X > 10) = 1 - \Phi\left(\frac{10 - 10.10}{2.29}\right) = 1 - \Phi\left(\frac{-0.10}{2.29}\right)$$

$$= 1 - \Phi(-0.044) = \Phi(0.044) = .517$$

Group B:

$$Pr(X > 10) = 1 - \Phi\left(\frac{10 - 8.36}{2.36}\right) = 1 - \Phi\left(\frac{1.64}{2.36}\right)$$

$$= 1 - \Phi(0.695) = 1 - .756 = .244$$

Group C:

$$Pr(X > 10) = 1 - \Phi\left(\frac{10 - 5.99}{1.20}\right) = 1 - \Phi\left(\frac{4.01}{1.20}\right)$$

$$= 1 - \Phi(3.342) = 1 - .9996 = .0004$$

5.50 From Figure 5.1, we see that there are 7 high values in Group A, 3 high values in group B, and 0 high values in group C. The expected number of high values if the distribution of mean plasma testosterone within a group is normal are as follows: group A: $12 \times .517 = 6.2$; group B: $12 \times .244 = 2.9$; group C: $12 \times .0004 = 0.005$. Thus, the observed and expected number of high values agree very closely, which implies that a normal distribution is fitting the data well within specific groups.

5.51 The coefficient of variation (CV) is defined as $100\% \times s/\bar{x}$. In this case, s refers to the within-person standard deviation over the 3 values for a given person. Also $\bar{x}$ = overall mean in the study = $(5.99 + 8.36 + 10.10)/3 = 8.15$. Thus, $CV = 100\% \times 1.32/8.15 = 16.2\%$.

REFERENCES

[1] Berson, E. L., Sandberg, M. A., Rosner, B., Birch, D. G., & Hanson, A. H. (1985). Natural course of retinitis pigmentosa over a three-year interval. *American Journal of Ophthalmology, 99,* 240–251.

[2] Boston Globe, April 25, 1980.

[3] Speizer, F. E., Doll, R., Heaf, P., & Strang, L. B. (1968, February 8). Investigation into use of drugs preceding death from asthma. *British Medical Journal, 1,* 339–343.

[4] Weissfeld, J. L., Johnson, E. H., Brock, B. M., Hawthorne, V. M. (1988, September). Sex and age interactions in the association between alcohol and blood pressure. *American Journal of Epidemiology, 128*(3), 559–569.

[5] Report of the Second Task Force on Blood Pressure Control in Children—1987. (1987, January). National Heart, Lung and Blood Institute, Bethesda, Maryland. *Pediatrics, 79*(1), 1–25.

[6] Schwartz, J., & Dockery, D. W. (1992). Particulate air pollution and daily mortality in Steubenville, Ohio. *American Journal of Epidemiology, 135*(1), 12–19.

[7] Ehrenkranz, J., Bliss, E., & Sheard, M. H. (1974). Plasma testosterone: Correlation with aggressive behavior and dominance in men. *Psychosomatic Medicine, 36*(6), 469–475.

CHAPTER SIX

ESTIMATION

SECTION 6.1 Relationship of Population to Sample

Reference population Collection of all sampling units that potentially might be under study (or collection of all sampling units to whom we wish to generalize our findings).

Sample A finite collection of sampling units from the reference population.

Random sample A sample chosen such that each member of the reference population has an equal chance of being chosen. This is actually a simple random sample; a random sample can allow for unequal probabilities of selection. In this text, however, we will use the term *random sample* to mean a simple random sample.

SECTION 6.2 Random-Number Tables

Suppose we had a registry of 5000 children with a rare cancer. How can we select a random sample of 100 children from this registry?

Random digit Each digit 0, 1, ... , 9 has a probability of 1/10 of being selected; successive digits are independent.

Table 4 (Appendix, text) contains a collection of 1000 random digits. To select a random sample of 100 children,

1. Start at row 10 of Table 4 and obtain the following groups of 5 consecutive random digits: *39446 01375 75264 51173 16638 04680 98617 90298.*

2. Use the first 4 digits from each group of 5 digits and obtain *3944 0137 7526 5117 1663 0468 9861 9029.*

3. Sort the original list of 5000 children according to alphabetical order, ID order, or some other prespecified order.

4. Select the children from the list according to the numbers obtained in step 2; e.g., the 1st child selected would be the 3944th on the list; the 2nd child selected would be the 137th on the list, ... , etc. If some of the numbers in step 2 are either > 5000 or are duplicates of previously used numbers, then ignore them and go on to the next number.

Randomized Clinical Trials

An important use of random numbers is in randomized clinical trials. Patients are assigned to treatment groups at random, using random numbers obtained from either a table or generated using a computer. The primary rationale for randomization is that it ensures approximate comparability of treatment groups on baseline characteristics.

6.3.1   Techniques of Study Design in Randomized Clinical Trials

(1) Blocking   An equal number of patients are assigned to each treatment group for every $2n$ patients. The purpose is to ensure that an equal number of patients will be assigned to each treatment group over the short run (i.e., every $2n$ patients), because study protocols sometimes change over time.

(2) Stratification   Separate random-number tables are used for each of a number of homogeneous strata (e.g., age groups). The purpose is to ensure balance between treatment groups within strata. The strata are usually constructed to have very different risk profiles for disease (e.g., different age groups).

(3) Blinding   the purpose of blinding is to avoid bias on the part of either the patient or the physician in the ascertainment of outcome. In **double-blind** clinical trials, neither the patient nor the physician is aware of the treatment assignment. In **single-blind** clinical trials, one of either the physician or the patient is aware of the treatment assignment, but not both. It is usually used when it is either impossible or impractical to conduct a double-blind study.

The gold standard of modern clinical research is the *randomized double-blind placebo-controlled* clinical trial. By placebo control, we mean that one treatment arm is an active treatment, while the other treatment arm is a *placebo*— a treatment that appears similar to active treatment to the patient, but actually has no biologic effect (e.g., sugar pills).

Sampling Distribution

A distribution of all possible values of a statistic of interest (e.g., $\bar{x}$) over all possible random samples that can be selected from the reference population.

Estimation of the Mean of a Distribution

The sample mean is used to estimate the underlying or population mean of a distribution. Why is the sample mean a good estimator of the population mean?

(1) It is unbiased (i.e., $E(\bar{x}) = \mu$).

(2) It has minimum variance among all unbiased estimators.

By an *unbiased estimator,* we mean that the average value of the sample mean over all possible random samples of size n that can be selected from the reference population = the population mean.

SECTION 6.6   Standard Error of the Mean

The standard error of the mean (*se*) reflects the variability of the sample mean in repeated samples of size n drawn from the reference population. It differs from the standard deviation, which reflects the variation of individual sample points. It is given by $\sigma/\sqrt{n}$, where σ = standard deviation and n = sample size. In situations where σ is unknown, the *se* is estimated by $s/\sqrt{n}$.

Example: Suppose we take blood pressure measurements on twenty 30–49-year-old males. The mean ± sd of diastolic blood pressure (DBP) is 78.5 ± 10.3 mm Hg. Our best estimate of μ is 78.5. The standard error (*se*) of this estimate is $10.3/\sqrt{20}$ = 2.3. Notice that the standard error of the mean (2.3) is much smaller than the standard deviation (10.3).   ■

The standard error is dependent on

(1) σ = the variability of individual sample points
(2) n = sample size

The following table gives the *se* for mean DBP for different sample sizes (n):

n	se
1	10.3
20	2.3
100	1.0

SECTION 6.7   Interval Estimation for the Mean-Known Variance

Suppose we assume the sd of the distribution of DBP in 30–49-year-old white males is known to be 10.9. Our best estimate of μ from the preceding data = 78.5. A 95% confidence interval (CI) for μ based on the sample data is given by

$$(\bar{x} - 1.96\sigma/\sqrt{n} < \mu < \bar{x} + 1.96\sigma/\sqrt{n})$$

This is a 95% CI (confidence interval) for μ. 95% of the CI's constructed in this manner from repeated samples of size n from the reference population (30–49-year-old males) will contain the true mean (μ).

Example: Compute a 95% CI for μ = true mean DBP for 30–49-year-old males. A 95% CI for μ is given by 78.5 ± 1.96(10.9)/$\sqrt{20}$ = 78.5 ± 4.8 = (73.7, 83.3).   ■

6.7.1   Use of Confidence Intervals for Decision-Making Purposes

Suppose the community we are studying is near a major highway and we wonder whether the noise level has increased BP level in this community relative to the national average. Suppose the national average SBP for age 65+-year-olds = 140.0 mm Hg, σ = 20 mm Hg, based on previous national health surveys. We draw a sample of size 50 and find $\bar{x} = 150$ mm Hg. Can we conclude that the average SBP for 65+-year-olds in our community is higher? We will construct a 95% CI for μ and use it to help us make the decision. We have

$$95\% \text{ CI for } \mu = \bar{x} \pm 1.96 \frac{\sigma}{\sqrt{n}}$$

$$= 150 \pm 1.96 \frac{20}{\sqrt{50}} = 150 \pm 5.5 = (144.5, 155.5)$$

Since this interval excludes the national (140) average, we conclude that the true mean SBP for 65+-year-olds in this community is above the national average.

In general, a $100\% \times (1-\alpha)$ CI for $\mu = \bar{x} \pm z_{1-\alpha/2}\sigma/\sqrt{n}$.

6.7.2 **Factors Influencing the Length of a Confidence Interval**

$$\text{Length} = 2z_{1-\alpha/2}\sigma/\sqrt{n}$$

(1) α: level of confidence; as α gets smaller, CI gets longer
(2) σ: as $\sigma \uparrow$ length $\uparrow$
(3) n: as $n \uparrow$ length $\downarrow$

SECTION 6.8 The Central-Limit Theorem

What if our distribution is not normal? For large enough n, $\bar{x} \sim N(\mu,\sigma^2/n)$ even if the distribution of the individual observations is not normal. This is crucial for the validity of many hypothesis-testing and confidence-interval methods that assume approximate normality for $\bar{x}$ in repeated samples of size n from the reference population. Based on the central-limit theorem, this will be true if n is large enough even if the original distribution is not normal.

SECTION 6.9 Interval Estimation for the Mean-Unknown Variance

If the variance is assumed unknown, then we estimate σ^2 by s^2 and obtain a 95% CI for μ from

$$\left(\bar{x} - t_{n-1,.975}\frac{s}{\sqrt{n}} < \mu < \bar{x} + t_{n-1,.975}\frac{s}{\sqrt{n}} \right)$$

where $t_{n-1,.975} = 97.5$ percentile of a t distribution with $n-1$ df. There are two important differences in CI estimation for μ when σ^2 is unknown versus when σ^2 is known:

(1) We estimate σ^2 by s^2
(2) We base the CI on percentiles of the t distribution rather than the normal distribution

Percentiles of the t distribution are given in Table 5 (in the Appendix of the main text) and are also available in many computer packages, including MINITAB. The t

distribution is symmetric about 0 but has slightly longer tails than the normal distribution. Because of this, CIs based on the t distribution tend to be slightly wider than those based on the normal distribution.

Example: Derive a 95% CI for true mean systolic blood pressure for 65+-year-old men in the noisy community if σ^2 is assumed unknown, and $s = 20$. A 95% CI for $\mu = 150 \pm t_{49,.975} 20/\sqrt{50} = 150 \pm 2.009(20)/\sqrt{50} = 150 \pm 5.7 = (144.3, 155.7)$. Notice that this interval is slightly wider than the CI for σ^2 known $= (144.5, 155.5)$. We still would conclude that this community is different from the national average, because the 95% CI does not include 140. ∎

6.9.1 Confidence Intervals Other Than 95%

In general, a $100\% \times (1 - \alpha)$ CI for $\mu = \bar{x} \pm t_{n-1,1-\alpha/2} \dfrac{s}{\sqrt{n}}$.

SECTION 6.10 Estimation of the Variance of a Distribution

When are we interested in estimating σ^2? In reproducibility studies, estimation of σ^2 is often of primary interest. Suppose we are planning a study of the role of alpha tocopherol (vitamin E) on heart disease. We want to assess the reproducibility of serum alpha-tocopherol levels because we are setting up a new laboratory to perform this test.

We recruit 15 volunteers who provide two blood samples one week apart. We compute $d_i = x_{1i} - x_{2i}$, where $x_{1i} = $ 1st sample ith subject, $x_{2i} = $ 2nd sample ith subject, and find a mean difference $= \bar{x}$ of 0 and a standard deviation $= s$ of $= 0.25$. In the literature, we find that $\sigma = 0.20$ for a similar experiment based on a much larger sample. Can we say that our lab is less reproducible than the literature laboratory? We need to obtain a 95% CI for σ^2. We will base our interval on s^2, which we use as a point estimate of σ^2. A 95% CI for σ^2 is given by the interval

$$[(n - 1)s^2/\chi^2_{n-1,.975} < \sigma^2 < (n - 1)s^2/\chi^2_{n-1,.025}]$$

where $\chi^2_{n-1,p} = $ the pth percentile of a chi-square distribution with $n - 1$ df.

The chi-square distribution is a family of distributions indexed by a parameter d known as the *degrees of freedom* (or df). The pth percentile of a chi-square distribution with d df is denoted by $\chi^2_{d,p}$ and is given in Table 6 (see the Appendix in the main text).

Example: Compute a 95% CI for σ^2 for the vitamin-E reproducibility study. Since $n = 15$, $s = 0.25$, the 95% CI is given by

$$\left[\frac{14(0.25)^2}{\chi^2_{14,.975}}, \frac{14(0.25)^2}{\chi^2_{14,.025}} \right] = \left[\frac{14(0.0625)}{26.12}, \frac{14(0.0625)}{5.63} \right] = (0.0335, 0.1554)$$

$$95\% \text{ CI for } \sigma = (\sqrt{0.0335}, \sqrt{0.1554}) = (0.183, 0.394)$$

This interval includes 0.20 and indicates that our lab is comparable in reproducibility to the laboratory in the literature. ∎

Estimation for the Binomial Distribution

6.11.1 **Large-Sample Method**

Some studies have suggested a relationship between exposure to anesthetic gases and incidence of breast cancer. To examine this relationship, a study was set up among 10,000 female operating-room nurses ages 40–59. Suppose that the 5-year risk of breast cancer in the general population in this age group $= p_0 = .005$.

We find that among 10,000 female operating-room nurses, 60 women have developed disease over 5 years. Is this a significant excess over the expected rate based on national rates? Our best estimate of the risk $= \hat{p} = \dfrac{60}{10,000} = .006$. This is a 20% excess over the national average. In this example,

$p =$ true rate for 40–59-year-old female operating-room nurses, which is unknown

$p_0 = .005 =$ general-population rate for 40–59-year-old women

$\hat{p} =$ estimate of $p = .006$

We will use $\hat{p}$ as an estimate of p. We need to obtain a 95% CI for p based on $\hat{p}$.

Based on the central-limit-theorem, an approximate 95% CI for p is given by the interval

$$\left(\hat{p} - 1.96\sqrt{\frac{\hat{p}\hat{q}}{n}} < p < \hat{p} + 1.96\sqrt{\frac{\hat{p}\hat{q}}{n}} \right)$$

We will only use this interval if $n\hat{p}\hat{q} \geq 5$ (this is the condition under which the normal approximation to the binomial distribution is valid).

In this example, the 95% CI for p is given by

$$(.006 - 1.96\sqrt{.006(.994)/10,000}, .006 + 1.96\sqrt{.006(.994)/10,000})$$
$$= (.006 - .0015, .006 + .0015)$$
$$= (.0045, .0075)$$

Since this interval includes $p_0 = .005$, we would conclude that there is not a significant excess risk of breast cancer among female operating-room nurses. It would still warrant setting up a better study based on a control group of unexposed nurses (e.g., female general-duty nurses who do not work in operating rooms).

6.11.2 **Small-Sample Method**

The drug erythromycin has been postulated to prevent infections in the last trimester of pregnancy and as a result prevent low-birthweight deliveries. A *matched-pair* study was set up comparing erythromycin (E) versus placebo (P). The patients were matched on age, parity, race, smoking, and miscarriages. 50 matched pairs of women were formed. One member of the matched pair randomly received (E) and the other member of the matched pair received (P). The results were as follows:

Outcome for E member of matched pair	+	+	−	−
Outcome for P member of matched pair	+	−	+	−
Number of matched pairs	1	2	6	41

where $+$ indicates a low-birthweight delivery (i.e., ≤ 88 oz) and $-$ indicates a normal birthweight delivery (i.e., > 88 oz). Thus, there are 2 $(+, -)$ pairs, or in other words for two matched pairs, the erythromycin-treated woman had a low-birthweight infant, while the placebo-treated woman had a normal birthweight infant, 6 $(-, +)$ pairs, ... , etc.

How can we compare the E treatment versus the P treatment for the prevention of low-birthweight deliveries? We will focus on the discordant pairs (i.e., the pairs of women who had a differential response). If E and P are equally effective, then the number of $(+, -)$ pairs should be about the same as the number of $(-, +)$ pairs. Another way of saying this is that p = proportion of discordant pairs that are $(+, -)$ should be $1/2$. In this case, there are 8 $(2 + 6)$ discordant pairs of which 2 are $(+, -)$ pairs. Therefore, our best estimate of p is $\hat{p} = 2/8 = .25$. We also wish to compute a 95% CI for p. Since $n\hat{p}\hat{q} - 8(1/4)(3/4) - 1.5 < 5$, we can't use the large-sample method. Instead, we need to compute exact binomial confidence limits for p.

To obtain exact binomial confidence limits requires a computer or use of a nomogram such as the one given in Table 7 (in the Appendix of the main text). Nomograms are provided in Tables 7a and 7b, giving 95 and 99% confidence limits, respectively. To use these nomograms,

If $\hat{p} \leq .50$,

(1) Find $\hat{p}$ on the lower horizontal axis.

(2) Draw a vertical line and note the points of intersection with the appropriate n curves.

(3) Draw a horizontal line across to the left vertical scale. The lower value $= p_1$ and the upper value $= p_2$.

If $\hat{p} > .50$,

(1) Find $\hat{p}$ on the upper horizontal axis.

(2) Same as step 2 in first list.

(3) Draw a horizontal line to the right vertical scale. The upper value $= p_1$, and the lower value $= p_2$.

In this case, we use Table 7a with $n = 8$ and $\hat{p} = 2/8 = .25$. The exact 95% CI for p is obtained from the left vertical axis and is $(.03, .65)$. Since this interval includes .50, we conclude that E and P are equivalent; that is, erythromycin is no better than placebo for the prevention of low-birthweight deliveries.

SECTION 6.12   Estimation for the Poisson Distribution

Suppose we consider the data in Example 4.36 (text, p. 94). In this example, we described the cancer-mortality experience from 161 white male employees of two plants in Texas and Michigan who were exposed to ethylene dibromide (EDB) during the time period 1940–1975. Seven deaths due to cancer were observed, while only 5.8 were expected based on U.S. white male cancer-mortality rates. Is the observed number of cases excessive?

We answered this question by calculating $Pr(X \geq 7 \mid \mu = 5.8)$ where $X =$ observed number of cancer deaths, which was assumed to follow a Poisson distribution with mean $= \mu = 5.8$. Under this assumption, the probability was .36 and we concluded that the number of cancer deaths was not excessive in this group.

Another approach to this problem is to compute a 95% CI for μ based on the observed number of deaths ($X = 7$). To obtain this CI, we will usually use exact methods based on Table 8 (see the Appendix in the text). We refer to the $X = 7$ row and find that the 95% CI for $\mu = (2.81, 14.42)$. Since this interval includes 5.8, we can again conclude that the observed number of deaths are not excessive in this group. We can also obtain a corresponding 95% CI for the standardized mortality ratio (SMR) given by $(2.81/5.8, 14.42/5.8) = (0.48, 2.49)$. The SMR is an index of (mortality rate in an exposed group)/(mortality rate in the general population).

SECTION 6.13   One-Sided Confidence Limits

Suppose it is hypothesized that anesthetic gases can only potentially be harmful and not protective with regard to breast cancer incidence. Instead of a CI of the form $Pr(c_1 < p < c_2) = 1 - \alpha$ we might want a CI for breast cancer incidence of the form $Pr(p > c_1) = 1 - \alpha$. That is, what is the minimum plausible value for the probability of breast cancer among exposed nurses? This is referred to as an *upper one-sided CI*. This interval is given by

$$p > \hat{p} - z_{1-\alpha}\sqrt{\frac{\hat{p}\hat{q}}{n}}$$

In this case, if we refer to Section 6.11.1, we have $\alpha = .05$, $z_{1-\alpha} = 1.645$, $\hat{p} = .006$, $\hat{q} = .994$, $n = 10,000$. The interval is

$$p > .006 - 1.645\sqrt{\frac{.006(.994)}{10,000}}$$

$$= .006 - 1.645(.0008) = .006 - .0013 = .0047$$

Thus, $p > .0047$ is an upper 95% CI. This includes $p = .005$ and we again conclude that breast cancer incidence is not significantly elevated among female operating room nurses.

Similarly, a lower one-sided CI is an interval of the form $Pr(p < c_2) = 1 - \alpha$.

This might be useful in the birthweight example (Section 6.11.2) if erythromycin is seen as only being possibly protective. Such an interval is given by

$$p < \hat{p} + z_{1-\alpha}\sqrt{\frac{\hat{p}\hat{q}}{n}}$$

Suppose in the birthweight example that there were actually 100 discordant pairs instead of 8, which would enable us to use the normal approximation to the binomial distribution. In this case, $\hat{p} = .25$, $n = 100$ and the interval would be

$$p < .25 + 1.645\sqrt{\frac{.25(.75)}{100}}$$

$$= .25 + 1.645(.0433) = .25 + .0712 = .321$$

Thus, $p < .321$ is a lower 95% CI for p, and we would conclude that erythromycin prevents low birthweight deliveries.

Similar methods can be used to obtain one-sided confidence limits for other parameters (e.g., the mean of a normal distribution).

6.1 What is the upper 10th percentile of a chi-square distribution with 5 *df*?

6.2 What is the upper 1st percentile of a chi-square distribution with 3 *df*?

Pathology

The data in Table 6.1 are measurements from a group of 10 normal males and 11 males with left-heart disease taken at autopsy at a particular hospital. Measurements were made on several variables at that time, and the table presents the measurements on total heart weight (THW) and total body weight (BW). Assume that the diagnosis of left-heart disease is made independently of these variables.

6.3 Using the data in Table 6.1 construct a 99% confidence interval for the variance of total heart weight for left-heart disease males.

6.4 Using the data in Table 6.1 answer Problem 6.3 for normal males.

Cancer

A case–control study of the effectiveness of the Pap test in preventing cervical cancer (by identifying precancerous lesions) was performed [1]. It was found that 28.1% of 153 cervical-cancer cases and 7.2% of 153 age-matched (within 5 years) controls had never had a Pap test prior to the time of the case's diagnosis.

6.5 Provide a 95% CI for the percentage of cervical-cancer cases who never had a Pap test.

6.6 Provide a 95% CI for the percentage of controls who never had a Pap test.

6.7 Do you think the Pap test is helpful in preventing cervical cancer?

Hypertension

Hypertensive patients are screened at a neighborhood health clinic and are given methyl dopa, a strong antihypertensive medication for their condition. They are asked to come back 1 week later and have their blood pressures measured again. Suppose the initial and follow-up systolic blood pressures of the patients are given in Table 6.2.

To test the effectiveness of the drug, we want to measure the difference (*D*) between initial and follow-up blood pressures for each person.

6.8 What is the mean and sd of *D*?

6.9 What is the standard error of the mean?

6.10 Assume that *D* is normally distributed with unknown mean and known standard deviation = 20; that is, $D \sim N(\mu, 400)$. Construct a 95% CI for μ.

6.11 Do you have any opinion on the effectiveness of methyl dopa from the results of these 10 patients?

TABLE 6.1 Autopsy data

Left-heart disease males			Normal males		
Observation number	THW (g)	BW (kg)	Observation number	THW (g)	BW (kg)
1	450	54.6	1	245	40.8
2	760	73.5	2	350	67.4
3	325	50.3	3	340	53.3
4	495	44.6	4	300	62.2
5	285	58.1	5	310	65.5
6	450	61.3	6	270	47.5
7	460	75.3	7	300	51.2
8	375	41.1	8	360	74.9
9	310	51.5	9	405	59.0
10	615	41.7	10	290	40.5
11	425	59.7			

TABLE 6.2 Initial and follow-up SBP (mm Hg) for hypertensive patients given methyl dopa

Patient number	Initial SBP	Follow-up SBP
1	200.0	188.0
2	194.0	212.0
3	236.0	186.0
4	163.0	150.0
5	240.0	200.0
6	225.0	222.0
7	203.0	190.0
8	180.0	154.0
9	177.0	180.0
10	240.0	225.0

Cancer

Data from U.S. cancer-tumor registries suggest that of all people with the type of lung cancer where surgery is the recommended therapy, 40% survive for 3 years from the time of diagnosis and 33% survive for 5 years.

6.12 Suppose that a group of patients who would have received standard surgery are assigned to a new type of surgery. Of 100 such patients, 55 survive for 3 years and 45 survive for 5 years. Can we say that the new form of surgery is better in any sense than the standard form of surgery?

Cardiovascular Disease

A recent hypothesis states that vigorous exercise is an effective preventive measure for subsequent cardiovascular death. To test this hypothesis, a sample of 750 men aged 50–75 who report that they jog at least 10 miles per week is ascertained. After 6 years, 64 have died of cardiovascular disease.

6.13 Compute a 95% CI for the incidence of cardiovascular death in this group.

6.14 If the expected death rate from cardiovascular disease over 6 years in 50–75-year-old men based on large samples is 10%, then can a conclusion be drawn concerning this hypothesis from these data?

Cardiovascular Disease

A group of 60 men under the age of 55 with a prior history of myocardial infarction are put on a strict vegetarian diet as part of an experimental program. After 5 years, 3 men from the group have died.

6.15 What is the best point estimate of the 5-year mortality rate in this group of men?

6.16 Derive a 95% confidence interval for the 5-year mortality rate.

6.17 Suppose that from a large sample of men under age 55 with a prior history of myocardial infarction, we know that the 5-year mortality rate is 18%. How does the observed mortality rate obtained in Problem 6.15 compare with the large sample rate of 18%?

TABLE 6.3 Estimation of flow rates (L/sec) at 50% of forced vital capacity by a manual and a digitizer method

Person	Manual method replicate 1	Manual method replicate 2	Digitizer replicate 1	Digitizer replicate 2
1	1.80	1.84	1.82	1.83
2	2.01	2.09	2.05	2.04
3	1.63	1.52	1.62	1.60
4	1.54	1.49	1.49	1.45
5	2.21	2.36	2.32	2.36
6	4.16	4.08	4.21	4.27
7	3.02	3.07	3.08	3.09
8	2.75	2.80	2.78	2.79
9	3.03	3.04	3.06	3.05
10	2.68	2.71	2.70	2.70

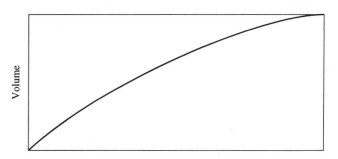

FIGURE 6.1
A typical spirometric
tracing

Pulmonary Disease

A spirometric tracing is a standard device used to measure pulmonary function. These tracings represent plots
of the volume of air expelled over a 6-second period and
tend to look like Figure 6.1. One quantity of interest is
the slope at various points along the curve. The slopes
are referred to as **flow rates.** A problem that arises is that
the flow rates cannot be accurately measured, and some
observer error is always introduced. To quantify the
observer error, an observer measures the flow at 50% of
forced vital capacity (volume as measured at 6 seconds)
twice on tracings from 10 different people. A machine
called a *digitizer* can trace the curves automatically and
can estimate the flow mechanically. Suppose the digitizer
is also used to measure the flow twice on these 10
tracings. The data are given in Table 6.3.

6.18 Find a 95% CI for the standard deviation of the
difference between the first and second replicates using
the manual method.

6.19 Answer Problem 6.18 for the difference between
the first and second replicates using the digitizer method.

Suppose we want to compare the variability of the
two methods within the same person. Let x_{i_1}, x_{i_2}
represent the 2 replicates on the ith person using the
manual methods, and let y_{i_1}, y_{i_2} represent the 2 replicates
on the ith person using the digitizer method. Let

$$d_i = \mid x_{i_1} - x_{i_2} \mid - \mid y_{i_1} - y_{i_2} \mid$$

Then, d_i is a measure of the difference in variability using
the two methods. Assume that d_i is normally distributed
with mean μ_d and variance σ_d^2.

6.20 Find a 95% CI for μ_d.

6.21 What is your opinion on the relative variability of
the two methods?

Turn to the table of random digits (Table 4 in the
Appendix of the main text). Start at the top left. Reading

across, record for each of the first 11 sets of 10 consecutive digits (1) the second digit, X_i, and (2) the mean of
the 10 digits, Y_i.

6.22 Derive the theoretical mean and variance of the
distribution of second digits.

6.23 How do the sample properties of the X_i compare
with the results of Problem 6.22?

6.24 From your answer to Problem 6.22, what should
the mean and variance of the Y_i be?

6.25 How do the actual Y_i's compare with the results of
Problem 6.24?

6.26 Do the actual Y_i's relate to the central-limit theorem
in any way? Elaborate.

TABLE 6.4 Survival time of mice after inoculation
with a specific type of bacteria

Survival time (days)	Number of mice
10	5
11	11
12	29
13	30
14	40
15	51
16	71
17	65
18	48
19	36
20	21
21	12
22	7
23	2
24	1

Bacteriology

Suppose a group of mice are inoculated with a uniform dose of a specific type of bacteria and all die within 24 days, with the distribution of survival times given in Table 6.4.

6.27 Assume that the underlying distribution of survival times is normal. Estimate the probability p that a mouse will survive for 20 or more days.

6.28 Suppose we are not willing to assume that the underlying distribution is normal. Estimate the probability p that a mouse will survive for 20 or more days.

6.29 Compute 95% confidence limits for the parameter estimated in Problem 6.28.

6.30 Compute 99% confidence limits for the parameter estimated in Problem 6.28.

Cardiovascular Disease

In Table 2.1 (p. 2) data on serum-cholesterol levels of 24 hospital employees before and after they adopted a vegetarian diet were provided.

6.31 What is your best estimate of the effect of adopting a vegetarian diet on change in serum-cholesterol levels?

6.32 What is the standard error of the estimate given in Problem 6.31?

6.33 Provide a 95% CI for the effect of adopting the diet.

6.34 What can you conclude from your results in Problem 6.33?

Some physicians consider only changes of at least 10 mg/dL (the same units as in Table 2.1) to be clinically significant.

6.35 Among people with a clinically significant change in either direction, what is the best estimate of the proportion whose cholesterol levels have declined?

6.36 Provide a 95% CI associated with the estimate in Problem 6.35.

6.37 What can you conclude from your results in Problem 6.36?

Pulmonary Disease

Wheezing is a common respiratory symptom reported by both children and adults. A study is conducted to assess whether either personal smoking or maternal smoking is associated with wheezing in children ages 6–19.

6.38 Suppose that 400 children in this age group whose mothers smoke are assessed in 1985, and it is found that 60 report symptoms of wheezing. Provide a point estimate and a 95% CI for the underlying rate of wheezing in the population of children whose mothers smoke.

6.39 Suppose the rate of wheezing in the general population is 10%. What can you conclude from the results in Problem 6.38?

6.40 Suppose we have a subgroup of 30 children where both the mother and the child are smokers. Six of these 30 children have wheezing symptoms. Provide a 95% CI for the true rate of wheezing in this population.

6.41 How do you assess the results in Problem 6.40 if the nationwide rate of wheezing is 10%?

6.42 Suppose 6 events are realized over a 1-year period for a random variable that is thought to be Poisson-distributed. Provide a 95% CI for the true expected number of events over 1 year.

SOLUTIONS

6.1 9.24

6.2 11.34

6.3 A 99% CI for the variance of total heart weight is given by

$$\left[\frac{(n-1)s^2}{\chi^2_{n-1,.995}}, \frac{(n-1)s^2}{\chi^2_{n-1,.005}}\right]$$

We have $s^2 = 19{,}415$, $n = 11$. Therefore, the 99% CI is

$$\left[\frac{10(19{,}415)}{\chi^2_{10,.995}}, \frac{10(19{,}415)}{\chi^2_{10,.005}}\right]$$

$$= \left(\frac{194{,}150}{25.19}, \frac{194{,}150}{2.16}\right)$$

$$= (7707, 89{,}884)$$

6.4 We have that $s^2 = 2{,}217.8$, $n = 10$. Thus, the 99% CI is

$$\left[\frac{9(2{,}217.8)}{\chi^2_{9,.995}}, \frac{9(2{,}217.8)}{\chi^2_{9,.005}}\right]$$

$$= \left(\frac{19{,}960}{23.59}, \frac{19{,}960}{1.73}\right) = (846, 11{,}538)$$

6.5 A 95% CI is given by $\hat{p} \pm 1.96\sqrt{\hat{p}\hat{q}/n} = .281 \pm 1.96\sqrt{.281(.719)/153} = .281 \pm .071 = (.210, .352)$.

6.6 A 95% CI is given by $\hat{p} \pm 1.96\sqrt{\hat{p}\hat{q}/n} = .072 \pm 1.96\sqrt{.072(.928)/153} = .072 \pm .041 = (.031, .113)$.

6.7 Since the confidence intervals do not overlap, it is clear that the true percentage of cervical-cancer cases who do not have a Pap smear is larger than the comparable percentage for controls.

6.8 We have the following data:

Patient number	d_i (I-F)	d_i^2
1	12.0	144
2	−18.0	324
3	50.0	2500
4	13.0	169
5	40.0	1600
6	3.0	9
7	13.0	169
8	26.0	676
9	−3.0	9
10	15.0	225
	151.0	5825

The summary statistics are obtained as follows:

$$\bar{d} = \frac{\Sigma d_i}{n} = \frac{151}{10} = 15.1$$

$$s = \sqrt{\frac{\Sigma d_i^2 - (\Sigma d_i)^2/n}{n-1}}$$

$$= \sqrt{\frac{5825 - 151^2/10}{9}}$$

$$= \sqrt{393.878} = 19.85$$

6.9 $se = \dfrac{s}{\sqrt{n}} = \dfrac{19.85}{\sqrt{10}} = 6.28$

6.10 A 95% CI for μ is provided by

$$\left(\bar{d} - \frac{1.96\sigma}{\sqrt{n}}, \bar{d} + \frac{1.96\sigma}{\sqrt{n}}\right)$$

$$= [15.1 - 1.96(20)/\sqrt{10},\ 15.1 + 1.96(20)/\sqrt{10}]$$

$$= (15.1 - 12.4,\ 15.1 + 12.4) = (2.7, 27.5)$$

6.11 Since the 95% CI in Problem 6.10 does not contain 0, the results allow us to state that the true mean drop in blood pressure is greater than 0. Thus, we conclude from the sample evidence that methyl dopa is effective.

6.12 We want to construct a one-sided confidence interval for the new-surgical-method survival rates. Since $n\hat{p}\hat{q} = 100 \times .55 \times .45 = 24.75$, we use the normal approximation to the binomial, whereby we obtain the following one-sided 95% CI:

$$p > \hat{p} - z_{1-\alpha}\sqrt{\hat{p}\hat{q}/n}$$

$$= .55 - 1.645\sqrt{.55(.45)/100}$$

$$= .55 - .082 = .468$$

Since this confidence interval does *not* include .40, we can say that the 3-year survival rate is improved with the new surgery.

For 5-year survival, we have $\hat{p} = .45$, $\hat{q} = .55$. Thus, the one-sided 95% CI is given by

$$p > \hat{p} - z_{1-\alpha}\sqrt{\hat{p}\hat{q}/n}$$

$$= .45 - 1.645\sqrt{.45(.55)/100}$$

$$= .368$$

Since this interval does *not* contain .33, we can say that the 5-year survival rate is also improved with the new method.

6.13 A 95% CI is given by

$$\hat{p} \pm 1.96\sqrt{\hat{p}\hat{q}/n}$$

$$= \frac{64}{750} \pm 1.96\sqrt{\left(\frac{64}{750}\right)\left(1 - \frac{64}{750}\right)/750}$$

$$= .085 \pm 1.96\sqrt{.085(.915)/750}$$

$$= .085 \pm .020 = (.065, .105)$$

6.14 The rate of .10 is compatible with these data, because it falls within the 95% CI in Problem 6.13. Thus, we *cannot* conclude from these data that jogging 10 miles per week prevents death from cardiovascular disease.

6.15 The best point estimate of the 5-year mortality rate is $3/60 = .05$.

6.16 We must use the exact method for obtaining the confidence interval because $n\hat{p}\hat{q} = 60 \times .05 \times .95 = 2.85 < 5$. We refer to Table 7 (Appendix, text), $\alpha = .05$ and note the two curves with $n = 60$. Since $\hat{p} = .05 < .5$, we refer to the lower horizontal axis and draw a vertical line at $\hat{p} = .05$. We then locate where the vertical line intersects the two $n = 60$ curves and refer to the left vertical axis for the confidence limits. The confidence limits are given by $(.01, .14)$.

6.17 Since 18% is not within the confidence interval obtained in Problem 6.16, we can exclude the possibility that the large-sample mortality rate holds for the group on a vegetarian diet and conclude that the vegetarian-diet group has a lower mortality rate.

6.18 The concern here is with the variability of two methods of measuring flow, a manual method and a digitizer method. A 95% CI must be constructed for

$$\sigma^2_{\text{manual differences}}$$

We have

$$s_{\text{diff}} = 0.0779, n = 10$$

The interval is given by

$$\left(\frac{(n-1)s^2}{\chi^2_{n-1, .975}}, \frac{(n-1)s^2}{\chi^2_{n-1, .025}} \right)$$

$$= \left(\frac{9(0.0779)^2}{\chi^2_{9, .975}}, \frac{9(0.0779)^2}{\chi^2_{9, .025}} \right)$$

$$= \left(\frac{9(0.0779)^2}{19.02}, \frac{9(0.0779)^2}{2.70} \right)$$

$$= (0.0029, 0.0202)$$

So, the 95% CI for

$$\sigma^2_{\text{manual}} = (0.0029, 0.0202)$$

and thus the 95% CI for

$$\sigma_{\text{manual}} = (0.054, 0.142)$$

6.19 The same method can be used for the digitizer differences. We have that $s_{\text{diff}} = 0.0288$. Thus, the 95% CI is given by

$$\left(\frac{(n-1)s^2}{\chi^2_{9, .975}}, \frac{(n-1)s^2}{\chi^2_{9, .025}} \right)$$

$$= \left(\frac{9(0.0288)^2}{19.02}, \frac{9(0.0288)^2}{2.70} \right)$$

$$= (0.00039, 0.00276)$$

Finally, the 95% CI for

$$\sigma_{\text{dig}} = (0.020, 0.053)$$

6.20 $d_i = |\text{manual diff}_i| - |\text{digitizer diff}_i|$. A 95% CI for μ_d comes from the usual t statistic formulation

$$\bar{d} \pm t_{9, .975}\left(\frac{s_d}{\sqrt{10}} \right)$$

$$= 0.0440 \pm 2.262\left(\frac{0.0353}{\sqrt{10}} \right)$$

$$= (0.0187, 0.0693)$$

6.21 The digitizer method appears to be significantly less variable than the manual method, as shown in Problem 6.20, since the 95% CI for μ_d does not include 0.

6.22 $Pr(X_i = i) = .1, i = 0, 1, \ldots, 9$. Therefore,

$$\mu = E(X) = \frac{0 + 1 + \ldots + 9}{10} = 4.5$$

$$\sigma^2 = E(X - \mu)^2 = E(X^2) - \mu^2$$

$$= \frac{0^2 + 1^2 + \ldots + 9^2}{10} - 4.5^2 = 8.25$$

6.23 Since $\bar{x}$ and s^2 are unbiased estimators of μ and σ^2, respectively, the calculated mean and variance of the 11 sample X_i's should approximate the theoretical values in Problem 6.22. Indeed, we have that $X_i = 2, 8, 4, 4, 8, 7, 1, 1, 0, 7, 0$. Thus, $\bar{x} = 3.82 \cong \mu, s^2_x = 10.36 \cong \sigma^2$.

6.24 We have

$$Y_i = \frac{\Sigma X_i}{10}$$

$$E(Y_i) = \frac{\Sigma E(X_i)}{10} = E(X_i) = 4.5$$

$$Var(Y_i) = \frac{Var(X_i)}{10} = \frac{8.25}{10} = 0.825$$

6.25 The calculated mean and variance of the actual Y_i should approximate the theoretical values in Problem 6.24. Since $Var(Y_i) < Var(X_i)$, we would expect the Y_i to be more closely clustered than the X_i about the common mean of 4.5. Indeed, we have that $Y_i = 3.3, 4.0, 4.3, 5.0, 4.0, 4.9, 3.4, 3.2, 6.2, 6.0, 4.0$. Furthermore, $\bar{y} = 4.39 \cong \mu, s^2_y = 1.05 \cong \sigma^2_y$.

6.26 Since the Y_i's are means of random samples selected from a population with mean $= \mu = 4.5$ and variance $= \sigma^2 = 8.25$, the central-limit theorem implies that their distribution will be approximately normal, with mean $\mu_y = 4.5$ and variance $\sigma^2_y = 8.25/10 = 0.825$. The number of samples of size 10(11) is too small to confirm whether the central-limit theorem is applicable to samples of size 10 drawn from this population of random digits.

6.27 We first estimate the parameters μ and σ^2 of the normal distribution. We use $\bar{x}$ as an estimate of μ and s^2 as an estimate of σ^2. We have $\bar{x} = 16.13, s^2 = 7.07$. If we assume that the underlying distribution of survival times is $N(16.13, 7.07)$, then an estimate of the probability that a mouse will survive for 20 or more days is given by

$$Pr(X \geq 19.5) = 1 - \Phi\left(\frac{19.5 - 16.13}{\sqrt{7.07}} \right)$$

$$= 1 - \Phi(1.27)$$

$$= 1 - .8980 = .102$$

6.28 We can treat this random variable as binomial; that is,

$$X_i = 1 \text{ if mouse } i \text{ survives for 20 or more days}$$

$$= 0 \text{ otherwise}$$

and

$$X_i = 1 \text{ with probability } p$$

$$= 0 \text{ with probability } 1 - p$$

Then an estimate of p is given by the sample proportion of successes, namely

$$\hat{p} = \frac{21 + 12 + 7 + 2 + 1}{5 + 11 + \ldots + 1} = \frac{43}{429} = .100$$

6.29 95% CI's for p are given by

$$\hat{p} \pm 1.96 \sqrt{\hat{p}\hat{q}/n}$$
$$= .100 \pm 1.96\sqrt{.100(.900)/429}$$
$$= .100 \pm .028 = (.072, .129)$$

6.30 99% CI's for p are given by

$$\hat{p} \pm 2.576 \sqrt{\hat{p}\hat{q}/n}$$
$$= .100 \pm 2.576\sqrt{.100(.900)/429}$$
$$= .100 \pm .037 = (.063, .138)$$

6.31 We compute $\bar{d} = \Sigma d_i/n = 469/24 = 19.54$ mg/dL, which is our best estimate of the mean decline in serum-cholesterol levels.

6.32 $sem = s/\sqrt{n} = 16.81/\sqrt{24} = 3.43$

6.33 A 95% CI for μ is provided by

$$\bar{d} \pm t_{n-1,.975}s/\sqrt{n}$$
$$= 19.54 \pm t_{23,.975}(3.43)$$
$$= 19.54 \pm 2.069(3.43)$$
$$= 19.54 \pm 7.10 = (12.4, 26.6)$$

6.34 We can conclude that cholesterol levels have declined after adopting the diet since 0 is not within the 95% CI in Problem 6.33.

6.35 Among people with an absolute change of at least 10 mg/dL in either direction, 18 have declined, while 2 have increased. Thus, the proportion who have declined $= 18/20 = .90$. Therefore, 90% of people with a clinically significant change have declined.

6.36 Since $n\hat{p}\hat{q} = 20 \times .9 \times .1 = 1.8 < 5$, we must use the exact method. We refer to Table 7 (Appendix, text),

$\alpha = .05$ and locate .9 on the upper horizontal axis. We then draw a vertical line and find the point of intersection with the $n = 20$ curves. Referring to the right vertical axis, we find that $c_1 = .68$, $c_2 = .99$. Thus, the 95% CI $= (.68, .99)$.

6.37 Since 50% is not within the 95% CI in Problem 6.36, we can conclude that more than 50% of people with a clinically significant change have declined. Thus, among people with a clinically significant change, a decline in cholesterol levels is more likely than an increase.

6.38 The best point estimate is $\hat{p} = 60/400 = .15$. Since $n\hat{p}\hat{q} = 400(.15)(.85) = 51.0$, we can use the normal approximation method to obtain an interval estimate. The estimate is given by

$$\hat{p} \pm 1.96 \sqrt{\hat{p}\hat{q}/n}$$
$$= .15 \pm 1.96 \sqrt{.15(.85)/400}$$
$$= .15 \pm .035 = (.115, .185)$$

6.39 Since 10% is not included in the preceding 95% CI, we can conclude that the true rate of wheezing among children whose mothers smoke is significantly higher than the nationwide rate of wheezing.

6.40 The best point estimate is $\hat{p} = 6/30 = .20$. Since $n\hat{p}\hat{q} = 30(.20)(.80) = 4.8 < 5$, we must use the exact binomial confidence limits. We refer to Table 7a (Appendix, text) for $n = 30$, $\hat{p} = .20$ and obtain the 95% CI $= (.08, .38)$.

6.41 Since the 95% CI in Problem 6.40 includes 10%, we cannot be certain that the true rate of wheezing is higher among smoking children whose mothers smoke than in the general population. We would need a larger population of smoking children whose mothers also smoke to make a more definitive assessment.

6.42 Referring to Table 8 (Appendix, text) with $X = 6$, we obtain the 95% CI $= (2.20, 13.06)$.

REFERENCE

[1] Celentano, D. D., Klassan, A. C., Weisman, C. S., & Rosenshein, N. S. (1988). Cervical-cancer screening practices among older women: Results from the Maryland cervical-cancer case–control study. *Journal of Clinical Epidemiology, 41*(6), 531–541.

CHAPTER SEVEN

HYPOTHESIS TESTING: ONE-SAMPLE INFERENCE

SECTION 7.1   Fundamentals of Hypothesis Testing

Let us focus on the birthweight example mentioned in section 6.11. Suppose instead of dichotomizing the outcome, we compute differences in birthweight $x_{i_1} - x_{i_2}$, i.e., (drug $-$ placebo) $= x_i$ for all 50 pairs, not just discordant pairs. Assume $x_i \sim N(\mu, \sigma^2)$, $\sigma = 30$ oz. If $\bar{x} = 8$ oz, then is this far enough from zero for us to be convinced that the drug is effective in increasing mean birthweight? We set up two hypotheses:

(1) H_0 = null hypothesis, $\mu = \mu_0 = 0$.

(2) H_1 = alternative hypothesis, $\mu = \mu_1 > \mu_0$.

Under H_0, the mean birthweight for the drug and placebo babies are the same (i.e., the drug is ineffective). Under H_1, the mean birthweight for the drug babies is greater than for placebo babies (i.e., the drug is effective in increasing mean birthweight). How can we use the data to decide which of these two hypotheses is likely to be correct?

What types of errors can we make? If we decide that H_1 is true and H_0 is actually true, then we call this a *type I error*. If we decide that H_0 is true and H_1 is actually true, then we call this a *type II error*. The probabilities of these two types of errors are called α and β, respectively. Also, the *power of a test* is defined as $1 - \beta =$ the probability of deciding that H_1 is true when H_1 is actually true.

We usually set $\alpha = .05$ by convention. Another name for the decision that H_1 is true is that we reject H_0; another name for the decision that H_0 is true is that we accept H_0. We will use hypothesis tests that for a given α error will minimize the type II error (or equivalently maximize the power).

SECTION 7.2   One-Sample z Test

For the birthweight example, we will base the test on $\bar{x}$ and will reject H_0 for large values of $\bar{x}$ (that is, if $\bar{x} > c$). How large should c be? This depends on the type I error.

For a type I error $= \alpha$, $c = \mu_0 + z_{1-\alpha}\dfrac{\sigma}{\sqrt{n}}$. Therefore, we

Reject H_0 if $\bar{x} > \mu_0 + z_{1-\alpha}\dfrac{\sigma}{\sqrt{n}}$

Accept H_0 if $\bar{x} \leq \mu_0 + z_{1-\alpha}\dfrac{\sigma}{\sqrt{n}}$

We refer to this method of hypothesis testing as the *critical-value method* and refer to the set of values of $\bar{x}$ for which we reject H_0 (that is, $\bar{x} > \mu_0 + z_{1-\alpha}\sigma/\sqrt{n}$) as the *critical region.*

Another way of implementing the critical-value method is by using a *test statistic.* In general, we reject H_0 if the test statistic is in the critical region and accept H_0 otherwise. For the birthweight example, the test statistic is

$$z = \frac{\bar{x} - \mu_0}{\sigma/\sqrt{n}}$$

and we reject H_0 if $z > z_{1-\alpha}$ and accept H_0 if $z \leq z_{1-\alpha}$. The name of this test is the *one-sample z test.* In our case, $\mu_0 = 0$, $\sigma = 30$, $n = 50$, $z_{1-\alpha} = z_{.95} = 1.645$. Therefore, we reject H_0 if

$$z = \frac{\bar{x} - 0}{30/\sqrt{50}} > 1.645$$

and accept H_0 otherwise. Since $\bar{x} = 8$ oz, $z = (8-0)/(30/\sqrt{50}) = 8/4.243 = 1.886 > 1.645$, we reject H_0.

What if we conducted our test at the 1% significance level? Since $z_{.99} = 2.326$ and $z = 1.886 < 2.326$, this implies that we would accept H_0 in this case. Instead of conducting significance tests at many different α levels, we can use a different method of hypothesis testing called the *p-value method.* A **p-value** is defined as the α level at which we would be indifferent between accepting and rejecting H_0 based on the observed data. In this case, *p-value* $= 1 - \Phi(z)$.

If we use an $\alpha \leq p$, then we accept H_0.

If we use an $\alpha > p$, then we reject H_0.

In our case,
$$p = 1 - \Phi(1.886) = 1 - .970 = .030 = p\text{-value}$$

SECTION 7.3   Guidelines for Assessing Statistical Significance

How can we interpret the *p*-value? If $p < .05$, then the results are statistically significant. Different levels of significance are sometimes noted. Specifically, if

   $.01 \leq p < .05$ significant

   $.001 \leq p < .01$ highly significant

   $p < .001$ very highly significant

If $p \geq .05$, then the results are not statistically significant. However, if $.05 \leq p < .10$ then a trend toward statistical significance is sometimes noted.

Statistical Significance Versus Clinical Significance

Suppose $\bar{x} = 1$ oz, $n = 10,000$; then

$$p = 1 - \Phi\left(\frac{1}{30/\sqrt{10,000}}\right) = 1 - \Phi(3.33) = 1 - .99957 \approx .0004 < .001$$

The results are statistically significant, but not very important.

Conversely, suppose that $\bar{x} = 4$ oz., $n = 50$; then $p = 1 - \Phi\left(\frac{4}{30/\sqrt{50}}\right)$ $= 1 - \Phi(0.943) = 1 - .827 = .173$.

The results are not significant but could possibly be important if confirmed by a larger study.

SECTION 7.4   Two-Sided Alternatives

A randomized study was conducted among patients with unstable angina (i.e., angina severe enough to require hospitalization). The study was based on a comparison of two drugs: Nifedipine and Propranolol. The patients were not matched.

Primary outcome measure—relief of pain within 48 hours

Secondary outcome measure(s)—change in heart rate and blood pressure from baseline to 48 hours.

If we focus on the secondary outcome measures, then we're interested in changes in these parameters within each drug group separately and also in a comparison of the change between the two drug groups. Let's look at each drug group separately now. We will compare the two drug groups in Chapter 8. The results are as follows:

Nifedipine group: mean change (heart rate 48 hours $-$ heart-rate baseline) in heart rate $= \bar{x} = +3.0$ beats per minute, $sd = 10.0$, $n = 61$

Propranolol group: mean change in heart rate $= \bar{x} = -2.2$ beats per minute, $sd = 10.0$, $n = 61$

Assume σ is known. We wish to test H_0: $\mu = 0$ is H_1: $\mu \neq 0$ and assume $X \sim N(\mu, \sigma^2)$. In this case, we will reject H_0 if $\bar{x}$ (or equivalently, the test statistic $z = (\bar{x} - \mu_0)/(\sigma/\sqrt{n})$ is either too small or too large. This type of alternative is called a *two-sided alternative*, because we are interested in deviations from the null hypothesis where either $\mu < \mu_0 = 0$ or $\mu > \mu_0 = 0$. In contrast, the alternative hypothesis for the birthweight example was a one-sided alternative (i.e., $\mu > \mu_0$).

The specific test criteria for a level of significance $= \alpha$ are

Reject H_0 if $z < z_{\alpha/2} = -z_{1-\alpha/2}$

or $z > z_{1-\alpha/2}$

Accept H_0 otherwise.

Alternatively, to compute a *p*-value

If $z < 0$, then $p = 2\Phi(z)$

If $z \geq 0$, then $p = 2 \times [1 - \Phi(z)]$

Nifedipine Group

Assess if mean heart rate has changed significantly among patients in the nifedipine group. Assume that the true value of $\sigma = 10$. In this case, our hypotheses are

$$H_0: \mu = 0, \text{ versus } H_1: \mu \neq 0$$

The test statistic is

$$z = \frac{3 - 0}{10/\sqrt{61}} = \frac{3}{1.280} = 2.343$$

Since $z > 0$,

$$p\text{-value} = 2 \times [1 - \Phi(2.343)] = 2 \times (1 - .9904) = .019$$

Thus, mean heart rate has significantly increased after 48 hours in the nifedipine group.

Propranolol Group

We test the same hypotheses as for the nifedipine group. The test statistic is

$$z = \frac{-2.2 - 0}{10/\sqrt{61}} = \frac{-2.2}{1.280} = -1.718$$

Since $z < 0$,

$$p\text{-value} = 2 \times \Phi(-1.718) = 2 \times [1 - \Phi(1.718)]$$
$$= 2 \times (1 - .9571) = 2 \times .0429 = .086$$

Thus, there is no significant change in mean heart rate over 48 hours for the propranolol group.

Notice that it is possible for results to be significant using a one-sided test, while they would not be significant using a two-sided test. For example, if we tested the hypothesis $H_0: \mu = 0$ versus $H_1: \mu < 0$, the one-tailed p-value for the propranolol group would be $p = .043$, which is statistically significant.

Generally, most research results are reported using two-sided p-values because the direction of the alternative hypothesis is not known in advance. A strong case needs to be made for using a one-sided p-value and should be written down in a protocol before a study begins. All tests for the remainder of this study guide will be two-sided.

SECTION 7.5   One-Sample t Test

Suppose we don't assume that σ is known to be 10 in the unstable angina example but instead estimate it from the sample standard deviation $= s$. How does the test procedure change in this case? The hypotheses are now $H_0: \mu = \mu_0$, σ^2 unknown versus $H_1: \mu \neq \mu_0$, σ^2 unknown.

We base our significance test on the test statistic $t = \dfrac{\bar{x} - \mu_0}{s/\sqrt{n}}$. Also, the critical region will be based on percentiles of the t distribution rather than the $N(0, 1)$ distribution that we used for the one-sample z test. Specifically, for a significance level of α, we reject H_0 if $t < c_1 = t_{n-1,\alpha/2} = -t_{n-1,1-\alpha/2} - t_{n-1,1-\alpha/2}$ or $t > c_2 = t_{n-1,1-\alpha/2}$ and accept H_0 otherwise.

To compute the *p*-value, if $t > 0$, *p*-value $= 2 \times Pr(t_{n-1} > t)$. If $t < 0$, *p*-value $= 2 \times Pr(t_{n-1} < t)$.

The name of this procedure is the *one-sample t test*.

Nifedipine Group

Assess if the mean heart rate has changed significantly among patients in the nifedipine group, assuming that the true variance is unknown. Our hypotheses are H_0: $\mu = 0$, σ^2 unknown versus H_1: $\mu \neq 0$, σ^2 unknown. The test statistic is

$$t = \frac{3.0}{10/\sqrt{61}} = \frac{3}{1.28} = 2.343 \sim t_{60}$$

Since $t_{60, .99} = 2.390$ and $t_{60, .975} = 2.000$, it follows that $.01 < p/2 < .025$, or $.02 < p < .05$. Thus, there is a significant increase in heart rate for nifedipine users. The exact *p*-value $= .022$ versus $.019$ for the one-sample *z* test.

Propranolol Group

Test the same hypotheses as for the nifedipine group. The test statistic is

$$t = \frac{-2.2}{10/\sqrt{61}} = \frac{-2.2}{1.28} = -1.718 \sim t_{60}$$

Since $t_{60, .95} = 1.671$ and $t_{60, .975} = 2.000$, it follows that $.025 < p/2 < .05$, or $.05 < p < .10$. Thus, there is no significant change in heart rate for propranolol users. The exact *p*-value $= .091$ versus $.086$ for the one-sample *z* test.

SECTION 7.6 The Power of a Test

Suppose we wish to test the hypotheses H_0: $\mu = \mu_0$ versus H_1: $\mu = \mu_1 > \mu_0$.

The power of the test

$$= \text{Prob(reject null hypothesis given } H_1 \text{ true)}$$

$$= \Phi\left[\frac{\sqrt{n}\,|\mu_1 - \mu_0|}{\sigma} - z_{1-\alpha} \right]$$

The power depends on

(1) n; as $n \uparrow$, power $\uparrow$

(2) $|\mu_1 - \mu_0|$; as $|\mu_1 - \mu_0| \uparrow$, power $\uparrow$

(3) σ; as $\sigma \uparrow$, power $\downarrow$

(4) α; as $\alpha \downarrow$, power $\downarrow$

It is important for a study to have adequate power to achieve its goals. Consider the birthweight example. Suppose $\mu_1 - \mu_0 = 4$ oz, $\alpha = .05$, $\sigma = 30$ oz, $n = 50$. What power does such a study have?

$$\text{Power} = \Phi\left(\frac{\sqrt{50}\,(4)}{30} - 1.645 \right) = \Phi(-0.702) = .24$$

Therefore we will be able to detect a significant difference 24% of the time based on a sample of size 50. To compute power for a two-sided alternative, substitute $z_{1-\alpha/2}$ for $z_{1-\alpha}$ in the preceding formula.

SECTION 7.7   Sample Size

Another way of looking at the same question is to ask how many subjects do we need to achieve a given level of power. Specifically, if we wish to use the one-sided alternative given in Section 7.6 with a type I error of α and a power of $1 - \beta$, then the required sample size is

$$n = \frac{\sigma^2(z_{1-\beta} + z_{1-\alpha})^2}{(\mu_1 - \mu_0)^2}$$

Sample size is related to

(1) α; as $\alpha \downarrow, n \uparrow$

(2) β; as $\beta \downarrow, n \uparrow$

(3) σ; as $\sigma \uparrow, n \uparrow$

(4) $|\mu_1 - \mu_0|$; as $|\mu_1 - \mu_0| \uparrow, n \downarrow$

Suppose $\mu_1 - \mu_0 = 8$ oz, $\alpha = .05$, $\beta = .20$, $\sigma = 30$. How many subjects are needed? In this case, $z_{1-\beta} = z_{.8} = 0.84$, $z_{1-\alpha} = z_{.95} = 1.645$ and

$$n = \frac{900(0.84 + 1.645)^2}{64} = \frac{900(6.175)}{64}$$

$$= 86.8 \rightarrow 87 \text{ matched pairs or } 174 \text{ subjects}$$

For a two-sided test, we substitute $z_{1-\alpha/2}$ for $z_{1-\alpha}$ in the preceding sample-size formula.

SECTION 7.8   Relationship Between Hypothesis Testing and Confidence Intervals

Notice that from Chapter 6, another method for making the decision about whether $\mu = \mu_0$ (that is, whether the null hypothesis is true), would be to construct a 95% CI for μ and note if μ_0 falls in this interval.

From a *CI approach,* we decide that μ_0 is a likely value for μ if

$$\bar{x} - z_{1-\alpha/2}\sigma/\sqrt{n} < \mu_0 < \bar{x} + z_{1-\alpha/2}\sigma/\sqrt{n} \qquad \text{(1)}$$

and decide it is an unlikely value, otherwise.

From a *hypothesis-testing approach,* we will accept H_0 if

$$\mu_0 - z_{1-\alpha/2}\sigma/\sqrt{n} < \bar{x} < \mu_0 + z_{1-\alpha/2}\sigma/\sqrt{n} \qquad \text{(2)}$$

and reject H_0 otherwise.

Note that Equations **(1)** and **(2)** are equivalent algebraically. Thus, the two criteria are the same. We will accept (reject) H_0 at level α using a two-sided test if and only if the two-sided $100\% \times (1 - \alpha)$ CI for μ does (does not) contain μ_0.

7.8.1   **Relative Advantages of Hypothesis-Testing Versus Confidence-Interval Approaches**

Hypothesis-testing approach	**Confidence-interval approach**
(1) We obtain an exact p-value.	**(1)** Only known if $p < .05$ or $p > .05$.
(2) We can distinguish between $\mu = \mu_0$ and $\mu \neq \mu_0$.	**(2)** We obtain an interval within which the parameter is likely to fall.

The two methods complement each other.

SECTION 7.9   **One-Sample χ^2 Test**

A new eye chart (the Ferris chart) has been used to measure visual acuity in several research studies. Different charts are used for the left and right eye of an individual. A measure of acuity is the number of letters read correctly. One measure of variability is the number of letters read with the left eye − the number of letters read with the right eye. The variability should increase with the new chart versus the standard eye chart (the Snellen chart), where the same chart is used for each eye. Let x_i = the number of letters read correctly with the left eye − the number of letters read correctly with the right eye for the ith subject. We assume $x_i \sim N(0, \sigma^2)$. Suppose σ is known to be 1.5 based on previous studies with the Snellen chart. We collect data on 20 subjects using the new chart and find $s = 2.0$. Is there significantly more variability with the new chart?

Let σ_0^2 = true variance for the Snellen chart (known), σ^2 = true variance for the Ferris chart (unknown). We wish to test the hypothesis H_0: $\sigma^2 = \sigma_0^2 = 2.25$ versus H_1: $\sigma^2 \neq \sigma_0^2$. We base our test on the test statistic

$$X^2 = \frac{(n-1)s^2}{\sigma_0^2} \sim \chi_{n-1}^2 \text{ under } H_0$$

Critical-Value Method

Reject H_0 if

$$X^2 < c_1 = \chi_{n-1, \alpha/2}^2 \quad \text{or} \quad X^2 > c_2 = \chi_{n-1, 1-\alpha/2}^2$$

Accept H_0 otherwise.

p-Value Method

$$p\text{-value} = 2 \times Pr[\chi_{n-1}^2 > X^2] \quad \text{if } s^2 > \sigma_0^2,$$
$$= 2 \times Pr[\chi_{n-1}^2 < X^2] \quad \text{if } s^2 < \sigma_0^2$$

In this case, $X^2 = 19(4.0)/2.25 = 33.78$. Since $c_2 = \chi_{19, .975}^2 = 32.85$, $c_1 = \chi_{19, .025}^2 = 8.91$ and $X^2 > c_2$, it follows that $p < .05$. The p-value $= 2 \times Pr(\chi_{19}^2 > 33.78)$. Since $\chi_{19, .99}^2 = 36.19$, $\chi_{19, .975}^2 = 32.85$ and $32.85 < 33.78 < 36.19$, it follows that $.02 < p < .05$. Thus, there is significantly more variability between left and right eyes using the Ferris chart compared with the Snellen chart. This may actually indicate that the Ferris chart is more valid, because we are obtaining independent assessments of acuity with the left and right eyes.

One-Sample Inference for the Binomial Distribution

The incidence of myocardial infarction (MI) in 50–59-year-old postmenopausal women is on the order of 2 events per 1000 women per year. Suppose in a large group-health practice, 1000 women are identified who have been taking post-menopausal hormones (PMH) for at least 1 year as of 1985, of whom 3 have an MI from 1985 to 1990. Is this an unusual event. We wish to test the hypothesis: H_0: $p = p_0$ versus H_1: $p \neq p_0$ where p_0 = overall population incidence rate, and p = true incidence rate for PMH users.

We use a normal-theory method if $np_0q_0 \geq 5$ and an exact method otherwise.

Normal-Theory Method

We use the test statistic

$$z = \frac{\hat{p} - p_0}{\sqrt{p_0q_0/n}} \sim N(0, 1) \text{ under } H_0$$

where $\hat{p}$ = sample proportion of events (or more generally, successes). If $z \leq 0$, then the p-value = $2\Phi(z)$; if $z > 0$, then the p-value = $2[1 - \Phi(z)]$.

In this case, $n = 1000$, $\hat{p} = 3/1000 = .003$, $p_0 = 5(2/1000) = .01$, $q_0 = .99$, $np_0q_0 = 9.9 > 5$. Thus, we can use the normal-theory test. The test statistic is

$$z = \frac{.003 - .01}{\sqrt{.01(.99)/1000}} = \frac{-.007}{.0031} = -2.225$$

Since $z < 0$, the p-value = $2 \times [1 - \Phi(2.225)] = 2 \times [1 - .98695] = .026$.

Thus, there are significantly fewer MI's than expected among postmenopausal estrogen users. This could form the basis for a better study using a control group of postmenopausal women who never used estrogens.

Exact Method

If $np_0q_0 < 5$, then we have to use the exact method. The exact method can also be used at any time, but is more difficult computationally than the normal-theory test.

$$\text{If } \hat{p} < p_0, \text{ then } p\text{-value} = 2 \times \sum_{k=0}^{x} \binom{n}{k} p_0^k q_0^{n-k}$$

$$\text{If } \hat{p} \geq p_0, \text{ then } p\text{-value} = 2 \times \sum_{k=x}^{n} \binom{n}{k} p_0^k q_0^{n-k}$$

where x = observed number of events (successes).

We now apply the exact method to the MI data for comparison with the normal-theory method. In our case, $\hat{p} = .003 < p_0 = .01$. Thus,

$$p\text{-value} = 2 \times \sum_{k=0}^{3} \binom{1000}{k}(.01)^k(.99)^{1000-k}$$

Using the recursion rule,

$$Pr(0) = (.99)^{1000} \qquad\qquad\qquad = .000043$$

$$Pr(1) = \frac{1000}{1} \times \frac{1}{99} \times .000043 = .000436$$

$$Pr(2) = \frac{999}{2} \times \frac{1}{99} \times .000436 \;\; = .002200$$

$$Pr(3) = \frac{998}{3} \times \frac{1}{99} \times .002200 \;\; = .007393$$

$$Pr(X \le 3) \qquad\qquad\qquad\qquad = .010073$$

$$2 \times Pr(X \le 3) \qquad\qquad\qquad = .020$$

Thus, the results are also significant using exact methods. We only need to use exact methods if $np_0 q_0 < 5$.

SECTION 7.11   One-Sample Inference for the Poisson Distribution

The usual annual death rate due to asthma in England over the period 1862–1962 for people aged 5–34 was approximately 1 death per 100,000 people. Suppose that in 1963, 20 deaths due to asthma were observed among 1 million people in this age group living in Britain. Is this number of deaths inconsistent with the preceding 100-year rate?

Since n is large (1 million) and $p =$ the probability of death due to asthma is small, we can model the number of asthma deaths in 1963 by a Poisson distribution with parameter $= \mu$. Specifically, we wish to test the hypothesis $H_0: \mu = \mu_0$ versus $H_1: \mu \ne \mu_0$, where

$$\mu = \text{true expected number of deaths due to asthma in 1963}$$

$$\mu_0 = \text{expected number of deaths due to asthma in 1963 if}$$
$$\text{the asthma death rate from 1862–1962 holds in 1963}$$

To obtain a p-value,

$$p = \min\left[2 \times \sum_{k=0}^{x} \frac{e^{-\mu_0}\mu_0^{k}}{k!}, 1\right] \quad \text{if } x < \mu_0$$

$$p = \min\left[2 \times \left(1 - \sum_{k=0}^{x-1} \frac{e^{-\mu_0}\mu_0^{k}}{k!}\right), 1\right] \quad \text{if } x \ge \mu_0$$

where $x =$ observed number of deaths (or in general, events).

In this example, $x = 20$, $\mu_0 = 1,000,000(1/100,000) = 10$. Therefore, the p-value is given by

$$p = 2 \times \left(1 - \sum_{k=0}^{19} \frac{e^{-10}10^{k}}{k!}\right) = 2 \times \left(\sum_{k=20}^{\infty} \frac{e^{-10}10^{k}}{k!}\right)$$

We refer to Table 2 (see the appendix in the text) under $\mu = 10$. Since $Pr(X = 25 \,|\, \mu = 10) = .0000$, we compute

$$p = 2 \times [Pr(20) + Pr(21) + Pr(22) + Pr(23) + Pr(24)]$$
$$= 2 \times (.0019 + .0009 + .0004 + .0002 + .0001)$$
$$= .007$$

Thus, there are significantly more asthma deaths in 1963 than would be expected based on the asthma death rate from 1862 to 1962.

PROBLEMS

Nutrition

As part of a dietary-instruction program, ten 25–34-year-old males adopted a vegetarian diet for 1 month. During the diet, the average daily intake of linoleic acid was 13 g.

7.1 If the average daily intake among 25–34-year-old males in the general population is 15 g with standard deviation 4 g, then, using a significance level of .05, test the hypothesis that the intake of linoleic acid in this group is lower than that in the general population.

7.2 Compute a p-value for the hypothesis test in Problem 7.1.

As part of the same program, eight 25–34-year-old females report an average daily intake of saturated fat of 11 g.

7.3 If the average daily intake of saturated fat among 25–34-year-old females in the general population is 24 g with standard deviation 11 g, then, using a significance level of .01, test the hypothesis that the intake of saturated fat in this group is lower than that in the general population.

7.4 Compute a p-value for the hypothesis test in Problem 7.3.

7.5 What is the relationship between your answers to Problems 7.3 and 7.4?

Suppose we are uncertain what effect a vegetarian diet will have on the level of linoleic-acid intake in Problem 7.1.

7.6 What are the null and alternative hypotheses in this case?

7.7 Compare the mean level of linoleic acid in the vegetarian population with that of the general population under the hypotheses in Problem 7.6. Report a p-value.

7.8 Suppose $s = 5$ based on a sample of 20 subjects. Test the null hypothesis H_0: $\sigma^2 = 16$ versus H_1: $\sigma^2 \neq 16$ using the critical-value method based on a significance level of .05.

7.9 Answer Problem 7.8 using the p-value method.

7.10 Suppose the sample standard deviation of linoleic acid is 6 g in Problem 7.7. Assume that the standard deviation for the vegetarian population is not known, and perform the hypothesis test for the hypotheses in Problem 7.6. Report a p-value.

7.11 Compute a lower one-sided 95% CI for the true mean intake of linoleic acid in the vegetarian population depicted in Problem 7.1. Assume that the standard deviations of the vegetarian population and the general population are the same.

7.12 How does your answer to Problem 7.11 relate to your answer to Problem 7.1?

Nutrition

A food-frequency questionnaire was mailed to 20 subjects to assess the intake of various food groups. The sample standard deviation of vitamin-C intake over the 20 subjects was 15 (exclusive of vitamin-C supplements). Suppose we know from using an in-person diet-interview method in a large previous study that the standard deviation is 20.

7.13 What hypotheses can be used to test if there are any differences between the standard deviations of the two methods?

7.14 Perform the test described in Problem 7.13 and report a p-value.

The sample standard deviation for ln (vitamin-A intake) exclusive of supplements based on the 20 subjects using the food-frequency questionnaire was 0.016. Suppose the standard deviation from the diet-interview method is known to be 0.020 based on a large previous study.

7.15 Test the hypothesis that the variances using the two methods are the same. Use the critical-value method with $\alpha = .05$.

7.16 Report a p-value corresponding to the test in Problem 7.15.

Occupational Health

Suppose that 28 cancer deaths are noted among workers exposed to asbestos in a building-materials plant from 1981–1985. Only 20.5 cancer deaths are expected from statewide cancer-mortality rates.

7.17 What is the estimated SMR for total cancer mortality?

7.18 Is there a significant excess or deficit of total cancer deaths among these workers?

In the same group of workers, 7 deaths due to leukemia are noted. Only 4.5 are expected from statewide rates.

7.19 What is the estimated SMR for leukemia?

7.20 Is there a significant excess or deficit of leukemia deaths among these workers?

Infectious Disease

Suppose the annual incidence of diarrhea (defined as 1+ episodes per year) in a Third-World country is 5% in children under the age of 2.

7.21 If 10 children out of 108 under the age of 2 in a poor rural community in the country have 1+ episodes of diarrhea in a year, then test if this represents a significant departure from the overall rate for the country using the critical-value method.

7.22 Report a *p*-value corresponding to your answer to Problem 7.21.

Ophthalmology

Suppose the distribution of systolic blood pressure in the general population is normal with a mean of 130 mm Hg and a standard deviation of 20 mm Hg. In a special subgroup of 85 people with glaucoma, we find that the mean systolic blood pressure is 135 mm Hg with a standard deviation of 22 mm Hg.

7.23 Assuming that the standard deviation of the glaucoma patients is the same as that of the general population, test for an association between glaucoma and high blood pressure.

7.24 Answer Problem 7.23 without making the assumption concerning the standard deviation.

Occupational Health

7.25 Suppose it is known that the average life expectancy of a 50-year-old man in 1945 was 18.5 years. Twenty men aged 50 who have been working for at least 20 years in a potentially hazardous industry were ascertained in 1945. On follow-up in 1985 all the men have died, with an average lifetime of 16.2 years and a standard deviation of 7.3 years since 1945. Assuming that life expectancy of 50-year-old men is approximately normally distributed, test if the underlying life expectancy for workers in this industry is shorter than for comparably aged men in the general population.

Cancer

7.26 An area of current interest in cancer epidemiology is the possible role of oral contraceptives (OC's) in the development of breast cancer. Suppose that in a group of 1000 premenopausal women ages 40–49 who are current users of OC's, 15 subsequently develop breast cancer over the next 5 years. If the expected 5-year incidence rate of breast cancer in this group is 1.2% based on national incidence rates, then test the hypothesis that there is an association between current OC use and the subsequent development of breast cancer.

Cancer

A group of investigators wishes to explore the relationship between the use of hair dyes and the development of breast cancer in females. A group of 1000 beauticians 30–39 years of age is identified and followed for 5 years. After 5 years, 20 new cases of breast cancer have occurred. Assume that breast-cancer incidence over this time period for an average woman in this age group is 7/1000. We wish to test the hypothesis that using hair dyes increases the risk of breast cancer.

7.27 Is a one-sided or two-sided test appropriate here?

7.28 Test the hypothesis.

Renal Disease

The level of serum creatinine in the blood is considered a good indicator of the presence or absence of kidney disease. Normal people generally have low concentrations of serum creatinine, whereas diseased people have high concentrations. Suppose we want to look at the relation between analgesic abuse and kidney disorder. In particular, suppose we look at 15 people working in a factory who are known to be "analgesic abusers" (i.e., they take more than 10 pills per day) and we measure their creatinine levels. The creatinine levels are

0.9, 1.1, 1.6, 2.0, 0.8,
0.7, 1.4, 1.2, 1.5, 0.8,
1.0, 1.1, 1.4, 2.2, 1.4

7.29 If we assume that creatinine levels for normal people are normally distributed with mean 1.0 and standard deviation 0.40, then can we make any comment about the levels for analgesic abusers via some statistical test?

7.30 Suppose we are skeptical about assuming that the standard deviation is known. Can we test the validity of this assumption?

7.31 If we do not want to assume that the standard deviation is known, then answer the question in Problem 7.29 in some other way.

Otolaryngology

Otitis media is an extremely common disease of the middle ear in children under 2 years of age. It can cause prolonged hearing loss during this period and may result in subsequent defects in speech and language. Suppose we wish to design a study to test the latter hypothesis, and we set up a study group consisting of children with 3 or more episodes of otitis media in the first 2 years of life. We have no idea what the size of the effect will be. Thus, we set up a pilot study with 20 cases, whereby we find that 5 of the cases have speech and language defects at age 3.

7.32 If we regard this experience as representative of what would occur in a large study and if we *know* that 15% of all normal children have speech and language defects by age 3, then how large a study group is needed to have an 80% chance of detecting a significant difference using a one-sided test at the 5% level?

7.33 Suppose only 50 cases can be recruited for the study group. How likely are we to find a significant difference if the true proportion of affected children with speech and language defects at age 3 is the same as that in the pilot study?

Epidemiology

Height and weight are often used in epidemiological studies as possible predictors of disease outcomes. If the people in the study are assessed in a clinic, then heights and weights are usually measured directly. However, if the people are interviewed at home or by mail, then a person's self-reported height and weight are often used instead. Suppose we conduct a study on 10 people to test the comparability of these two methods. The data for weight are given in Table 7.1.

7.34 Should a one-sided or two-sided test be used here?

7.35 Which test procedure should be used to test the preceding hypothesis?

7.36 Conduct the test in Problem 7.35 using the critical-value method with $\alpha = .05$.

7.37 Compute the *p*-value for the test in Problem 7.35.

7.38 Is there evidence of digit preference among the self-reported weights? Specifically, compare the observed proportion of reported weights whose last digit is 0 or 5 with the expected proportion based on chance and report a *p*-value.

TABLE 7.1 A comparison of self-reported and measured weight (lb) for 10 subjects

Subject number	Self-reported weight	Measured weight	Difference
1	120	125	−5
2	120	118	+2
3	135	139	−4
4	118	120	−2
5	120	125	−5
6	190	198	−8
7	124	128	−4
8	175	176	−1
9	133	131	+2
10	125	125	0

Hypertension

Several studies have been performed relating urinary-potassium excretion to blood-pressure level. These studies have tended to show an inverse relationship between these two variables, with the higher the level of potassium excretion, the lower the BP level. Therefore, a treatment trial is planned to look at the effect of potassium intake in the form of supplement capsules on changes in DBP level. Suppose that in a pilot study, 20 people are given potassium supplements for 1 month. The data are as follows:

Mean change	−3.2
(1 month–baseline)	
sd change	8.5
n	20

7.39 What test should be used to assess if the potassium supplements have any effect on DBP level?

7.40 Perform the test in Problem 7.39 using a two-sided test and report the *p*-value.

7.41 Derive a 95% CI for the true mean change based on the preceding data. What is the relationship of your results here and in Problem 7.40?

7.42 How many subjects need to be studied to have an 80% chance of detecting a significant treatment effect using a two-sided test with an α level of .05 if the mean and sd of the pilot study are assumed to be the population mean and sd?

Hospital Epidemiology

A study was conducted to identify characteristics that would predict 1-year survival for patients admitted to the medical service at New York Hospital [1]. One factor that was considered was the physician's estimate of the patients' severity of illness at the time of admission.

Suppose that it is expected, based on previous studies in this hospital, that 2/3 of admitted patients will survive for at least 1 year.

7.43 If 47% of 136 patients deemed severely ill survive at least 1 year, then what test can be used to test the hypothesis that the severity-of-illness rating is predictive of 1-year mortality?

7.44 Perform the test in Problem 7.43 using the critical-value method.

7.45 Provide a 95% CI for the 1-year survival rate among severely ill patients.

Suppose that it is expected, based on previous studies, that among *all patients who survive hospitalization,* 75% will survive for 1 year. Furthermore, of the 136 severely ill patients, 33 die during hospitalization, and an additional 39 will die during the 1st year, but after hospitalization.

7.46 Test the hypothesis that severity of illness is predictive of 1-year mortality among patients who are discharged from the hospital. Report a *p*-value.

Cancer

Radiotherapy is a common treatment for breast cancer, with about 25% of women receiving this form of treatment. Assume that the figure 25% is based on a very large sample and is known *without error.* One hypothesis is that radiotherapy applied to the contralateral breast may be a risk factor for development of breast cancer in the opposite breast 5 or more years after the initial cancer. Suppose that 655 women are identified who developed breast cancer in the opposite breast 5 or more years after the initial cancer [2].

7.47 If 206 of the women received radiotherapy after their initial diagnosis, then test the hypothesis that radiotherapy is associated with the development of breast cancer in the opposite breast. Please report a *p*-value.

7.48 Provide a 95% CI for the true proportion of women with contralateral breast cancer who received radiotherapy treatment.

7.49 Suppose that our *p*-value in Problem 7.47 = .03 (this is not necessarily the actual *p*-value in Problem 7.47). If we conduct a test using the critical-value method with $\alpha = .05$, then would we accept or reject H_0 and why? (Do not actually conduct the test.)

7.50 Suppose we wanted a 99% CI instead of the 95% CI in Problem 7.48. Would the length of the 99% CI be narrower, wider, or the same as the CI in Problem 7.48? (Do not actually construct the interval.)

Renal Disease

To compare two methods for the analysis of uric acid, 23 blood samples from 23 individuals were divided and analyzed both by the colorimetric method and the uricase method. Suppose the sample means of the colorimetric and uricase assessments were 6.26 and 6.20 mg/dL, respectively, the sample standard deviation of the paired difference between repeated assessments is .50 mg/ dL, and it is reasonable to assume that this paired difference has a normal distribution.

7.51 What is the best estimate for the mean difference between approaches?

7.52 Construct a 95% CI for the mean difference.

7.53 What can be concluded about the difference between assessments by the two methods?

Cardiovascular Disease

The relationship between serum cholesterol and coronary-heart-disease mortality has been a subject of much debate over the past 30 years. Some data relevant to this question were presented in the Whitehall study [3]. It was shown there that among 3615 men 40–64 years of age who were in the top quintile of serum cholesterol at baseline, 194 died from coronary heart disease over the next 10 years. Suppose the incidence rate of coronary-heart-disease mortality over 10 years among men in this age group in Great Britain is 4%.

7.54 What test can be performed to compare the incidence rate of coronary-heart-disease mortality among men in the top cholesterol quintile with general-population incidence rates?

7.55 Implement the test in Problem 7.54 and report a *p*-value.

7.56 Construct a 95% CI for the true incidence rate in the group of 40–64-year-old men in the top quintile of serum cholesterol.

Cancer

It is well established that exposure to ionizing radiation at or after puberty increases a woman's risk of breast cancer. However, it is uncertain whether such exposure early in life is also carcinogenic. A study was performed [4] in a cohort of 1201 women who received x-ray treatment in infancy for an enlarged thymus gland and were followed prospectively for 36 years. It was found that 22 breast cancers occurred over a 36-year period among the 1201 women.

7.57 If the expected incidence rate of breast cancer in this age group over this time period is 1 event per 200 women based on New York State cancer-incidence rates, then test the hypothesis that irradiation has an effect on breast-cancer incidence. Please report a p-value and construct a 95% CI for the incidence rate among the exposed group.

It was also found that 6 breast cancers occurred over a 36-year period among a subgroup of 138 women who were exposed to a high radioactive dose (0.50–1.99 gray, where a gray is a unit of radiation absorbed by the breast tissue).

7.58 Test the hypothesis that the high-risk subgroup is at excess risk for breast cancer, and report a p-value.

SOLUTIONS

7.1 Test the hypothesis H_0: $\mu = 15 = \mu_0$ versus H_1: $\mu < 15$. Reject H_0 if $z = (\bar{x} - \mu_0)/(\sigma/\sqrt{n}) < -1.645$ and accept H_0 if $z \geq -1.645$. Since $z = (13 - 15)/(4/\sqrt{10}) = -1.581 > -1.645$, we accept H_0 at the 5% level.

7.2 The p-value is given by $\Phi(z) = \Phi(-1.581) = 1 - \Phi(1.581) = 1 - .9431 = .057$.

7.3 Test the hypothesis H_0: $\mu = 24 = \mu_0$ versus H_1: $\mu < 24$. Reject H_0 if $z = (\bar{x} - \mu_0)/(\sigma/\sqrt{n}) < -2.326$ and accept H_0 otherwise. Since $z = (11 - 24)/(11/\sqrt{8}) = -3.343 < -2.326$, reject H_0 at the 1% level.

7.4 The p-value is given by $\Phi(z) = \Phi(-3.343) = 1 - \Phi(3.343) = 1 - .9996 = .0004$.

7.5 Since H_0 was rejected at the 1% level in Problem 7.3, p must be $< .01$, which is the case.

7.6 The hypotheses to be tested are H_0: $\mu = 15$ versus H_1: $\mu \neq 15$. This alternative is two-sided in contrast to the one-sided alternative in Problem 7.1.

7.7 The rejection region is given by $z < z_{.025} = -1.96$ or $z > z_{.975} = 1.96$. We have that $z = -1.581$. Since $-1.96 < -1.581 < 1.96$, we accept H_0 using a two-sided test at the 5% level. The exact p-value is given by $2 \times \Phi(z) = 2 \times \Phi(-1.581) = 2 \times [1 - \Phi(1.581)] = 2 \times (1 - .9431) = .114$.

7.8 We have the test statistic

$$X^2 = \frac{(n-1)s^2}{\sigma^2_0} \sim \chi^2_{n-1} \text{ under } H_0$$

We reject H_0 if $X^2 > \chi^2_{n-1, 1-\alpha/2}$ or $X^2 < \chi^2_{n-1, \alpha/2}$. In this case,

$$X^2 = \frac{19(25)}{16} = 29.69 \sim \chi^2_{19} \text{ under } H_0$$

The critical values are $\chi^2_{19, .025} = 8.91$ and $\chi^2_{19, .975} = 32.85$. Since $8.91 < 29.69 < 32.85$, it follows that we accept H_0.

7.9 The two-sided p-value is given by

$$p = 2 \times Pr(\chi^2_{19} > 29.69)$$

Since $\chi^2_{19, .90} = 27.20$, $\chi^2_{19, .95} = 30.14$ and $27.20 < 29.69 < 30.14$, it follows that $1 - .95 < p/2 < 1 - .90$ or $.05 < p/2 < .10$ or $.10 < p < .20$. Using a computer, the exact p-value $= 2 \times Pr(\chi^2_{19} > 29.69) = .112$.

7.10 Use a one-sample t test. We have the test statistic $t = (\bar{x} - \mu_0)/(s/\sqrt{n}) = (13 - 15)/(6/\sqrt{10}) = -2/1.897 = -1.054 \sim t_9$ under H_0. Since a two-sided test is being performed, reject H_0 if $t < t_{9, .025} = -2.262$ or $t > t_{9, .975} = 2.262$ and accept H_0 otherwise. Since $-2.262 \leq -1.054 \leq 2.262$, it follows that H_0 is accepted. The exact p-value is given by $2 \times Pr(t_9 < [(\bar{x} - \mu_0)/(s/\sqrt{n})] = 2 \times Pr(t_9 < -1.054) = 2 \times Pr(t_9 > 1.054)$. Since $t_{9, .8} = 0.883$, $t_{9, .85} = 1.100$, and $0.883 < 1.054 < 1.100$, it follows that $2 \times (1 - .85) < p < 2 \times (1 - .8)$, or $.3 < p < .4$. The exact p-value $= .32$.

7.11 A lower one-sided 95% CI for the true mean intake of linoleic acid is given by

$$\mu < \bar{x} + z_{.95}\sigma/\sqrt{n}$$
$$= 13 + 1.645(4)/\sqrt{10}$$
$$= 13 + 2.08 = 15.08$$

7.12 This interval contains 15.0 g = mean of the general population. Thus, this answer corresponds to the answer to Problem 7.1, where we accepted H_0 using a one-sided significance test at the 5% level.

7.13 Test the hypothesis H_0: $\sigma^2 = \sigma^2_0$ versus H_1: $\sigma^2 \neq \sigma^2_0$. Use the one-sample χ^2 test for the variance of a normal distribution to test these hypotheses.

7.14 We have the test statistic

$$X^2 = \frac{(n-1)s^2}{\sigma^2_0} = \frac{19(15)^2}{20^2}$$
$$= 10.69 \sim \chi^2_{19} \text{ under } H_0$$

Since $\chi^2_{19,.05} = 10.12$, $\chi^2_{19,.10} = 11.65$, and $10.12 < 10.69 < 11.65$, it follows that $2 \times .05 < p < 2 \times .10$, or $.10 < p < .20$. Thus, there is no significant difference between the variances using the two methods. The exact p-value $= .13$.

7.15 Test the hypothesis H_0: $\sigma^2 = \sigma_0^2 = (0.020)^2$ versus H_1: $\sigma^2 \neq \sigma_0^2$. Reject H_0 if $X^2 = (n-1)s^2/\sigma_0^2 > \chi^2_{n-1,1-\alpha/2}$ or $X^2 < \chi^2_{n-1,\alpha/2}$ and accept H_0 otherwise. We have the test statistic

$$X^2 = \frac{19(.016)^2}{(.020)^2} = 12.16 \sim \chi^2_{19} \text{ under } H_0$$

Since $\chi^2_{19,.025} = 8.91$, $\chi^2_{19,.975} = 32.85$ and $8.91 < 12.16 < 32.85$ it follows that H_0 is accepted at the 5% level and we conclude that the variances of the two methods are not significantly different.

7.16 We have the test statistic $X^2 = 12.16 \sim \chi^2_{19}$ under H_0. Since $\chi^2_{19,.10} = 11.65$, $\chi^2_{19,.25} = 14.56$, and $11.65 < 12.16 < 14.56$, it follows that $2 \times .10 < p < 2 \times .25$, or $.20 < p < .50$. The exact p-value $= .24$.

7.17 SMR $= 100\% \times 28/20.5 = 136.6$

7.18 Since the expected number of cancer deaths ≥ 10, we can use the large-sample test procedure in **(7.43)**, see text, p. 241. We have the test statistic

$$X^2 = \frac{(28 - 20.5)^2}{20.5} = \frac{7.5^2}{20.5} = \frac{56.25}{20.5}$$

$$= 2.74 \sim \chi^2_1 \text{ under } H_0$$

Since $\chi^2_{1,.90} = 2.71$, $\chi^2_{1,.95} = 3.84$ and $2.71 < 2.74 < 3.84$, it follows that $1 - .95 < p < 1 - .90$ or $.05 < p < .10$. The exact p-value, obtained by computer, is $p = Pr(\chi^2_1 > 2.74) = .098$.

7.19 SMR $= 100\% \times 7/4.5 = 155.6$

7.20 Since $E = 4.5 < 10$, we must use the small-sample method in **(7.42)**, see text, p. 239, to test these hypotheses. Since O > E, the p-value is given by

$$p = 2 \times \left[1 - \sum_{k=0}^{6} \frac{e^{-4.5}(4.5)^k}{k!} \right]$$

We refer to the exact Poisson tables (Table 2 in the Appendix of the text) under $\mu = 4.5$ and obtain $Pr(0) = .0111$, $Pr(1) = .0500$, ... , $Pr(6) = .1281$. Therefore,

$$p = 2 \times [1 - (.0111 + .0500 + ... + .1281)]$$
$$= 2 \times (1 - .8310) = .338$$

There is no significant excess or deficit of leukemia deaths in this group of workers.

7.21 We wish to test the hypothesis H_0: $p = p_0 = .05$ versus H_1: $p \neq p_0$. Since $np_0q_0 = 108(.05)(.95) = 5.13 \geq 5$, we can use the normal-theory test. If we use the critical-value method with a 5% significance level, then the critical values are $z_{.025} = -1.96$ and $z_{.975} = 1.96$. We reject H_0 if the test statistic z is ≤ -1.96 or ≥ 1.96. The test statistic is given by

$$z = \frac{\hat{p} - p_0}{\sqrt{p_0 q_0/n}} = \frac{10/108 - .05}{\sqrt{.05/(.95)/108}}$$

$$= \frac{.0426}{.0210} = 2.031$$

Since $z > 1.96$, it follows that we reject H_0 and conclude that there is a significant excess rate of diarrhea in the rural community.

7.22 Since $\hat{p} > p_0$, the p-value $= 2 \times [1 - \Phi(z)] = 2 \times [1 - \Phi(2.031)] = .042$.

7.23 Test the hypothesis H_0: $\mu = 130 = \mu_0$ versus H_1: $\mu \neq 130$, where σ is assumed to be 20 mm Hg. Use the test statistic

$$z = \frac{\bar{x} - \mu_0}{\sigma/\sqrt{n}} = \frac{135 - 130}{20/\sqrt{85}} = 2.305 \sim N(0, 1) \text{ under } H_0$$

Thus, the two-tailed p-value equals $2 \times [1 - \Phi(2.305)] = 2 \times (1 - .9894) = .021$. Therefore, there is a significant association between glaucoma and high blood pressure.

7.24 Perform the one-sample t test here, since the standard deviation is not assumed known. Use the test statistic

$$t = \frac{\bar{x} - \mu_0}{s/\sqrt{n}} = \frac{135 - 130}{22/\sqrt{85}} = 2.095 \sim t_{84} \text{ under } H_0$$

Since $t_{60,.975} = 2.000$, $t_{60,.99} = 2.390$, $t_{120,.975} = 1.980$, $t_{120,.99} = 2.358$

it follows that if there were 60 or 120 df, then

$$2 \times (1 - .99) < p < 2 \times (1 - .975)$$

or $.02 < p < .05$

Since there are 84 df, we must also have $.02 < p < .05$. Thus, there is a significant association between glaucoma and high blood pressure in this case as well.

7.25 We wish to test the hypothesis H_0: $\mu = \mu_0 = 18.5$ versus H_1: $\mu < \mu_0$. We will use a one-sample t test. We have the test statistic

$$t = \frac{\bar{x} - \mu_0}{s/\sqrt{n}} = \frac{16.2 - 18.5}{7.3/\sqrt{20}} = \frac{-2.3}{1.632}$$

$$= -1.409 \sim t_{19} \text{ under } H_0$$

Since $t_{19,.9} = 1.328$, $t_{19,.95} = 1.729$, it follows that $1 - .95 < p < 1 - .90$ or $.05 < p < .1$. Thus, the results are not statistically significant, and we cannot conclude that workers in this industry have significantly shorter lifetimes.

7.26 We wish to test the hypothesis $H_0: p = p_0 = .012$ versus $H_1: p \neq p_0$. We use the normal-theory test since $np_0q_0 = 1000(.012)(.988) = 11.86 \geq 5$. We have the test statistic

$$z = \frac{\hat{p} - p_0}{\sqrt{p_0q_0/n}} = \frac{.015 - .012}{\sqrt{.012(.988)/1000}}$$

$$= \frac{.003}{.00344} = 0.87 \sim N(0, 1) \text{ under } H_0$$

Thus, $p = 2[1 - \Phi(0.87)]$

$$= 2(1 - .8082) = .38$$

and there is no significant association between current OC use and the subsequent development of breast cancer.

7.27 Test the hypothesis $H_0: p = p_0$ versus $H_1: p > p_0$. A one-sided alternative is more appropriate here, since we wish to detect only if hair dyes increase the risk of breast cancer rather than decrease the risk.

7.28 Since $np_0q_0 = 1000(.007)(.993) = 6.95 \geq 5$, use the large-sample test. We have the test statistic

$$z = \frac{\hat{p} - p_0}{\sqrt{p_0q_0/n}} = \frac{20/1000 - .007}{\sqrt{.007 \times .993/1000}}$$

$$= \frac{.013}{.0026} = 4.93 \sim N(0, 1) \text{ under } H_0$$

The p-value is given by $p = 1 - \Phi(z) = 1 - \Phi(4.93) < .001$. Thus, the results are very highly statistically significant, and we conclude that extensive occupational exposure to hair dyes significantly increases the risk of breast cancer.

7.29 We have that $\bar{x} = 1.273$, $s = 0.435$, and $n = 15$. We must test the hypothesis $H_0: \mu = 1.0$, $\sigma = 0.40$ versus $H_1: \mu \neq 1.0$, $\sigma = 0.40$. We perform the one-sample z test using the test statistic

$$z = \frac{\bar{x} - 1.0}{0.40/\sqrt{n}} \sim N(0, 1) \text{ under } H_0$$

Hence, we have

$$z = \frac{1.273 - 1.0}{0.40/\sqrt{15}} = \frac{0.273}{0.103} = 2.647$$

Thus, the p-value $= 2 \times Pr(z > 2.647) = 2 \times [1 - \Phi(2.647)] = .008$. Therefore, we reject the null hypothesis at a significance level of $p = .008$.

7.30 We can test for whether $\sigma = 0.40$ by performing a one-sample χ^2 test. In particular, we test the hypothesis $H_0: \sigma = 0.40$ versus $H_1: \sigma \neq 0.40$. We have the test statistic

$$X^2 = \frac{(n-1)s^2}{\sigma^2} \sim \chi^2_{n-1} \text{ under } H_0$$

Thus, $X^2 = \frac{14(0.435)^2}{(0.40)^2} = 16.56 \sim \chi^2_{14}$ under H_0

Since $\chi^2_{14,.975} = 26.12$, $\chi^2_{14,.025} = 5.63$ and $5.63 < 16.56 < 26.12$ it follows that $p > .05$. Thus, we can accept the null hypothesis that $\sigma = 0.40$.

7.31 If we don't want to assume the standard deviation is known, then we can use a one-sample t test instead of a one-sample z test. We still are testing the hypothesis $H_0: \mu = 1.0$ versus $H_1: \mu \neq 1.0$, but are not assuming σ is known. Under H_0 we have

$$t = \frac{\bar{x} - \mu_0}{s/\sqrt{n}} \sim t_{n-1} = \frac{1.273 - 1.0}{0.435/\sqrt{15}}$$

$$= 2.434 \sim t_{14}$$

Since $t_{14,.975} = 2.145$ and $t_{14,.99} = 2.624$, we have that the p-value is given by $.02 < p < .05$. The exact two-sided p-value, obtained by computer, is $p = 2 \times Pr(t_{14} > 2.434) = .029$. Thus, we again reject H_0 as we did in Problem 7.29, but with a higher significance level.

7.32 We use the sample-size formula in **(7.40)** (text, p. 237) using $z_{1-\alpha}$ instead of $z_{1-\alpha/2}$ for a one-sided test. In this case, $p_0 = .15$, $q_0 = .85$, $p_1 = .25$, $q_1 = .75$, $\alpha = .05$. Therefore

$$n = \frac{p_0q_0[z_{1-\alpha} + z_{1-\beta}\sqrt{p_1q_1/(p_0q_0)}]^2}{(p_1 - p_0)^2}$$

$$= \frac{.15(.85)\left[z_{.95} + z_{.80}\sqrt{\dfrac{.25(.75)}{.15(.85)}}\right]^2}{(.25 - .15)^2}$$

$$= \frac{.1275[1.645 + 0.84(1.213)]^2}{.01}$$

$$= 12.75(7.095) = 90.5$$

Therefore, we need to study 91 children with 3+ episodes of otitis media in the first two years of life in order to achieve an 80% power.

7.33 We must compute the power of the study. We refer to the power formula in **(7.39)** (text, p. 236) using z_α instead of $z_{\alpha/2}$ for a one-sided test. We have

$$\text{Power} = \Phi\left[\sqrt{\frac{p_0q_0}{p_1q_1}}\left(z_\alpha + \frac{|p_1 - p_0|\sqrt{n}}{\sqrt{p_0q_0}}\right)\right]$$

where $n = 50$, $\alpha = .05$, $p_0 = .15$, $q_0 = .85$, $p_1 = .25$, $q_1 = .75$. Thus

$$\text{Power} = \Phi\left[\sqrt{\frac{.15(.85)}{.25(.75)}}\left(z_{.05} + \frac{|.25 - .15|\sqrt{50}}{\sqrt{.15(.85)}}\right)\right]$$

$$= \Phi\left[\sqrt{\frac{.1275}{.1875}}\left(-1.645 + \frac{.10\sqrt{50}}{\sqrt{.1275}}\right)\right]$$

$$= \Phi[0.8246(-1.645 + 1.9803)]$$

$$= \Phi(0.276) = .61$$

Thus, there is a 61% chance of finding a significant difference if we enroll 50 children in the study.

7.34 Use a two-sided test, because there is no reason to expect self-reported weights to be higher or lower than actual weights.

7.35 Use a one-sample t test to test the hypothesis H_0: $\mu = 0$ versus H_1: $\mu \neq 0$, where μ is the true mean for self-reported weight $-$ measured weight.

7.36 We have $\bar{x} = -2.5$, $s = 3.27$. The rejection region is given by $t > t_{9,.975} = 2.262$ or $t < t_{9,.025} = -2.262$. We have the test statistic

$$t = \frac{\bar{x} - \mu_0}{s/\sqrt{n}} = \frac{-2.5 - 0}{3.27/\sqrt{10}}$$

$$= \frac{-2.5}{1.035} = -2.414$$

Since $t = -2.414 < -2.262$, we reject H_0 at the 5% level.

7.37 Since $\bar{x} < \mu_0 = 0$, the p-value is obtained from

$$2 \times Pr(t_9 < -2.414)$$

$$= 2 \times Pr(t_9 > 2.414)$$

Since $t_{9,.975} = 2.262$, $t_{9,.99} = 2.821$, and $2.262 < 2.414 < 2.821$, it follows that $2 \times (1 - .99) < p < 2 \times (1 - .975)$, or $.02 < p < .05$. The exact p-value, obtained by computer, is $p = 2 \times Pr(t_9 < -2.414) = .039$.

7.38 There are 7 reported weights out of 10 that end in 0 or 5 ($\hat{p} = .7$) compared with 2 expected ($p_0 = .2$). Use a one-sided hypothesis test of the form H_0: $p = p_0$ versus H_1: $p > p_0$, since p would only be expected to be larger than p_0 due to digit preference. Since $np_0q_0 = 10 \times .2 \times .8 = 1.6 < 5$, use the exact method. The p-value is given by

$$p = \sum_{k=7}^{10} {}_{10}C_k(.2)^k(.8)^{10-k}$$

Refer to Table 1 (Appendix in the text) under $n = 10$, $p = .2$ and obtain $p = .0008 + .0001 + .0000 + .0000 = .0009$. Thus, there is a significant excess of 0's and 5's among the self-reported weights.

7.39 Use the one-sample t test to test the hypothesis H_0: $\mu = 0$ versus H_1: $\mu \neq 0$ where μ is the true mean change in DBP after taking potassium supplements.

7.40 We have the test statistic

$$t = \frac{\bar{x} - \mu_0}{s/\sqrt{n}} = \frac{-3.2}{8.5/\sqrt{20}}$$

$$= \frac{-3.2}{1.90} = -1.684 \sim t_{19} \text{ under } H_0$$

Note from the t tables that $t_{19,.95} = 1.729$, $t_{19,.90} = 1.328$. Since $1.328 < 1.684 < 1.729$, it follows that $2 \times (1 - .95) < p < 2 \times (1 - .90)$, or $.1 < p < .2$. The exact p-value, obtained by computer, is $p = 2 \times Pr(t_{19} < -1.684) = .11$. Thus, there is not a significant change in diastolic blood pressure based on the pilot-study results.

7.41 A two-sided 95% CI is given by

$$\bar{x} \pm t_{19,.975}s/\sqrt{n} = -3.2 \pm 2.093(8.5/\sqrt{20})$$

$$= -3.2 \pm 2.093(1.90)$$

$$= -3.2 \pm 4.0 = (-7.2, 0.8)$$

The 95% CI includes the null value for μ (0), which is consistent with the nonsignificant results in Problem 7.40.

7.42 Use the sample-size formula

$$n = \frac{\sigma^2(z_{1-\alpha/2} + z_{1-\beta})^2}{\Delta^2}$$

In this case, $\sigma = 8.5$, $\alpha = .05$, $\beta = 1 - .80 = .20$, $\Delta = 3.2$. Therefore,

$$n = \frac{8.5^2(z_{.975} + z_{.80})^2}{3.2^2} = \frac{8.5^2(1.96 + 0.84)^2}{3.2^2}$$

$$= 55.3$$

Thus, 56 subjects are needed in order to have an 80% chance of detecting a significant difference using a two-sided test with α level = .05.

7.43 We wish to test the hypothesis H_0: $p = p_0$ versus H_1: $p \neq p_0$, where p = the true one-year survival rate among patients deemed to be severely ill at admission to the hospital and $p_0 = 2/3$. We use the one-sample binomial test using the test statistic $z = (\hat{p} - p_0)/\sqrt{p_0q_0/n} \sim N(0, 1)$ under H_0.

7.44 If we use an α level of .05, then the critical values for this test are given by $z_{.025} = -1.96$ and $z_{.975} = 1.96$, where we reject H_0 if $z < -1.96$ or $z > 1.96$ and accept H_0 otherwise. We have

$$z = (.47 - 2/3)/\sqrt{(2/3)(1/3)/136}$$

$$= \frac{-.1967}{.0404} = -4.865$$

Since $-4.865 < -1.96$, we reject H_0 at the 5% level.

7.45 The 95% CI is given by $\hat{p} \pm 1.96\sqrt{\hat{p}\hat{q}/n} = .47 \pm 1.96 \times \sqrt{.47(.53)/136} = .47 \pm .084 = (.386, .554)$.

7.46 We use the same test as in Problem 7.43. In this case, the one-year survival rate posthospitalization $= 1 - 39/(136 - 33) = 1 - 39/103 = .621 = \hat{p}$, while $p_0 = .75$. We have the test statistic

$$z = \frac{.621 - .75}{\sqrt{.75(.25)/103}}$$

$$= \frac{-.129}{.0427} = -3.015 \sim N(0, 1) \text{ under } H_0$$

We have that $p = 2 \times \Phi(-3.015) = 2 \times [1 - \Phi(3.015)] = 2 \times (1 - .9987) = .003$. Thus, severely ill patients who survive the hospitalization are significantly less likely to survive the remainder of the first year after hospitalization compared with the expected survival rate of .75 for all hospitalized patients.

7.47 We wish to test the hypothesis $H_0: p = .25 = p_0$ versus $H_1: p \neq .25$, where $p =$ proportion of women who received radiotherapy among women with breast cancer who subsequently develop contralateral breast cancer. We have that $\hat{p} = 206/655 = .315$. Since $np_0q_0 = 655(.25)(.75) = 122.8 \geq 5$ we can use the normal-theory test. We have the test statistic

$$z = \frac{\hat{p} - p_0}{\sqrt{p_0q_0/n}} = \frac{.315 - .25}{\sqrt{.25(.75)/655}}$$

$$= \frac{.065}{.0169} = 3.81 \sim N(0, 1) \text{ under } H_0$$

The p-value $= 2 \times [1 - \Phi(3.81)] < .001$. Thus, women with breast cancer who subsequently develop contralateral breast cancer were significantly more likely to have received radiotherapy previously than the average breast-cancer patient.

7.48 A 95% CI for p is given by $\hat{p} \pm 1.96\sqrt{\hat{p}\hat{q}/n} = .315 \pm 1.96\sqrt{.315(.685)/655} = .315 \pm .036 = (.279, .350)$. This interval excludes $p_0 = .25$.

7.49 We would reject H_0 because the p-value represents the α level at which we would be indifferent between accepting and rejecting H_0. For all α levels $> p$, we reject H_0; for all α levels $\leq p$, we accept H_0. Since $\alpha = .05 > .03 = p$, we would reject H_0.

7.50 The 99% CI would be wider than the 95% CI in Problem 7.48. As the level of confidence increases, the width of the CI must increase.

7.51 The best estimate for the mean difference $= 6.26 - 6.20 = 0.06$ mg/dL.

7.52 A 95% CI for the mean difference is given by

$$0.06 \pm t_{22,.975}(0.50/\sqrt{23})$$

$$= 0.06 \pm 2.074(0.50/\sqrt{23})$$

$$= 0.06 \pm 0.22 = (-0.16, 0.28)$$

7.53 Since the 95% CI includes zero, we conclude that there is no significant difference between the methods; i.e., *on average* we obtain the same value using either method.

7.54 We can use the one-sample binomial test. The hypotheses are $H_0: p = p_0$ versus $H_1: p \neq p_0$, where $p_0 =$ incidence rate of coronary heart disease among 40–64-year-old men in the general population and $p =$ incidence rate of coronary heart disease among 40–64-year-old men in the top quintile of serum cholesterol.

7.55 In this example, $n = 3615, \hat{p} = 194/3615 = .0537$. Since $np_0q_0 = 3615(.04)(.96) = 138.8 \geq .5$, we can use the normal-theory test. We have the test statistic

$$z = \frac{\hat{p} - p_0}{\sqrt{p_0q_0/n}} = \frac{194/3615 - .04}{\sqrt{.04(.96)/3615}}$$

$$= \frac{.0137}{.0033} = 4.19 \sim N(0, 1) \text{ under } H_0$$

The p-value $= 2 \times [1 - \Phi(4.19)] < .001$. Thus, men in the top quintile of serum cholesterol at baseline have a significantly higher incidence rate of coronary-heart disease mortality over 10 years than comparably aged men in the general population.

7.56 A 95% CI for the true incidence rate of coronary-heart-disease mortality among 40–64-year-old men in the top quintile of serum cholesterol is

$\hat{p} \pm 1.96\sqrt{\hat{p}\hat{q}/n} = .0537 \pm 1.96\sqrt{.0537(.9463)/3615}$

$$= .0537 \pm .0073 = (.046, .061)$$

7.57 We wish to test the hypothesis $H_0: p = p_0$ versus $H_1: p \neq p_0$ where $p =$ incidence rate of breast cancer among irradiated women, $p_0 =$ incidence rate of breast cancer from New York State incidence rates $= 1/200 = .005$. Since $np_0q_0 = 1201(.005)(.995) = 5.97 \geq 5$, we can use the normal-theory test. We have the test statistic

$$z = \frac{\hat{p} - p_0}{\sqrt{p_0q_0/n}}$$

where $\hat{p} = 22/1201 = .0183, n = 1201$. Thus,

$$z = \frac{.0183 - .005}{\sqrt{.005(.995)/1201}}$$

$$= \frac{.0133}{.0020} = 6.54 \sim N(0, 1)$$

under H_0. The p-value $= 2 \times [1 - \Phi(6.54)] < .001$.

Thus, there is a highly significant excess risk among the irradiated group. A 95% CI for p is given by

$$\hat{p} \pm 1.96 \sqrt{\hat{p}\hat{q}/n}$$

$$= .0183 \pm 1.96\sqrt{.0183(.9817)/1201}$$

$$= .0183 \pm 1.96(.0039)$$

$$= .0183 \pm .0076 = (.011, .026)$$

7.58 In this case, $np_0q_0 = 138(.005)(.995) = 0.69 < 5$. Thus, we must use the small sample test based on exact binomial probabilities. The p-value is given by

$$p = 2 \times \sum_{k=6}^{138} \binom{138}{k} p_0^k q_0^{138-k}$$

$$= 2 \times \sum_{k=6}^{138} \binom{138}{k} (.005)^k (.995)^{138-k}$$

$$= 2 \times \left[1 - \sum_{k=0}^{5} \binom{138}{k} (.005)^k (.995)^{138-k} \right]$$

We use the recursion rule to evaluate the p-value. We have

$$Pr(0) = .995^{138} = .5007$$

$$Pr(1) = \frac{138}{1} \times \frac{.005}{.995}(.5007) = .3472$$

$$Pr(2) = \frac{137}{2} \times \frac{.005}{.995}(.3472) = .1195$$

$$Pr(3) = \frac{136}{3} \times \frac{.005}{.995}(.1195) = .0272$$

$$Pr(4) = \frac{135}{4} \times \frac{.005}{.995}(.0272) = .0046$$

$$Pr(5) = \frac{134}{5} \times \frac{.005}{.995}(.0046) = .0006$$

Thus, the p-value $= 2 \times [1 - (.5007 + \ldots + .0006)] = 2 \times (1 - .9999) < .001$. Thus, there is a significant excess risk in this subgroup as well.

REFERENCES

[1] Pompei, P., Charlson, M. E., & Douglas, R. G., Jr. (1988). Clinical assessments as predictors of one-year survival after hospitalization: Implications for prognostic stratification. *Journal of Clinical Epidemiology, 41*(3), 275–284.

[2] Boice, J. D., Jr., Harvey, E. B., Blettner, M., Stovall, M., Flannery, J. (1992). Cancer in the contralateral breast after radiotherapy for breast cancer. *New England Journal of Medicine, 326*(12), 781–785.

[3] Rose, G. & Shipley, M. (1986). Plasma cholesterol concentration and death from coronary heart disease: 10-year results of the Whitehall study. *British Medical Journal* (Clinical Research Edition), *293*(6542), 306–307.

[4] Hildreth, N. G., Shore, R. E., & Dvoretsky, P. M. (1989). The risk of breast cancer after irradiation of the thymus in infancy. *New England Journal of Medicine, 321*(19), 1281–1284.

CHAPTER EIGHT

HYPOTHESIS TESTING: TWO-SAMPLE INFERENCE

SECTION 8.1   Paired *t* Test

A study was performed to compare blood-pressure readings obtained from an automated machine versus standard blood-pressure recordings from a mercury sphygmomanometer obtained by a human observer. This study was performed using four different machines, with results given as follows:

Machine	n	Machine Mean	Machine sd	Human Mean	Human sd	Difference Mean	Difference sd
A	98	142.5	21.0	142.0	18.1	0.5	11.2
B	84	134.1	22.5	133.6	23.2	0.5	12.1
C	98	147.9	20.3	133.9	18.3	14.0	11.7
D	62	135.4	16.7	128.5	19.0	6.9	13.6

To analyze these data, we let x_{1i} = machine reading for the ith person, x_{2i} = human reading for the ith person, $d_i = x_{1i} - x_{2i} \sim N(\mu_d, \sigma_d^2)$. We wish to test the hypothesis $H_0 : \mu_d = 0$ versus $H_1 : \mu_d \neq 0$.

We use the *paired t test*. The test statistic is

$$t = \frac{\bar{d} - 0}{} \sim t_{n-1} \text{ under } H_0, \text{ where}$$

$$s_d^2 = \sum_{i=1}^{n}(d_i - \bar{d})^2/(n - 1)$$

We reject H_0 if $t > t_{n-1,1-\alpha/2}$ or $t < t_{n-1,\alpha/2}$ and accept H_0 otherwise.

Alternatively,

$$p\text{-value} = 2 \times Pr(t_{n-1} > t) \text{ if } t \geq 0$$
$$= 2 \times Pr(t_{n-1} < t) \text{ if } t < 0$$

The results are as follows for each of the four machines:

(A) $t = 0.5/(11.2/\sqrt{98}) = 0.5/1.13 = 0.44 \sim t_{97}$, since $t_{60,.75} = 0.679$ and $t_{120,.75} = 0.677$, $p > 2 \times (1 - .75) = .50$

(B) $t = 0.5/(12.1/\sqrt{84}) = 0.5/1.32 = 0.38 \sim t_{83}$, $p > .50$

(C) $t = 14.0/(11.7/\sqrt{98}) = 14.0/1.18 = 11.85 \sim t_{97}$; since $t > t_{60,.9995} = 3.460 > t_{97,.9995}$, $p < 2 \times .0005 = .001$

(D) $t = 6.9/(13.6/\sqrt{62}) = 6.9/1.73 = 3.99$; since $t > 3.460 > t_{61,.9995}$, $p < .001$

For Machines A and B, the machine and human readings are comparable; for Machines C and D, the machine reads higher.

In each case, one can develop a 95% CI for μ_d given by

$$\bar{d} \pm t_{n-1,.975} s_d / \sqrt{n}$$

For example, for Machine D this is given by

$$6.9 \pm t_{61,.975}(1.73)$$

$$= 6.9 \pm 1.999(1.73) = 6.9 \pm 3.5 = (3.4, 10.4)$$

SECTION 8.2 Two-Sample *t* test for Independent Samples— Equal Variances

Infant BP's were collected during the first week of life in the newborn nursery and again at one week, 1 month, 6 months, 12 months, 18 months, 24 months. One question is, How does the level of consciousness of the infant affect BP? We have the following results for 1 month SBP (systolic blood pressure):

	Mean	sd	n
Quiet sleep	81.9	9.8	64
Awake and quiet	86.1	10.3	175

Let us assume that σ is the same in both groups. We will check this later. We wish to test the hypothesis H_0: $\mu_1 = \mu_2$, $\sigma_1^2 = \sigma_2^2 = \sigma^2$, versus H_1: $\mu_1 \neq \mu_2$, $\sigma_1^2 = \sigma_2^2 = \sigma^2$. We assume

$$x_i \sim N(\mu_1, \sigma_1^2) \qquad i = 1, \dots, n_1$$
$$y_j \sim N(\mu_2, \sigma_2^2) \qquad j = 1, \dots, n_2$$

where

x_i = SBP for the ith infant in the quiet sleep group

y_j = SBP for the jth infant in the awake and quiet group

n_1 does not have to be equal to n_2

We base our test on $\bar{x} - \bar{y}$ and use the test statistic

$$t = \frac{\bar{x} - \bar{y}}{s\sqrt{\dfrac{1}{n_1} + \dfrac{1}{n_2}}} \sim t_{n_1+n_2-2}, \text{ where}$$

$$s^2 = \frac{(n_1 - 1) s_1^2 + (n_2 - 1)s_2^2}{n_1 + n_2 - 2}$$

$$= \text{pooled variance estimate over the 2 samples}$$

We reject H_0 if $t > t_{n_1+n_2-2,1-\alpha/2}$ or $t < t_{n_1+n_2-2,\alpha/2}$ and accept H_0 otherwise.
    Alternatively,

$$p\text{-value} = 2 \times Pr(t_{n_1+n_2-2} > t) \text{ if } t > 0$$
$$p\text{-value} = 2 \times Pr(t_{n_1+n_2-2} < t) \text{ if } t \leq 0$$

This test is called the *two-sample t test for independent samples with equal variances.*
    In our case,

$$s^2 = \frac{63(9.8)^2 + 174(10.3)^2}{237} = 103.42$$

$$t = \frac{81.9 - 86.1}{\sqrt{103.42\left(\dfrac{1}{64} + \dfrac{1}{175}\right)}} = \frac{-4.2}{1.49}$$

$$= -2.83 \sim t_{239-2} = t_{237}$$

We have $t_{120,.995} = 2.617$, $t_{120,.9995} = 3.373$. Also, $t_{\infty,.995} = 2.576$, $t_{\infty,.9995} = 3.291$.
Since $2.617 < |t| < 3.373$, and $2.576 < |t| < 3.291$, it follows that $.001 < p < .01$.
Thus, there is a significant effect of sleep status on SBP.
    A $100\% \times (1 - \alpha)$ CI for $\mu_1 - \mu_2$ is given by

$$\bar{x} - \bar{y} \pm t_{n_1+n_2-2,1-\alpha/2}s\sqrt{1/n_1 + 1/n_2}$$

In this example, the 95% CI is given by

$$-4.2 \pm t_{237,.975}(1.49)$$

Using MINITAB, we estimate $t_{237,.975}$ by 1.97. Thus, the 95% CI is

$$-4.2 \pm 1.97(1.49) = -4.2 \pm 2.9 = (-7.1, -1.3)$$

SECTION 8.3   *F* Test for the Equality of Two Variances

How can we verify that the underlying sd's are in fact the same in the 2 groups? We
wish to test the hypothesis H_0: $\sigma_1^2 = \sigma_2^2$, $\mu_1 \neq \mu_2$ versus H_1: $\sigma_1^2 \neq \sigma_2^2$, $\mu_1 \neq \mu_2$.
    We use the variance ratio $F = s_1^2/s_2^2$ as a test statistic. Under H_0, s_1^2/s_2^2 follows an
F distribution with $n_1 - 1$ and $n_2 - 1$ df. The percentiles of the F distribution are
given in Table 9 (in the Appendix of the text). The pth percentile of an F distribution
with d_1 and d_2 df is denoted by $F_{d_1, d_2, p}$.

8.3.1   Characteristics of the *F* Distribution

(1) It is a skewed distribution.

(2)
$$F_{n_1-1, n_2-1, p} = \frac{1}{F_{n_2-1, n_1-1, 1-p}}$$

This enables us to obtain lower percentiles of an F distribution indirectly from the inverse of the corresponding upper percentile of an F distribution with the df reversed. This is important because F tables usually only provide upper percentiles.

Find the upper 2.5% point of an F distribution with 3 and 8 df. From Table 9 (Appendix, text),

$$F_{3,8,.975} = 5.42$$

Find the lower 2.5% point of an F distribution with 3 and 8 df. From Table 9,

$$F_{3,8,.025} = \frac{1}{F_{8,3,.975}} = \frac{1}{14.54} = 0.069$$

We will use the F distribution to test hypotheses concerning the equality of two variances. We reject H_0 if $F = s_1^2/s_2^2$ is either too large or too small. We reject H_0 if $F > F_{n_1-1,n_2-1,1-\alpha/2}$ or $F < F_{n_1-1,n_2-1,\alpha/2}$ and accept H_0 otherwise.

This test is called the *F test for the equality of two variances.*

It is usually easier to put the larger variance in the numerator so that $F > 1$. In our example, from Section 8.2,

$$F = \left(\frac{10.3}{9.8}\right)^2 = 1.105 \sim F_{174,63} \text{ under } H_0$$

From the F table, $F_{\infty,120,.975} = 1.31$. Since the percentiles of the F distribution decrease as either the numerator or denominator df increases, it follows that $F = 1.105 < 1.31 = F_{\infty,120,.975} < F_{174,63,.975}$. Also, since $F > 1$, F is also $> F_{174,63,.025}$. Therefore, $p > .05$, and there is no significant difference between the variances of the two groups.

SECTION 8.4   *t Test for Independent Samples—Unequal Variances*

In the same infant BP study referred to in Section 8.2, the following question was asked at 6 months of life: Do you add salt to the baby food? The mean SBPs of the infants were obtained in the subgroups of mothers who answered yes or no to this question with results as follows:

	Mean SBP	sd	n
Yes	95.0	13.0	23
No	91.3	9.5	503

We first perform the F test to test the hypothesis H_0: $\sigma_1^2 = \sigma_2^2$ versus H_1: $\sigma_1^2 \neq \sigma_2^2$. The test statistic is $F = (13.0/9.5)^2 = 1.873 \sim F_{22,502}$ under H_0. The upper critical value, obtained using MINITAB, is $F_{22,502,.975} = 1.70$. Since $F = 1.873 > 1.70$, this implies that $p < .05$ and the variances are significantly different. Therefore, we cannot use the equal-variance t test. Instead, we wish to test the hypothesis

$$H_0: \mu_1 = \mu_2, \sigma_1^2 \neq \sigma_2^2 \text{ versus } H_1: \mu_1 \neq \mu_2, \sigma_1^2 \neq \sigma_2^2$$

We will use the *two-sample t test with unequal variances.* The test statistic is

$$t = \frac{\bar{x} - \bar{y}}{\sqrt{s_1^2/n_1 + s_2^2/n_2}} \sim t_d$$

The key issue is what degrees of freedom to use for d.

There is no exact solution. The most common approach is to use the *Satterthwaite approximation* for d given by

$$d' = \frac{(s_1^2/n_1 + s_2^2/n_2)^2}{\dfrac{(s_1^2/n_1)^2}{n_1 - 1} + \dfrac{(s_2^2/n_2)^2}{n_2 - 1}}$$

We reject H_0 if $t > t_{d'',1-\alpha/2}$ or $t < t_{d'',\alpha/2}$, and accept H_0 otherwise, where d'' is the largest integer $\le d'$. The p-value is given by

$$p\text{-value} = 2 \times Pr(t_{d''} > t) \text{ if } t > 0$$
$$= 2 \times Pr(t_{d''} < t) \text{ if } t \le 0$$

This test is called the *two-sample t test for independent samples with unequal variances.* In this case,

$$s_1^2/n_1 = 13^2/23 = 7.35$$
$$s_2^2/n_2 = 9.5^2/503 = 0.18$$

$$d' = \frac{7.53^2}{\dfrac{7.35^2}{22} + \dfrac{0.18^2}{502}} = \frac{56.66}{2.45} = 23.1 \ df$$

We round *down* to the nearest integer $df = 23 = d''$. Therefore,

$$t = \frac{95.0 - 91.3}{\sqrt{13.0^2/23 + 9.5^2/503}}$$

$$= \frac{3.7}{2.74} = 1.349 \sim t_{23}$$

Since $t_{23,.90} = 1.319$, $t_{23,.95} = 1.714$ and $1.319 < t < 1.714$, it follows that $.05 < p/2 < .10$ or $.10 < p < .20$ (i.e., no significant difference in mean SBP between the two groups of babies). The exact p-value obtained from MINITAB $= 2 \times Pr(t_{23} > 1.349) = .191$.

The general strategy is to first perform the F test for the equality of two variances. If the variances are significantly different, then use the two-sample t test for independent samples with unequal variances; otherwise, use the two-sample t test for independent samples with equal variances.

SECTION 8.5   Sample-Size Determination for Comparing Two Means from Independent Samples

Suppose we wish to test the hypothesis H_0: $\mu_1 = \mu_2$, $\sigma_1^2 \ne \sigma_2^2$ versus H_1: $\mu_1 \ne \mu_2$, $\sigma_1^2 \ne \sigma_2^2$. We intend to use a test with a two-sided significance level $= \alpha$, power $= 1 - \beta$, and projected difference between the two means under $H_1 = \mu_1 - \mu_2 = \Delta$. How many subjects do we need in each group?

If the sample sizes are expected to be the same in the two groups, then

$$n_1 = n_2 = n = \frac{(z_{1-\alpha/2} + z_{1-\beta})^2(\sigma_1^2 + \sigma_2^2)}{\Delta^2}$$

$$= \text{sample size in each group}$$

If the sample sizes are expected to be different, and $n_2 = kn_1$, where k is the projected ratio of sample sizes in group 2 versus group 1, then

$$n_1 = \frac{(z_{1-\alpha/2} + z_{1-\beta})^2(\sigma_1^2 + \sigma_2^2/k)}{\Delta^2}$$

$$n_2 = kn_1$$

Example: Blood flow is known to be very reduced in certain parts of the brain in people with stroke. Suppose we plan an observational study to compare cerebral blood flow in "stroke-prone" people (with transient ischemic attacks, or TIAs) and stroke-age normal controls (age 50–79). We will use mean blood flow over the entire brain as an outcome variable. Suppose we expect

	Mean	sd
Stroke prone	45	10
Normal	50	10

How many subjects do we need per group to have a 90% chance of finding a significant difference, if we plan to enroll an equal number of subjects per group? In this case,

$$\alpha = .05, \ \beta = .10, \ \sigma_1 = \sigma_2 = 10, \ \Delta = 5$$

$$z_{1-\alpha/2} = z_{.975} = 1.96$$

$$z_{1-\beta} = z_{.90} = 1.28$$

$$n = \frac{(1.96 + 1.28)^2(200)}{25} = (10.50)(8) = 83.98 \text{ or } 84 \text{ subjects}$$

We need 84 subjects per group.

We can also derive power as a function of sample size as follows:

$$\text{Power} = 1 - \beta = \Phi\left(-z_{1-\alpha/2} + \frac{\sqrt{n_1}\Delta}{\sqrt{\sigma_1^2 + \sigma_2^2/k}}\right)$$

where (μ_1, σ_1^2), (μ_2, σ_2^2) are the means and variances of the two respective samples, $\Delta = |\mu_1 - \mu_2|$, $n_2 = kn_1$, $\alpha = $ type I error. ∎

Suppose 50 people are enrolled per group in the preceding study. How much power will we have?

Using the power formula,

$$\text{Power} = \Phi\left[-1.96 + \frac{\sqrt{50}(5)}{\sqrt{100 + 100/1}}\right]$$

$$= \Phi\left(-1.96 + \frac{35.36}{14.14}\right) = \Phi(0.540) = .705$$

Thus, we will have 71% power.

PROBLEMS

Ophthalmology

In a study of the natural history of retinitis pigmentosa (RP), 94 RP patients were followed for 3 years [1]. Among 90 patients with complete follow-up, the mean ± 1 *se* of ln (visual-field loss) over 1, 2, and 3 years was 0.02 ± 0.04, 0.08 ± 0.05, and 0.14 ± 0.07, respectively.

8.1 What test procedure can be used to test for changes in ln (visual field) over any given time period?

8.2 Implement the procedure in Problem 8.1 to test for significant changes in visual field over 1 year. Report a *p*-value.

8.3 Answer Problem 8.2 for changes over 2 years.

8.4 Answer Problem 8.2 for changes over 3 years.

8.5 Find the upper 5th percentile of an F distribution with 24 and 30 *df*.

8.6 Suppose we have two normally distributed samples of sizes 9 and 15 with sample standard deviations of 13.7 and 7.2, respectively. Test if the variances are significantly different in the two samples.

Pathology

Refer to the heart-weight data in Table 6.1 (page 69).

8.7 Test for a significant difference in total heart weight between the diseased and normal groups.

8.8 Test for a significant difference in body weight between the diseased and normal groups.

Psychiatry, Renal Disease

Severe anxiety often occurs in patients who must undergo chronic hemodialysis. A set of progressive relaxation exercises was shown on videotape to a group of 38 experimental subjects, while a set of neutral videotapes was shown to a control group of 23 patients who were also on chronic hemodialysis [2]. The results of a

TABLE 8.1 Pretest and posttest State-Trait Anxiety means and standard deviations for the experimental and control groups of hemodialysis patients

	Pretest			Posttest		
	Mean	sd	n	Mean	sd	n
Experimental	37.51	10.66	38	33.42	10.18	38
Control	36.42	8.59	23	39.71	9.16	23

Note: A lower score on the test corresponds to less anxiety.
Source: Reprinted with permission of the *Journal of Chronic Diseases,* 35(10), 797–802.

psychiatric questionnaire (the State-Trait Anxiety Inventory) are presented in Table 8.1.

8.9 Perform a statistical test to compare the experimental and control groups' pretest scores.

8.10 Perform a statistical test to compare the experimental and control groups' posttest scores.

Hypertension

Blood-pressure measurements taken on the left and right arms of a person are assumed to be comparable. To test this assumption, 10 volunteers are obtained and systolic blood-pressure readings are taken simultaneously on both arms by two different observers, Ms. Jones for the left arm and Mr. Smith for the right arm. The data are given in Table 8.2.

TABLE 8.2 Effect of arm on level of blood pressure (mm Hg)

Patient	Left arm	Right arm
1	130	126
2	120	124
3	135	127
4	100	95
5	98	102
6	110	109
7	123	124
8	136	132
9	140	137
10	155	156

8.11 Assuming that the two observers are comparable, test whether or not the two arms give comparable readings.

8.12 Suppose we do not assume that the two observers are comparable. Can the experiment, as it is defined, detect differences between the two arms? If not, can you suggest an alternative experimental design so as to achieve this aim?

Ophthalmology

A topic of current interest in ophthalmology is whether or not spherical refraction is different between the left and right eyes. For this purpose refraction is measured in both eyes of 17 people. The data are given in Table 8.3.

TABLE 8.3 Spherical refraction in the right and left eye

(i) Person	(x_i) Spherical refraction OD (right eye) (diopters)	(y_i) Spherical refraction OS (left eye) (diopters)	($d_i = x_i - y_i$) Difference (OD - OS)	(i) Person	(x_i) Spherical refraction OD	(y_i) Spherical refraction OS	($d_i = x_i - y_i$) Difference (OD - OS)
1	+1.75	+2.00	−0.25	9	0	0.50	−0.50
2	−4.00	−4.00	0	10	−1.00	−1.25	+0.25
3	−1.25	−1.00	−0.25	11	+0.50	−1.75	+2.25
4	+1.00	+1.00	0	12	−8.50	−5.00	−3.50
5	−1.00	−1.00	0	13	+0.50	+0.50	0
6	−0.75	+0.25	−1.00	14	−5.25	−4.75	−0.50
7	−2.25	−2.25	0	15	−2.25	−2.50	+0.25
8	+0.25	+0.25	0	16	−6.50	−6.25	−0.25
				17	+1.75	+1.75	0

$$\sum_{i=1}^{17} x_i = -27.00 \quad \sum_{i=1}^{17} x_i^2 = 180.00 \quad \sum_{i=1}^{17} y_i = -23.50 \quad \sum_{i=1}^{17} y_i^2 = 129.25 \quad \sum_{i=1}^{17} d_i = -3.50 \quad \sum_{i=1}^{17} d_i^2 = 19.125$$

8.13 Is a one-sample or two-sample test needed here?

8.14 Is a one-sided or two-sided test needed here?

8.15 Which of the following test procedures is appropriate to use on these data? (More than one may be necessary.)

(a) Paired t test
(b) Two-sample t test for independent samples with equal variances
(c) Two-sample t test for independent samples with unequal variances
(d) One-sample t test

8.16 Carry out the hypothesis test(s) in Problem 8.15 and report a p-value.

8.17 Estimate a 90% CI for the mean difference in spherical refraction between the two eyes.

Cardiovascular Disease
A study was performed in 1976 to relate the use of oral contraceptives with the levels of various lipid fractions in a group of 163 nonpregnant, premenopausal women ages 21–39. The mean serum cholesterol among 66 current users of oral contraceptives was 201 ± 37 (mg/dL) (mean ± sd), whereas for 97 nonusers it was 193 ± 37 (mg/dL).

8.18 Test for significant differences in mean cholesterol levels between the two groups.

8.19 Report a p-value based on your hypothesis test in Problem 8.18.

8.20 Derive a 95% CI for the true mean difference in cholesterol levels between the groups.

8.21 Suppose the two-tailed p-value in Problem 8.19 = .03 and the two-sided 95% CI in Problem 8.20 = (−0.6, 7.3). Do these two results contradict each other? Why or why not? (*Note:* These values are not necessarily the actual results in Problems 8.19 and 8.20.)

Hypertension
A 1982 study by the Lipid Research Clinics looked at the relationship between alcohol consumption and level of systolic blood pressure in women not using oral contraceptives [3]. Alcohol consumption was categorized as follows: no alcohol use; ≤ 10 oz/week alcohol consumption; > 10 oz/week alcohol consumption. The results for women 30–39 years of age are given in Table 8.4.

Suppose we wish to compare the levels of systolic blood pressure of groups A and B and we have no prior information regarding which group has higher blood pressure.

8.22 Should a one-sample or two-sample test be used here?

8.23 Should a one-sided or two-sided test be used here?

8.24 Which test procedure(s) should be used to test the preceding hypotheses?

8.25 Carry out the test in problem 8.24 and report a p-value.

8.26 Compute a 95% CI for the mean difference in blood pressure between the two groups.

TABLE 8.4 Relationship between systolic blood pressure and alcohol consumption in 30–39-year-old women not using oral contraceptives

	Systolic blood pressure (mm Hg)		
	Mean	sd	n
A. No alcohol use	110.5	13.3	357
B. ≤ 10 oz per week alcohol consumption	109.1	13.4	440
C. > 10 oz per week alcohol consumption	114.5	14.9	23

Source: Reprinted with permission of the *Journal of Chronic Diseases, 35*(4), 251–257.

8.27 Answer Problem 8.24 for a comparison of groups A and C.

8.28 Answer Problem 8.25 for a comparison of groups A and C.

8.29 Answer Problem 8.26 for a comparison of groups A and C.

Pulmonary Disease

Forced expiratory volume (FEV) is a standard measure of pulmonary function representing the volume of air expelled in 1 second. Suppose we enroll 10 nonsmoking males ages 35–39, heights 68–72 inches in a longitudinal study and measure their FEV (L) initially (year 0) and 2 years later (year 2). The data in Table 8.5 are obtained.

8.30 What are the appropriate null and alternative hypotheses in this case to test if pulmonary function has decreased over 2 years?

TABLE 8.5 Pulmonary function in nonsmokers at two points in time

Person	FEV year 0 (1)	FEV year 2 (1)	Person	FEV year 0 (1)	FEV year 2 (1)
1	3.22	2.95	6	3.25	3.20
2	4.06	3.75	7	4.20	3.90
3	3.85	4.00	8	3.05	2.76
4	3.50	3.42	9	2.86	2.75
5	2.80	2.77	10	3.50	3.32
			Mean	3.43	3.28
			sd	0.485	0.480

8.31 In words, what is the meaning of a type I and a type II error here?

8.32 Carry out the test in Problem 8.30. What are your conclusions?

Another aspect of the preceding study involves looking at the effect of smoking on baseline pulmonary function and on change in pulmonary function over time. We must be careful, since FEV depends on many factors, particularly age and height. Suppose we have a comparable group of 15 men in the same age and height group as in Table 8.5 who are smokers, and we measure their FEV at year 0 and year 2. The data are given in Table 8.6.

TABLE 8.6 Pulmonary function in smokers at two points in time

Person	FEV year 0 (1)	FEV year 2 (1)	Person	FEV year 0 (1)	FEV year 2 (1)
1	2.85	2.88	9	2.76	3.02
2	3.32	3.40	10	3.00	3.08
3	3.01	3.02	11	3.26	3.00
4	2.95	2.84	12	2.84	3.40
5	2.78	2.75	13	2.50	2.59
6	2.86	3.20	14	3.59	3.29
7	2.78	2.96	15	3.30	3.32
8	2.90	2.74			
			Mean	2.98	3.03
			sd	0.279	0.250

8.33 What are the appropriate null and alternative hypotheses to compare the FEV of smokers and nonsmokers at baseline?

8.34 Carry out the procedure(s) necessary to conduct the test in Problem 8.33.

8.35 Suggest a procedure for testing whether or not the *change* in pulmonary function over 2 years is the same in the two groups.

Hypertension

A study of the relationship between salt intake and blood pressure of infants is in the planning stages. A pilot study is done, comparing five 1-year-old infants on a high-salt diet with five 1-year-old infants on a low-salt diet. The results are given in Table 8.7.

8.36 If the means and standard deviations in Table 8.7 are considered to be true population parameters, then,

TABLE 8.7 Relationship between salt intake and level of systolic blood pressure (SBP)

	High-salt diet	**Low-salt diet**
Mean SBP	90.8	87.2
sd SBP	10.3	9.2
n	5	5

using a one-sided test with significance level = .05, how many infants are needed in each group to have an 80% chance of detecting a significant difference?

8.37 Suppose it is easier to recruit high-salt-diet infants and the investigators decide to enroll twice as many high-salt-diet infants as low-salt-diet infants. How many infants are needed in each group to have a 80% chance of detecting a significant difference using a one-sided test with an α level of .05?

8.38 Suppose the budget will only allow for recruiting 50 high-salt-diet and 50 low-salt-diet infants into the study. How much power would such a study have of detecting a significant difference using a one-sided test with significance level = .05 if the true difference between the groups is 5 mm Hg?

8.39 Answer Problem 8.38 for a true difference of 2 mm Hg.

8.40 Answer Problem 8.38 if a two-sided test is used and the true difference is 5 mm Hg.

8.41 Answer Problem 8.40 if the true difference is 2 mm Hg.

Hypertension

The effect of sodium restriction on blood pressure remains a controversial subject. To test this hypothesis a group of 83 individuals participated in a study of restricted sodium intake ($\leq$ 75 mEq/24 hrs) for a period of 12 weeks. The effect on diastolic blood pressure (DBP) is reported in Table 8.8 [4].

8.42 What is the appropriate procedure to test for whether sodium restriction has had an impact on DBP?

8.43 Implement the procedure in Problem 8.42 for people age < 40 using the critical-value method.

One of the interesting findings is the difference in response to dietary therapy between people in the two age groups.

8.44 Test the hypothesis that the response to sodium restriction is different in the two groups and report a p-value.

8.45 Obtain a 95% CI for the response to dietary therapy in each age group separately.

8.46 Obtain a 95% CI for the difference in response between the 2 age groups.

Suppose the results of this study are to be used to plan a larger study on the effects of sodium restriction on DBP.

8.47 How many subjects need to be enrolled in the larger study to test if sodium restriction results in lower DBP if the mean and sd of decline in DBP over the total group of 83 subjects are used for planning purposes and a 90% chance of detecting a significant difference using a two-sided test with a 5% level of significance is desired?

8.48 Suppose 200 patients are enrolled in the larger study. How much power would the larger study have if a two-sided test with significance level = .05 is used?

Environmental Health

A study was conducted relating lead level in umbilical-cord blood and cognitive development [5]. Three groups were identified at birth (high/medium/low) cord blood lead. One issue is the consistency of the differences in blood-lead levels over time between these three groups. The data in Table 8.9 were presented.

8.49 What test procedure can be used to test if there are

TABLE 8.8 Effect of sodium restriction on diastolic blood pressure

	Baseline period			**Diet period**			**Change from control (diet–baseline)**		
	Mean	sd	n	Mean	sd	n	Mean	sd	n
Age < 40	69.7	8.5	61	69.1	8.5	61	−0.7	6.2	61
Age ≥ 40	77.0	8.0	22	71.9	7.5	22	−5.0	4.7	22
Total							−1.8	6.1	83

TABLE 8.9 Reproducibility of blood-lead level in the first two years of life

Cord blood-lead group		Blood-lead level (mean ± sd)				
		Birth	6 months	12 months	18 months	24 months
Low	Mean ± sd	1.8 ± 0.6	4.6 ± 3.9	5.8 ± 5.1	6.7 ± 5.5	5.4 ± 4.8
	n	85	70	69	65	61
Medium	Mean ± sd	6.5 ± 0.3	7.0 ± 7.8	8.5 ± 7.6	8.3 ± 5.8	7.2 ± 5.0
	n	88	70	70	65	63
High	Mean ± sd	14.6 ± 3.0	7.0 ± 8.7	8.8 ± 6.4	7.6 ± 5.8	7.7 ± 8.5
	n	76	61	60	57	58

differences between the observed blood-lead levels at 24 months between the low and high groups (as defined at baseline)?

8.50 Implement the test procedure in Problem 8.49 and report a p-value.

8.51 Provide a 95% CI for the difference in mean blood-lead levels between the low and high groups at 24 months.

Ophthalmology

Retinitis pigmentosa (RP) is the name given to a family of inherited retinal degenerative diseases that may be transmitted through various modes of inheritance. The most common features include a history of night blindness, loss of visual fields, and pigment clumping in the retina. Several reports of lipid abnormalities have been reported in RP patients. However, a consistent trend as regards either excesses or deficiencies in lipid levels has not been apparent. In one study, fatty-acid levels were measured in a group of RP patients and normal controls [6]. The data in Table 8.10 were reported on one particular fatty acid (docosahexaenoic acid) (labeled 22:6w3) in individuals with dominant disease and normal controls.

8.52 What is an appropriate procedure to test if the mean level of 22:6w3 differs between dominant affected

TABLE 8.10 Mean levels of plasma 22:6w3 (adjusted for age)(units are nmol/mL plasma)

	Mean	sd	n
Dominant affected individuals	34.8	20.8	36
Normal controls	47.8	30.3	68

individuals and normal controls? State the hypotheses being tested. Is a one-sided or a two-sided test appropriate here?

8.53 Perform the hypothesis test in Problem 8.52 and report a p-value.

8.54 Provide a 95% CI for the difference in means between the two groups. What does it mean in words?

Neurology

An article published in 1986 describes physical, social, and psychological problems in patients with multiple sclerosis [7]. Patients were classified as to mild, moderate, and severe disease and were graded on physical health using the McMaster physical health index, and mental health using the Rand mental health index scale. The data are shown in Table 8.11.

8.55 What test procedure should be used to test if there

TABLE 8.11 Physical and mental health indices for patients with multiple sclerosis

	Mild			Severe		
	Mean	sd	n	Mean	sd	n
McMaster physical health index	0.76	0.24	65	0.23	0.12	82
Rand mental health index	157.3	26.2	65	149.1	34.4	82

are differences in physical health between the two groups?

8.56 Perform the test in Problem 8.55 with the critical-value method using a two-sided test with an α level of .05.

8.57 What test procedure should be used to test if there are differences in mental health between the two groups?

8.58 Perform the test in Problem 8.57 with the critical-value method using an α level of .05.

8.59 What is the most accurate p-value corresponding to your answer for Problem 8.58?

Hypertension

As one component of a study of blood pressure (BP) response to nonpharmacologic interventions, subjects with high normal BP (DBP of 80–89) who were moderately overweight were randomized to either an active weight-loss intervention or a control group. Blood-pressure measurements were taken both at baseline and after 18 months of intervention and follow-up. Using change in blood pressure over this period as the outcome, the data in Table 8.12 were obtained for diastolic blood pressure (DBP).

TABLE 8.12 Comparison of change in DBP between the weight loss versus the control group

Treatment Group	Change in DBP (mm Hg)		
	n	Mean	sd
Weight loss	308	−6.16	5.88
Control	246	−3.91	6.12

8.60 Assuming that the variances are equal in the two groups, test whether there was a significant difference in mean DBP change between the weight-loss versus the control group.

Researchers were also interested in the amount of weight loss attributable to the intervention. Mean weight change for each group is given in Table 8.13.

8.61 Test whether the variances are the same in the weight-loss and control groups.

8.62 Test whether the intervention was effective in reducing weight in the weight-loss versus the control group.

TABLE 8.13 Comparison of change in weight between the weight-loss group versus the control group

Treatment Group	Change in weight (kg)		
	n	Mean	sd
Weight loss	293	−3.83	6.12
Control	235	0.07	4.01

Cardiology

A study is planned of the effect of drug therapy with cholesterol-lowering drugs on the coronary arteries of patients with severe angina. A pilot study was performed whereby each patient received the placebo treatment and had an angiogram at baseline and again three years later at follow-up. The average diameter of three coronary arteries was measured at baseline and follow-up. The results are given in Table 8.14.

TABLE 8.14 Change in diameter of coronary arteries in patients with severe angina

Mean change[a]	−0.049
sd	0.237
n	8

[a]Follow-up − baseline.

Suppose we wish to test the hypothesis that there has been significant change over 3 years in the placebo group.

8.63 What test should be used to test this hypothesis?

8.64 Implement the test in Problem 8.63 and report a p-value.

8.65 Suppose it is assumed that the mean change over 3 years would be 0 for an active drug group and −0.049 for a placebo group. We intend to randomize subjects to placebo and active drug and compare the mean change in the two groups. How much power would the proposed study have if 500 subjects in each of the drug and placebo groups are available to participate in the study, a two-sided test is used with $\alpha = .05$ and the variances in each group are assumed to be the same?

Cardiovascular Disease, Pediatrics

A family-based behavior-change program was used to modify cardiovascular risk factors among teenage children of patients with ischemic heart disease. The mean baseline level and change in HDL-cholesterol level after a 6-month period in the program is given in Table 8.15.

TABLE 8.15 Mean baseline level and change in HDL-cholesterol level (mmol/L) among teenage children over a 6-month period

	Mean	sd	n
Baseline	1.20	0.32	44
6-months	1.12	0.35	44
Difference (6 months–baseline)	0.08	0.23	44

8.66 What test procedure can be used to test if there has been a change in mean HDL-cholesterol levels among teenage children who undergo such a program?

8.67 Implement the test procedure mentioned in Problem 8.66 and report a p-value (two-sided) as precisely as possible using the appropriate tables.

8.68 Provide a 95% CI for the true mean change over 6 months among teenage children exposed to the program.

8.69 Suppose the results in Problem 8.67 are statistically significant (they may or may not be). Does this necessarily mean that the education program per se is the reason why the HDL-cholesterol levels have changed? If not, is there some way to change the design to allow us to be more confident about the specific effects of the education program?

Endocrinology

A study was performed comparing the rate of bone formation between black and white adults. The data in Table 8.16 were presented [8].

8.70 What method can be used to compare bone-formation rate between blacks and whites?

TABLE 8.16 Comparison of bone-formation rate between black versus white adults

	Blacks (n = 12) Mean ± se	Whites (n =13) Mean ± se
Bone-formation rate	0.033 ± 0.007	0.095 ± 0.012

8.71 Implement the method in Problem 8.70 and report a p-value.

8.72 Obtain a 95% CI for the mean difference in bone-formation rate between the two groups.

SOLUTIONS

8.1 Use the paired t test because each person is used as his or her own control.

8.2 We have the test statistic

$$t = \frac{\bar{d}}{s_d/\sqrt{n}} = \frac{0.02}{0.04} = 0.50 \sim t_{89} \text{ under } H_0$$

Since $0.50 < t_{120,.975} = 1.980 < t_{89,.975}$, it follows that H_0, that there is no mean change in visual-field loss over 1 year, is accepted. Furthermore, since $0.50 < t_{120,.75} = 0.677 < t_{89,.75}$, it follows that $p > 2 \times (1 - .75) = .50$.

8.3 We have the test statistic $t = 0.08/0.05 = 1.60 \sim t_{89}$ under H_0. Since $1.60 < 1.980$, H_0, that there is no mean change in visual-field loss over 2 years, is accepted. Furthermore, if there were 60 df, then since $t_{60,.9} = 1.296 < 1.60 < t_{60,.95} = 1.671$, it follows that $2 \times (1 - .95) < p < 2 \times (1 - .90)$, or $.10 < p < .20$. Similarly, if there were 120 df, then since $t_{120,.9} = 1.289 < 1.60 < t_{120,.95} = 1.658$, it would also follow that $.10 < p < .20$.

Finally, since there are 89 df and we reached the same decision with 60 or 120 df, then it follows that $.10 < p < .20$.

8.4 We have the test statistic $t = 0.14/0.07 = 2.000$. The critical value $= t_{89,.975} < t_{60,.975} = 2.000$. Since $t >$ critical value, H_0 is rejected at the 5% level. Since $2.000 < 2.358 = t_{120,.99} < t_{89,.99}$, it follows that $p > 2 \times (1 - 0.99) = .02$. Therefore, $.02 < p < .05$. The exact p-value, obtained by computer, is $p = 2 \times Pr(t_{89} > 2.000) = .049$.

8.5 $F_{24,30,.95} = 1.89$

8.6 We test the hypothesis H_0: $\sigma_1^2 = \sigma_2^2$ versus H_1: $\sigma_1^2 \neq \sigma_2^2$. We compute the F statistic $F = s_1^2/s_2^2 = 13.7^2/7.2^2 = 3.62 \sim F_{8,14}$ under H_0. Since $F_{8,14,.975} = 3.29$, $F_{8,14,.99} = 4.14$ and $3.29 < 3.62 < 4.14$, it follows that $(1 - .99) \times 2 < p < (1 - .975) \times 2$ or $.02 < p < .05$ and the variances are significantly different.

8.7 The basic statistics for total heart weight (THW) and body weight (BW) in the two groups are given as follows:

Basic statistics for total heart weight and body weight
in diseased and normal males

	Left-heart disease			Normals		
	Mean	**sd**	**n**	**Mean**	**sd**	**n**
THW (g)	450.0	139.34	11	317.0	47.09	10
BW (kg)	55.61	11.55	11	56.23	11.54	10

THW Perform the F test for the equality of two variances as follows:

$$F = s_1^2/s_2^2 = (139.34/47.09)^2 = 8.75 \sim F_{10,9} \text{ under } H_0$$

We have

$$F_{10,9,.995} < F_{8,9,.995} = 6.69 < F$$

Thus, the p-value for F is $<2(.005) = .01$, and the variances are significantly different. The t test with unequal variances must be used. We have the following test statistic:

$$t = \frac{\bar{x}_1 - \bar{x}_2}{\sqrt{\dfrac{s_1^2}{n_1} + \dfrac{s_2^2}{n_2}}} = \frac{450.0 - 317.0}{\sqrt{\dfrac{139.34^2}{11} + \dfrac{47.09^2}{10}}}$$

$$= \frac{133.0}{44.57} = 2.98$$

The appropriate df (d') must be computed.

$$d' = \frac{\left(\dfrac{139.34^2}{11} + \dfrac{47.09^2}{10}\right)^2}{\left(\dfrac{139.34^2}{11}\right)^2 \bigg/ 10 + \left(\dfrac{47.09^2}{10}\right)^2 \bigg/ 9}$$

$$= \frac{3,947,286}{316,988} = 12.45$$

Thus, there are 12 df. Since $|t| > t_{12,.975} = 2.179$, it follows that H_0 is rejected at the 5% level.

8.8 BW Again perform the F test for the equality of two variances.

$$F = s_1^2/s_2^2 = (11.55/11.54)^2 = 1.002 \sim F_{10,9}$$

which is clearly not statistically significant. The t test with equal variances is therefore used. We have

$$s^2 = \frac{(n_1 - 1)s_1^2 + (n_2 - 1)s_2^2}{n_1 + n_2 - 2}$$

$$= \frac{10(11.55)^2 + 9(11.54)^2}{19} = 133.30$$

$$t = \frac{\bar{x}_1 - \bar{x}_2}{\sqrt{s^2\left(\dfrac{1}{n_1} + \dfrac{1}{n_2}\right)}} = \frac{55.61 - 56.23}{\sqrt{133.30\left(\dfrac{1}{11} + \dfrac{1}{10}\right)}}$$

$$= \frac{-0.62}{5.045} = -0.12 \sim t_{19}$$

which is not statistically significant. Thus, there is no significant difference in body weight between the two groups.

8.9 We first compare the variances of the two groups to decide which t test to use. We have the test statistic $F = s_1^2/s_2^2 = 10.66^2/8.59^2 = 1.54 \sim F_{37,22}$ under H_0. We note that $F_{37,22,.975} > F_{\infty,30,.975} = 1.79 > F$. Thus, we accept H_0 and conclude that there is no significant difference between the two variances. Therefore, we use a two-sample t test with equal variances. We have the pooled-variance estimate

$$s^2 = \frac{(n_1 - 1)s_1^2 + (n_2 - 1)s_2^2}{n_1 + n_2 - 2}$$

$$= \frac{37(10.66)^2 + 22(8.59)^2}{59} = 98.78$$

We have the test statistic

$$t = \frac{\bar{x}_1 - \bar{x}_2}{\sqrt{s^2(1/n_1 + 1/n_2)}} = \frac{37.51 - 36.42}{\sqrt{98.78(1/38 + 1/23)}}$$

$$= \frac{1.09}{2.626} = 0.42 \sim t_{59} \text{ under } H_0$$

Since $t_{59,.975} > t_{60,.975} = 2.000 > 0.42$, we accept H_0 at the 5% level.

8.10 We again perform the variance-ratio test. We have $F = s_1^2/s_2^2 = 10.18^2/9.16^2 = 1.24 \sim F_{37,22}$. Since $F_{37,22,.975} > F_{\infty,30,.975} = 1.79 > F$, we accept H_0 that the variances are the same. The pooled-variance estimate is given by

$$s^2 = \frac{37(10.18)^2 + 22(9.16)^2}{59} = 96.28$$

The test statistic for the t test with equal variances is obtained as follows:

$$t = \frac{\bar{x}_1 - \bar{x}_2}{\sqrt{s^2(1/n_1 + 1/n_2)}} = \frac{33.42 - 39.71}{\sqrt{96.28(1/38 + 1/23)}}$$

$$= \frac{-6.29}{2.592} = -2.43 \sim t_{59} \text{ under } H_0$$

Since $|t| > t_{40,.975} = 2.021 > t_{59,.975}$, we reject H_0 at the 5% level. Furthermore, since $|t| > t_{40,.99} = 2.423 > t_{59,.99}$, we have that $p < .02$, while since $|t| < t_{60,.995} = 2.660 < t_{59,.995}$, we have that $p > .01$. Thus, $.01 < p < .02$.

8.11 This is a classical situation for the use of a paired t test where each person is his or her own control. Thus, in this case we test the hypothesis $H_0: \mu_{1i} = \mu_{2i}$ versus H_1: $\mu_{1i} - \mu_{2i} = \Delta \neq 0$. Therefore, we do *not* assume that each person has the same mean blood pressure for a particular arm. We perform this test by computing the ten differences $d_i = x_{1i} - x_{2i}$ as follows:

i	d_i	i	d_i
1	4	6	1
2	-4	7	-1
3	8	8	4
4	5	9	3
5	-4	10	-1

We have $\bar{d} = 1.50$, $s_d = 3.979$, $s_d/\sqrt{10} = 1.258$. Thus, we compute the paired t statistic

$$t = \frac{\bar{d}}{s_d/\sqrt{n}} = \frac{1.50}{1.258} = 1.19$$

This should follow a t distribution with 9 df under H_0. Since $t_{9,.975} = 2.262$, it follows that $p > .05$ and we conclude that blood-pressure readings taken on the left and right arms are not significantly different from each other. Indeed, a 95% CI for $\Delta = \mu_{1i} - \mu_{2i}$ is given by

$$\bar{d} \pm 2.262(s_d/\sqrt{n}) = 1.50 \pm 2.262(1.258)$$

$$= (-1.35, 4.35)$$

Thus, we can be reasonably certain that the true mean difference, if any, between the two arms is less than 5 mm Hg.

8.12 If the 2 observers are not comparable, then we cannot separate the differences between the 2 arms from the differences between the 2 observers. We would require that a second measurement be obtained from each person, with Mr. Smith observing the left arm and Ms. Jones observing the right arm. We could then compute

$$d = .5 \times [(\text{BP left arm}_{\text{Jones}} - \text{BP right arm}_{\text{Jones}})$$

$$+ (\text{BP left arm}_{\text{Smith}} - \text{BP right arm}_{\text{Smith}})]$$

to obtain a difference between arms that is not influenced by observers. We could analyze the 10 d's again using the paired t test.

8.13 A two-sample test is needed here because we are comparing refractive error in the right and left eyes.

8.14 A two-sided test is needed here because we have no prior idea as to whether the right or left eye has the larger mean refraction.

8.15 A paired t test is appropriate here because we are comparing two paired samples consisting of the right and left eyes from the same group of people.

8.16 We first compute the following sample statistics:

$$\bar{d} = \frac{\Sigma d_i}{17} = \frac{-3.50}{17} = -0.206$$

$$s_d^2 = \frac{\Sigma d_i^2 - (\Sigma d_i)^2/17}{16}$$

$$= \frac{19.25 - (-3.5)^2/17}{16} = 1.150$$

We wish to test the hypothesis H_0: $\mu_d = 0$ versus H_1: $\mu_d \neq 0$, where μ_d = underlying mean difference in spherical refractive error between right and left eyes. We have the test statistic

$$t = \frac{\bar{d}}{s_d/\sqrt{n}} = \frac{-0.206}{\sqrt{1.150/17}}$$

$$= \frac{-0.206}{0.260} = -0.79 \sim t_{16} \text{ under } H_0$$

Since $t_{16,.975} = 2.120$, it follows that $p > .05$ and there is no significant difference in mean spherical refractive error between the right and left eyes.

8.17 A two-sided 90% CI for the mean difference in spherical refractive error between right and left eyes is given by

$$\bar{d} \pm t_{n-1,.95}s_d/\sqrt{n}$$

$$= -0.206 \pm t_{16,.95}(0.260)$$

$$= -0.206 \pm 1.746(0.260)$$

$$= -0.206 \pm 0.454 = (-0.660, 0.248)$$

8.18 Clearly, there are no differences between the variances, so we use the two-sample t test with equal variances. We use the test statistic

$$t = \frac{\bar{x}_1 - \bar{x}_2}{\sqrt{s^2(1/n_1 + 1/n_2)}} \sim t_{n_1+n_2-2} \text{ under } H_0$$

where s^2 is the pooled-variance estimate over the 2 samples. In this case, since $s_1^2 = s_2^2$, we have that $s^2 = s_1^2 = s_2^2 = 37^2$. Thus,

$$t = \frac{201 - 193}{\sqrt{37^2(1/66 + 1/97)}}$$

$$= \frac{8}{5.904} = 1.36 \sim t_{161} \text{ under } H_0$$

Since $t < t_{\infty,.975} = 1.960 < t_{161,.975}$, we accept H_0 at the 5% level.

8.19 Since $t_{120,.9} = 1.289$, $t_{120,.95} = 1.658$, $t_{\infty,.9} = 1.282$, $t_{\infty,.95} = 1.645$, it follows that if we had either 120 or ∞ df, then $2 \times (1 - .95) < p < 2 \times (1 - .9)$ or $.1 < p < .2$. The exact p-value obtained by computer is $p = 2 \times Pr(t_{161} > 1.36) = .18$.

8.20 The 95% CI is given by $\bar{x}_1 - \bar{x}_2 \pm t_{161,.975} \sqrt{s^2(1/n_1 + 1/n_2)}$. We use MINITAB to obtain $t_{161,.975} = 1.975$. Thus, the 95% CI = $8 \pm 1.975(5.904) = 8 \pm 11.66$ or $(-3.66, 19.66)$.

8.21 These results contradict each other. If the p-value = $.03 < .05$, then the 95% CI must not contain 0, which it does in this case.

8.22 A two-sample test should be used here, because we are comparing 2 finite samples.

8.23 A two-sided test should be used here, because we have no prior conception as to which group has the higher blood pressure.

8.24 The samples being considered are independent and thus we should perform a two-sample t test for independent samples with either equal or unequal variances. We need to perform the F test for the equality of 2 variances in order to decide which of these 2 tests to use. We proceed as follows with the F test. We have the test statistic $F = s_B^2/s_A^2 = (13.4/13.3)^2 = 1.015 \sim F_{439,356}$. Using a computer program, we find that $p = 2 \times Pr(F_{439,356} > 1.015) = .89$. Thus there are no significant differences between the variances, and we should use the two-sample t test for independent samples with equal variances.

8.25 We have the test statistic

$$t = \frac{\bar{x}_1 - \bar{x}_2}{\sqrt{s^2(1/n_1 + 1/n_2)}} \sim t_{n_1+n_2-2} \text{ under } H_0$$

We must compute the pooled-variance estimate as follows:

$$s^2 = \frac{356(13.3)^2 + 439(13.4)^2}{795} = 178.36$$

Therefore,

$$t = \frac{110.5 - 109.1}{\sqrt{178.36(1/357 + 1/440)}}$$

$$= \frac{1.4}{0.951} = 1.47 \sim t_{795} \text{ under } H_0$$

Since $t_{120,.90} = 1.289$, $t_{120,.95} = 1.658$, it follows that if we had 120 df, then $.1 < p < .2$. Similarly, $t_{\infty,.90} = 1.282$, $t_{\infty,.95} = 1.645$ and if we had ∞ df, then $.1 < p < .2$. Therefore, since we obtain the same result for either 120 or ∞ df, it follows that $.1 < p < .2$ and there is no significant difference between the means of the two groups. The exact p-value obtained by computer is $p = 2 \times Pr(t_{795} > 1.47) = .14$.

8.26 The 95% CI for the true mean difference $(\mu_1 - \mu_2)$ is given by

$$\bar{x}_1 - \bar{x}_2 \pm t_{795,.975} \sqrt{s^2(1/n_1 + 1/n_2)}$$

$$= 1.4 \pm t_{795,.975}(0.951)$$

We use the computer to obtain $t_{795,.975} = 1.963$. Therefore, the 95% CI = $1.4 \pm 1.963(0.951) = 1.4 \pm 1.9 = (-0.5, 3.3)$.

8.27 The variance ratio = $(14.9/13.3)^2 = 1.26 \sim F_{22,356}$ under H_0. Since $F < F_{24,\infty,.975} = 1.64 < F_{22,356,.975}$, it follows that $p > .05$. Therefore, the variances are not significantly different and we must use the two-sample t test with equal variances.

8.28 We have the test statistic

$$t = \frac{114.5 - 110.5}{\sqrt{s^2(1/23 + 1/357)}}$$

$$= \frac{4.0}{\sqrt{s^2(1/23 + 1/357)}}$$

The pooled-variance estimate is given by

$$s^2 = \frac{356(13.3)^2 + 22(14.9)^2}{378} = 179.52$$

Therefore,

$$t = \frac{4.0}{\sqrt{179.52(1/23 + 1/357)}}$$

$$= \frac{4.0}{2.88} = 1.39 \sim t_{378} \text{ under } H_0$$

Since $1.289 = t_{120,.90} < t < 1.658 = t_{120,.95}$ and $1.282 = t_{\infty,.90} < t < 1.645 = t_{\infty,.95}$, it follows that if we had either 120 df or ∞ df, then $.10 < p < .20$. Therefore, we obtain $.10 < p < .20$.

8.29 The 95% CI is given by $4.0 \pm t_{378,.975}(2.88)$. Using the computer, we obtain $t_{378,.975} = 1.966$. Therefore, the 95% CI = $4.0 \pm 1.966(2.88) = 4.0 \pm 5.7 = (-1.7, 9.7)$.

8.30 We must use a paired t test formulation here because the same people are being compared at two different points in time. Our hypotheses are $H_0: \mu_d = 0$ versus $H_1: \mu_d > 0$, where μ_d = underlying mean for (FEV year 0–FEV year 2). We perform a one-sided test because we expect FEV to decline over time.

8.31 A type I error is the probability that we detect a significant decline in FEV level over time given that a true decline does not actually exist. A type II error is the probability that we do not detect a significant decline in FEV level given that a true decline does exist.

8.32 We must compute the 10 paired differences in FEV as follows:

i	d_i	i	d_i
1	0.27	6	0.05
2	0.31	7	0.30
3	−0.15	8	0.29
4	0.08	9	0.11
5	0.03	10	0.18

We have that $\bar{d} = 0.147$, $s_d = 0.150$. The test statistic is given as follows:

$$t = \frac{\bar{d}}{s_d/\sqrt{n}}$$

$$= \frac{0.147}{0.150/\sqrt{10}} = 3.09 \sim t_9 \text{ under } H_0$$

We find that $t_{9,.99} = 2.821$, $t_{9,.995} = 3.250$, and thus, $.005 < p < .01$ (one-sided test). Therefore, there is a significant decline in FEV level over time.

8.33 We will use a two-sample t test for independent samples. The appropriate hypotheses are H_0: $\mu_1 = \mu_2$ versus H_1: $\mu_1 \neq \mu_2$, where μ_1 = mean FEV at baseline for smokers and μ_2 = mean FEV at baseline for nonsmokers.

8.34 We first will test whether the variances in the two groups are significantly different, using the F test for the equality of two variances. We have the test statistic

$$F = \frac{s_1^2}{s_2^2} = \frac{0.485^2}{0.279^2} = 3.03 \sim F_{9,14} \text{ under } H_0$$

Since $F < F_{12,14,.975} = 3.05 < F_{9,14,.975}$ we have that $p > .05$ and thus the variances are not significantly different. Therefore, we use a two-sample t test with equal variances. We have the pooled-variance estimate

$$s^2 = \frac{9s_1^2 + 14s_2^2}{23}$$

$$= \frac{9(0.485)^2 + 14(0.279)^2}{23} = 0.139$$

Thus we have the test statistic

$$t = \frac{\bar{x}_1 - \bar{x}_2}{\sqrt{s^2(1/n_1 + 1/n_2)}}$$

$$= \frac{3.43 - 2.98}{\sqrt{0.139(1/10 + 1/15)}}$$

$$= \frac{0.45}{0.152} = 2.95 \sim t_{23} \text{ under } H_0$$

Since $t_{23,.995} = 2.807$ and $t_{23,.9995} = 3.767$, it follows that $.001 < p < .01$. Thus, there is a significant difference in the mean FEV level between smokers and nonsmokers at baseline, with smokers having lower FEVs.

8.35 The simplest approach would be to take the difference between year 0 and year 2 FEV for each person in each group and test whether the mean differences are significantly different in the two groups using a two-sample t test for independent samples. However, it is well known that adults with high initial levels of pulmonary function tend to decrease more than those with low initial levels (this is known as the "regression to the mean" phenomenon). Thus, if the true rate of decline was the same in smokers and nonsmokers, the nonsmokers might appear to decline faster (as is the case in this data set) because they had higher initial FEVs. We would have to correct the change in FEV for initial level of FEV before looking at differences in change between the two groups.

8.36 We use the sample-size formula in (**8.27**) (text, p. 283) changing $\alpha/2$ to α for a one-sided test. We have

$$n = \frac{(\sigma_1^2 + \sigma_2^2)(z_{1-\alpha} + z_{1-\beta})^2}{\Delta^2} = \text{sample size in each group}$$

In this application, $\sigma_1 = 10.3$, $\sigma_2 = 9.2$, $\Delta = 90.8 - 87.2 = 3.6$, $\alpha = .05$, $\beta = 1 - .80 = .20$. Thus, we have

$$n = \frac{(10.3^2 + 9.2^2)(z_{.95} + z_{.80})^2}{3.6^2}$$

$$= \frac{(106.09 + 84.64)(1.645 + 0.84)^2}{12.96} = 90.9$$

Therefore, we need to study 91 infants in each group in order to have an 80% probability of finding a significant difference using a one-sided significance test with $\alpha = .05$.

8.37 We refer to (**8.28**) (text, p. 284) substituting α for $\alpha/2$ and setting $k = 0.5$. We have

$$n_1 = \frac{(\sigma_1^2 + \sigma_2^2/k)(z_{1-\alpha} + z_{1-\beta})^2}{\Delta^2}$$

$$= \frac{(10.3^2 + 9.2^2/0.5)(1.645 + 0.84)^2}{3.6^2}$$

$$= \frac{275.37(6.175)}{12.96} = 131.2 \text{ or } 132 = \text{number of high salt infants}$$

$n_2 = 0.5 \times n_1 = 66 = \text{number of low salt infants}$

8.38 We use the power formula given in (**8.29**) (text, p. 285) substituting α for $\alpha/2$ because we are performing a one-sided test. We have

$$\text{Power} = \Phi\left(\frac{\sqrt{n}|\Delta|}{\sqrt{\sigma_1^2 + \sigma_2^2}} - z_{1-\alpha}\right)$$

where $n = 50$, $\Delta = 5$, $\sigma_1 = 10.3$, $\sigma_2 = 9.2$, $\alpha = .05$. We have

$$\text{Power} = \left(\frac{\sqrt{50}(5)}{\sqrt{10.3^2 + 9.2^2}} - z_{.95} \right)$$

$$= \Phi\left(\frac{35.36}{13.81} - 1.645 \right)$$

$$= \Phi(0.915) = .820$$

8.39 If $\Delta = 2$ mm Hg, then power $= \Phi[\sqrt{50}(2)/13.81 - 1.645] = \Phi(1.024 - 1.645) = \Phi(-0.621) = 1 - \Phi(0.621) = 1 - .7327 = .267$.

8.40 If $\Delta = 5$ mm Hg and a two-sided test is used, then from **(8.29)** (text, p. 285),

$$\text{Power} = \Phi\left(\frac{\sqrt{n}|\Delta|}{\sqrt{\sigma_1^2 + \sigma_2^2}} - z_{1-\alpha/2} \right)$$

$$= \Phi\left(\frac{35.36}{13.81} - z_{.975} \right)$$

$$= \Phi(0.600) = .726$$

Note that the two-sided test here has less power (.726) than the one-sided test using the same parameters in Problem 8.38 (.820).

8.41 If $\Delta = 2$ mm Hg, then Power $= \Phi[\sqrt{50}(2)/13.81 - 1.96] = \Phi(1.024 - 1.96) = \Phi(-0.936) = 1 - \Phi(0.936) = 1 - .825 = .175$.

8.42 Since each person is being used as his or her own control, we will use the paired t test to assess whether sodium restriction has had an impact on blood pressure.

8.43 We wish to test the hypothesis $H_0: \mu_d = 0$ versus $H_1: \mu_d \neq 0$. We use the test statistic

$$t = \frac{\bar{d}}{s_d/\sqrt{n}} \sim t_{n-1} \text{ under } H_0$$

If we perform the hypothesis test using an α level of .05, then we will reject H_0 if $t < t_{n-1,.025}$ or $t > t_{n-1,.975}$. In this case,

$$t = \frac{-0.7}{6.2/\sqrt{61}} = -0.882$$

The critical values are $t_{60,.025} = -2.000$ and $t_{60,.975} = 2.000$. Since $-2.000 < -0.882 < 2.000$, it follows that we accept H_0 at the 5% level and conclude that sodium restriction does not have a significant impact on DBP in the under-40 age group.

8.44 Since we are comparing two independent samples, we must use the two-sample t test for independent samples with either equal or unequal variances. To decide between these two tests, we first perform the F test for the equality of two variances. We have the test statistic $F = s_1^2/s_2^2 = 6.2^2/4.7^2 = 1.74 \sim F_{60,21}$ under H_0. We refer to the F tables and note that $F < F_{\infty,30,.975} = 1.79 < F_{60,21,.975}$. It follows that $p > .05$. Thus, there is

no significant difference between the variances and we use the t test with equal variances. We must first compute the pooled-variance estimate as follows:

$$s^2 = \frac{(n_1 - 1)s_1^2 + (n_2 - 1)s_2^2}{n_1 + n_2 - 2}$$

$$= \frac{60(6.2)^2 + 21(4.7)^2}{81}$$

$$= \frac{2770.29}{81} = 34.20$$

We now have the test statistic

$$t = \frac{\bar{x}_1 - \bar{x}_2}{\sqrt{s^2(1/n_1 + 1/n_2)}}$$

$$= \frac{-0.7 + 5.0}{\sqrt{34.20(1/61 + 1/22)}}$$

$$= \frac{4.3}{1.454} = 2.96 \sim t_{81} \text{ under } H_0$$

To evaluate the p-value, we must compute $p = 2 \times Pr(t_{81} > 2.96)$. Since $t_{60,.995} = 2.660$, $t_{60,.9995} = 3.460$, and $2.660 < 2.96 < 3.460$, it follows that if we had 60 df that $2 \times (1 - .9995) < p < 2 \times (1 - .995)$ or $.001 < p < .01$. Similarly, since $t_{120,.995} = 2.617$, $t_{120,.9995} = 3.373$ and $2.617 < 2.96 < 3.373$, it follows that if we had 120 df that $.001 < p < .01$. Thus, the response to sodium restriction is significantly different in the two age groups $(.001 < p < .01)$.

8.45 A 95% CI for true change in DBP for the age group < 40 is given by $-0.7 \pm t_{60,.975}(6.2)/\sqrt{61} = -0.7 \pm 2.0(6.2)/\sqrt{61} = -0.7 \pm 1.59 = (-2.3, 0.9)$. For the age group ≥ 40, we have the 95% CI $= -5.0 \pm t_{21,.975}(4.7)/\sqrt{22} = -5.0 \pm 2.080(4.7)/\sqrt{22} = -5.0 \pm 2.08 = (-7.1, -2.9)$.

8.46 A 95% CI for the true difference in response between the two groups is given by $4.3 \pm t_{81,.975}(1.454)$. We compute $t_{81,.975}$ using a computer and obtain $t_{81,.975} = 1.990$. Thus, the 95% CI $= 4.3 \pm 1.990(1.454) = 4.3 \pm 2.89 = (1.4, 7.2)$.

8.47 We use the sample-size formula

$$n = \frac{\sigma^2(z_{1-\beta} + z_{1-\alpha/2})^2}{\Delta^2}$$

In this case, $\Delta = -1.8$, $\sigma = 6.1$, $\alpha = .05$, $\beta = .1$. Therefore, we have that

$$n = \frac{6.1^2(z_{.90} + z_{.975})^2}{(-1.8)^2}$$

$$= \frac{6.1^2(1.28 + 1.96)^2}{(-1.8)^2} = \frac{390.62}{3.24} = 120.6$$

Therefore, we need to study 121 subjects to have a 90% chance of finding a significant difference using a two-sided test with $\alpha = .05$.

8.48 We use the following power formula:

$$\text{Power} = \Phi\left(\frac{\sqrt{n}\,|\Delta|}{\sigma} - z_{1-\alpha/2}\right)$$

$$= \Phi\left[\frac{\sqrt{200}(1.8)}{6.1} - 1.96\right] = \Phi(4.173 - 1.96)$$

$$= \Phi(2.213) = .987$$

8.49 We wish to test the hypothesis $H_0: \mu_1 = \mu_2$ versus $H_1. \mu_1 \neq \mu_2$. We will use the F test for the equality of two variances ($H_0: \sigma_1^2 = \sigma_2^2$ versus $H_1: \sigma_1^2 \neq \sigma_2^2$) to decide whether to use a two-sample t test with equal or unequal variances. We have the test statistic $F = (8.5/4.8)^2 = 3.14 \sim F_{57,60}$ under H_0. Referring to Table 9 in the text appendix, we note that $F > F_{24,60,.975} = 1.88 > F_{57,60,.975}$. Therefore, $p < .05$ and we must use the two-sample t test with unequal variances.

8.50 We have the test statistic

$$t = \frac{\bar{x} - \bar{y}}{\sqrt{s_1^2/n_1 + s_2^2/n_2}}$$

$$= \frac{5.4 - 7.7}{\sqrt{4.8^2/61 + 8.5^2/58}}$$

$$= \frac{-2.3}{1.274} = -1.81 \sim t_{d''} \text{ under } H_0$$

We must interpolate to compute the appropriate df. We have that

$$d' = \frac{(s_1^2/n_1 + s_2^2/n_2)^2}{(s_1^2/n_1)^2/(n_1 - 1) + (s_2^2/n_2)^2/(n_2 - 1)}$$

$$= \frac{(0.378 + 1.246)^2}{0.378^2/60 + 1.246^2/57}$$

$$= \frac{2.635}{0.0296} = 89.0 \quad \text{or} \quad d'' = 89 \, df$$

Referring to the t table, we note that $t_{60,.95} = 1.671$, $t_{60,.975} = 2.000$, $t_{120,.95} = 1.658$, $t_{120,.975} = 1.980$. Since $1.671 < 1.81 < 2.000$ and $1.658 < 1.81 < 1.980$, we reach the same conclusion with either 60 or 120 df. Thus, $2 \times (1 - .975) < p < 2 \times (1 - .95)$ or $.05 < p < .10$. Thus, there is not a significant difference in blood-lead level at 24 months between the two groups.

8.51 A 95% CI for the difference in mean blood-lead levels is given by

$$\bar{x}_1 - \bar{x}_2 \pm t_{89,.975}\sqrt{s_1^2/n_1 + s_2^2/n_2}$$

$$= 5.4 - 7.7 \pm t_{89,.975}(1.274)$$

$$= -2.3 \pm t_{89,.975}(1.274)$$

We compute $t_{89,.975}$ using the computer and obtain $t_{89,.975} = 1.987$. Therefore, the 95% CI is given by $-2.3 \pm 1.987(1.274) = -2.3 \pm 2.53 = (-4.83, 0.23)$.

8.52 We first test the hypothesis $H_0: \sigma_1^2 = \sigma_2^2$ versus $H_1: \sigma_1^2 \neq \sigma_2^2$ to decide on the appropriate t test. We have the test statistic $F = s_2^2/s_1^2 = (30.3/20.8)^2 = 2.12 \sim F_{67,35}$ under H_0. To evaluate significance, we must use a computer and obtain the p-value $= 2 \times Pr(F_{67,35} > 2.12) = .008$. Thus we must use the two-sample t test with unequal variances to test the hypothesis $H_0: \mu_1 = \mu_2, \sigma_1^2 \neq \sigma_2^2$ versus $H_1: \mu_1 \neq \mu_2, \sigma_1^2 \neq \sigma_2^2$.

8.53 We have the test statistic

$$t = \frac{\bar{x}_1 - \bar{x}_2}{\sqrt{s_1^2/n_1 + s_2^2/n_2}}$$

$$= \frac{34.8 - 47.8}{\sqrt{20.8^2/36 + 30.3^2/68}}$$

$$= \frac{-13.0}{5.052} = -2.57 \sim t_{d''}$$

The df is obtained from the Satterthwaite approximation as follows:

$$d' = \frac{(20.8^2/36 + 30.3^2/68)^2}{(20.8^2/36)^2/35 + (30.3^2/68)^2/67}$$

$$= \frac{651.22}{6.85} = 95.1$$

Thus, $d'' = 95 \, df$. Since $t_{60,.99} = 2.390 < 2.57 < 2.660 = t_{60,.995}$ and $t_{120,.99} = 2.358 < 2.57 < 2.617 = t_{120,.995}$, it follows that if we had either 60 or 120 df, then $2 \times (1 - .995) < p < 2 \times (1 - .99)$ or $.01 < p < .02$.

8.54 A 95% CI for the true difference in means is given by $-13.0 \pm t_{95,.975}(5.052)$. We use the computer to obtain $t_{95,.975} = 1.985$. Thus, the 95% CI $= -13.0 \pm 1.985(5.052) = -13.0 \pm 10.0 = (-23.0, -3.0)$. It means that 95% of all possible 95% CIs that could be constructed from repeated random samples of 36 dominants and 68 normals will contain the true difference in means $\mu_1 - \mu_2$.

8.55 We use a two-sample t test with equal or unequal variances. To decide on the appropriate procedures, we first perform the F test for the equality of two variances. We have the test statistic $F = s_1^2/s_2^2 = (0.24/0.12)^2 = 4.00 \sim F_{64,81}$ under H_0. Since $F_{64,81,.975} < F_{24,60,.975} = 1.88 < F$, it follows that $p < .05$ and the variances are significantly different. Therefore, we must use the two-sample t test with unequal variances.

8.56 We have the test statistic

$$t = \frac{\bar{x} - \bar{y}}{\sqrt{s_1^2/n_1 + s_2^2/n_2}}$$

$$= \frac{0.76 - 0.23}{\sqrt{0.24^2/65 + 0.12^2/82}}$$

$$= \frac{0.53}{0.033} = 16.3$$

To assess significance, we note that $d'' \geq 64$ and $t = 16.3 > t_{60,.975} = 2.000 > t_{64,.975}$. Therefore, it follows that $p < .05$ and there are significant differences in physical health between the two groups.

8.57 We again perform the F test for the equality of two variances. We have the test statistic $F = s_2^2/s_1^2 = (34.4/26.2)^2 = 1.72 \sim F_{81,64}$ under H_0. To determine the significance of the results, we use the computer to obtain the p-value $= 2 \times Pr(F_{81,64} > 1.72) = .002$. Thus, the variances are significantly different, and we must use the two-sample t test with unequal variances.

8.58 We have the test statistic

$$t = \frac{\bar{x} - \bar{y}}{\sqrt{s_1^2/n_1 + s_2^2/n_2}}$$

$$= \frac{157.3 - 149.1}{\sqrt{26.2^2/65 + 34.4^2/82}}$$

$$= \frac{8.2}{5.00} = 1.640$$

Since $t = 1.64 < t_{\infty,.975} = 1.96 < t_{d'',.975}$ (where d'' is the appropriate df), it follows that $p > .05$. Therefore, there is no significant difference in mental health between the two groups.

8.59 To determine the p-value, we compute the df as follows:

$$d' = \frac{(26.2^2/65 + 34.4^2/82)^2}{(26.2^2/65)^2/64 + (34.4^2/82)^2/81}$$

$$= \frac{624.59}{4.31} = 144.8$$

Thus, $d'' = 144$. Since $t_{120,.9} = 1.289 < 1.64 < t_{120,.95} = 1.658$ and $t_{\infty,.9} = 1.282 < 1.64 < t_{\infty,.95} = 1.645$, it follows that $2 \times (1 - .95) < p < 2 \times (1 - .9)$ or $.1 < p < .2$.

8.60 We wish to test the hypothesis H_0: $\mu_1 = \mu_2$, $\sigma_1^2 = \sigma_2^2$ versus H_1: $\mu_1 \neq \mu_2$, $\sigma_1^2 = \sigma_2^2$. We have the test statistic

$$t = \frac{\bar{x} - \bar{y}}{\sqrt{s^2(1/n_1 + 1/n_2)}} \sim t_{n_1+n_2-2} \text{ under } H_0.$$

The pooled-variance estimate is given by

$$s^2 = \frac{(n_1 - 1)s_1^2 + (n_2 - 1)s_2^2}{n_1 + n_2 - 2}$$

$$= \frac{307(5.88)^2 + 245(6.12)^2}{552} = 35.85$$

Thus,

$$t = \frac{-6.16 + 3.91}{\sqrt{35.85(1/308 + 1/246)}}$$

$$= \frac{-2.25}{0.512} = -4.39 \sim t_{552} \text{ under } H_0$$

Since $t_{552,.9995} < t_{120,.9995} = 3.373 < 4.39$, it follows that $p < .001$. Thus, there is a highly significant difference in mean DBP change between the weight-loss and control groups.

8.61 We wish to test the hypothesis H_0: $\sigma_1^2 = \sigma_2^2$ versus H_1: $\sigma_1^2 \neq \sigma_2^2$. We have the test statistic $F = (6.12/4.01)^2 = 2.33 \sim F_{292,234}$ under H_0. Since $2.33 > F_{24,120,.975} = 1.76 > F_{292,234,.975}$ it follows that $p < .05$ and there is a significant difference between the variances.

8.62 We wish to test the hypothesis H_0: $\mu_1 = \mu_2$, $\sigma_1^2 \neq \sigma_2^2$ versus H_1: $\mu_1 \neq \mu_2$, $\sigma_1^2 \neq \sigma_2^2$. We have the test statistic

$$t = \frac{\bar{x} - \bar{y}}{\sqrt{s_1^2/n_1 + s_2^2/n_2}}$$

$$= \frac{-3.83 - 0.07}{\sqrt{6.12^2/293 + 4.01^2/235}}$$

$$= \frac{-3.90}{0.443} = -8.80 \sim t_{d''} \text{ under } H_0$$

The degrees of freedom will be at least as large as 234. Clearly, since $8.80 > t_{120,.9995} = 3.373 > t_{234,.9995}$, it follows that $p < .001$. Thus, there is a significantly greater mean weight loss in the weight-loss group versus the control group.

8.63 We wish to test the hypothesis H_0: $\mu_d = 0$ versus H_1: $\mu_d \neq 0$ where $\mu_d =$ true mean change in diameter of the average of three coronary arteries. We will use the paired t test to test this hypothesis.

8.64 We have the test statistic

$$t = \frac{\bar{d}}{s_d/\sqrt{n}}$$

$$= \frac{-0.049}{0.237/\sqrt{8}} = \frac{-0.049}{0.084}$$

$$= -0.58 \sim t_7 \text{ under } H_0$$

Since $t_{7,.75} = 0.711$, and $|t| < 0.711$, it follows that $p > 2 \times (1 - .75) = .50$. Thus, there is no significant change in this study.

8.65 We use the following power formula for comparing two means:

$$\text{Power} = \Phi\left(\frac{-z_{1-\alpha/2} + \sqrt{n}\,|\Delta|}{\sqrt{\sigma_1^2 + \sigma_2^2}}\right)$$

where $\alpha = .05$, $n = 500$, $\Delta = 0.049$, $\sigma_1 = \sigma_2 = 0.237$. Thus, the power is given by

$$\text{Power} = \Phi\left[-z_{.975} + \frac{\sqrt{500}(0.049)}{\sqrt{2(0.237)^2}}\right]$$

$$= \Phi\left(-1.96 + \frac{1.0957}{0.3352}\right)$$

$$= \Phi(-1.96 + 3.269)$$

$$= \Phi(1.309) = .90$$

Thus, the study should have 90% power to detect this difference.

8.66 Since each person was used as his or her own control, we use the paired t test. The hypotheses to be tested are $H_0: \mu_d = 0$ versus $H_1: \mu_d \neq 0$, where $\mu_d =$ true mean change (6 months–baseline) in HDL cholesterol for teenage children exposed to this behavior-change program.

8.67 We use the test statistic

$$t = \frac{\bar{d}}{s_d/\sqrt{n}}$$

$$= \frac{0.08}{0.23/\sqrt{44}}$$

$$= \frac{0.08}{0.0347} = 2.307 \sim t_{43} \text{ under } H_0$$

We note that $t_{40,\,.975} = 2.021$, $t_{40,\,.99} = 2.423$. Since $2.021 < t < 2.423$, it follows that if we had 40 df, then $1 - .99 < p/2 < 1 - .975$ or $.01 < p/2 < .025$ or $.02 < p < .05$. Similarly, since $t_{60,\,.975} = 2.000$, $t_{60,\,.99} = 2.390$ and $2.000 < t < 2.390$, it follows that if we had 60 df, then $.02 < p < .05$. Therefore, since we have 43 df and we reach the same conclusion with either 40 or 60 df, it follows that $.02 < p < .05$ and there has been a significant decline in mean HDL cholesterol over the 6-month period.

8.68 A 95% CI for the true mean change over 6 months is given by

$$\bar{d} \pm t_{n-1,\,.975}s_d/\sqrt{n}$$

$$= 0.08 \pm t_{43,\,.975}(0.0347)$$

We use the computer to obtain $t_{43,\,.975}$, whereby $t_{43,\,.975} = 2.016$. Thus, the 95% CI $= 0.08 \pm 2.016(0.0347) = 0.08 \pm 0.070 = (0.01, 0.15)$.

8.69 We need to establish a control group that is also followed for 6 months and whose change in HDL cholesterol is assessed over a 6-month period. We could compare the change in the active treatment group with the change in the control group. This would allow us to be more confident of the specific effects of the education program.

8.70 We will assume that the bone-formation rate for blacks is normally distributed with mean μ_1 and variance σ_1^2, and the bone-formation rate for whites is normally distributed with mean μ_2 and variance σ_2^2. We wish to test the hypothesis $H_0: \mu_1 = \mu_2$ versus $H_1: \mu_1 \neq \mu_2$. We will use a two-sample t test with either equal or unequal variances. To decide on the appropriate t test, we will first perform the F test for the equality of 2 variances to test the hypotheses $H_0: \sigma_1^2 = \sigma_2^2$ versus $H_1: \sigma_1^2 \neq \sigma_2^2$. We have the test statistic $F = s_2^2/s_1^2 \sim F_{12,\,11}$ under H_0. We have $s_1 = 0.007\sqrt{12} = 0.024$, $s_2 = 0.012\sqrt{13} = 0.043$ and $F = (0.043/0.024)^2 = 3.18 \sim F_{12,\,11}$ under H_0. Since $F < F_{12,\,12,\,.975} = 3.28 < F_{12,\,11,\,.975}$, it follows that $p > .05$ and there is no significant difference between the variances. Thus, we will use the two-sample t test with equal variances.

8.71 We must compute the pooled-variance estimate:

$$s^2 = \frac{(n_1 - 1)s_1^2 + (n_2 - 1)s_2^2}{n_1 + n_2 - 2}$$

$$= \frac{11(0.024)^2 + 12(0.043)^2}{23} = .00126$$

We have the test statistic

$$t = \frac{\bar{x}_1 - \bar{x}_2}{\sqrt{s^2(1/n_1 + 1/n_2)}} \sim t_{23} \text{ under } H_0$$

$$= \frac{0.033 - 0.095}{\sqrt{0.00126(1/12 + 1/13)}}$$

$$= \frac{-0.062}{0.014} = -4.37$$

Since $|t| = 4.37 > t_{23,\,.9995} = 3.767$, it follows that $p < 2(1 - .9995) = .001$. Thus, there is a highly significant difference in the mean bone-formation rate between blacks and whites.

8.72 A 95% CI for $\mu_1 - \mu_2$ is given by

$$\bar{x}_1 - \bar{x}_2 \pm t_{23,\,.975}\sqrt{s^2(1/n_1 + 1/n_2)}$$

$$= -0.062 \pm 2.069(0.014)$$

$$= -0.062 \pm 0.029$$

$$= (-0.091, -0.033)$$

REFERENCES

[1] Berson, E. L., Sandberg, M. A., Rosner, B., Birch, D. G., & Hanson, A. H. (1985). Progression of retinitis pigmentosa over a three-year interval. *American Journal of Ophthalmology, 99,* 246–251.

[2] Alarcon, R. D., Jenkins, C. S., Heestand, D. E., Scott, L. K., & Cantor, L. (1982). The effectiveness of progressive relaxation in chronic hemodialysis patients. *Journal of Chronic Diseases, 35*(10), 797–802.

[3] Wallace, R. B., Barrett-Connor, E., Criqui, M., Wahl, P., Hoover, J., Hunninghake, D., & Heiss, G. (1982). Alteration in blood pressures associated with combined alcohol and oral contraceptive use—The Lipid Research Clinics prevalence study. *Journal of Chronic Diseases, 35*(4), 251–257.

[4] Miller, J. Z., Weinberger, M. H., Daugherty, S. A., Fineberg, N. S., Christian, J. C., & Grim, C. E. (1987). Heterogeneity of blood pressure response to dietary sodium restriction in normotensive adults. *Journal of Chronic Disease, 40*(3), 245–250.

[5] Bellinger, D., Leviton, A., Waternaux, C., Needleman, H., & Rabinowitz, M. (1987). Longitudinal analyses of prenatal and postnatal lead exposure and early cognitive development. *New England Journal of Medicine, 316*(17), 1037–1043.

[6] Anderson, R. E., Maude, M. B., Lewis, R. A., Newsome, D. A., & Fishman, G. A. (1987). Abnormal plasma levels of polyunsaturated fatty acid in autosomal dominant retinitis pigmentosa. *Experimental Eye Research, 44,* 155–159.

[7] Harper, A. C., Harper, D. A., Chambers, L. W., Cino, P. M., & Singer, J. (1986). An epidemiological description of physical, social and psychological problems in multiple sclerosis. *Journal of Chronic Disease, 39*(4), 305–310.

[8] Weinstein, R. S., & Bell, N. H. (1988). Diminished rates of bone formation in normal black adults. *New England Journal of Medicine, 319*(26), 1698–1701.

CHAPTER NINE

MULTISAMPLE INFERENCE

SECTION 9.1 General Concepts

In Chapter 8, we studied how to compare the means of two normally distributed samples using t test methods. In this chapter, we extend these methods to allow us to compare the means of more than 2 normally distributed samples.

Consider the data given in Table 8.4, p. 103, where we looked at the relationship between systolic blood pressure and alcohol consumption in a group of 30–39-year-old women not using oral contraceptives. The data are reproduced here:

	Systolic blood pressure (mm Hg)		
	Mean	sd	n
A. No alcohol use	110.5	13.3	357
B. $\leq$ 10 oz per week alcohol consumption	109.1	13.4	440
C. > 10 oz per week alcohol consumption	114.5	14.9	23

We could compare the means of each pair of groups, as we did in Problems 8.25–8.29. There are two important reasons for not doing so. If we have many ($k > 2$) groups and we compare the means of each pair of groups, then we will make a total of $_kC_2$ comparisons. Even for moderate k, $_kC_2$ is quite large and there is a high probability that some of the comparisons will be significant by chance alone. One approach to this problem is to use *one-way analysis-of-variance methods*. Using these methods, we will test the hypothesis $H_0: \mu_1 = \ldots = \mu_k$ versus H_1: at least two of the μ_i's are different, where μ_i is the underlying mean of the ith group. Only if we are able to reject H_0 and conclude that at least some of the means are different will we perform t tests comparing pairs of groups.

A second difference from the methods in Chapter 8 is that the t tests conducted in the context of the analysis of variance are different from ordinary two-sample t tests. First, the variance estimate used is based on a pooled-variance estimate over all groups, rather than simply the two groups being compared. Second, an explicit adjustment for the number of comparisons is sometimes made using what are known as *multiple-comparisons procedures*. We will study one popular method for adjusting for multiple comparisons known as the *Bonferroni method*.

Finally, there are two types of analysis-of-variance models, called *fixed-effects* and *random-effects models*. Under a fixed-effects model, we are primarily interested in comparing specific groups as in Table 8.4. Under a random-effects model, we are interested in determining what proportion of the total variation is attributable to between-group versus within-group variability. For example, in a dietary reproducibility study, where the same dietary instrument is administered on more than one occasion, we might be interested in comparing the degree of between-person versus within-person variation for a specific nutrient (e.g., dietary cholesterol), but not in comparing dietary cholesterol between specific people.

SECTION 9.2   One-Way Analysis of Variance—Fixed-Effects Model

Let $y_{ij} = j$th observation in the ith group; $i = 1, \ldots, k, j = 1, \ldots, n_i$.

The one-way analysis of variance fixed effects model is given by:

$$y_{ij} = \mu + \alpha_i + e_{ij}$$

where μ = overall mean

α_i = difference between mean of ith group and overall mean

$$e_{ij} \sim N(0, \sigma^2)$$

Differences among the α_i represent between-group variation. The e_{ij} represent within-group variation.

9.2.1   Overall Test of Significance

We with to test the hypothesis H_0: all $\alpha_i = 0$ versus H_1: some $\alpha_i \neq 0$. To test this hypothesis, we separate the total variation into a between-visit and within-visit component of variation. This can be written algebraically as

$$\sum_{ij}(y_{ij} - \bar{\bar{y}})^2 = \sum_{ij}(y_{ij} - \bar{y}_i)^2 + \sum_i n_i(\bar{y}_i - \bar{\bar{y}})^2$$

The terms in this equation are called *sums of squares (SS)*, and the equation can be written in words as follows:

Total SS = Within SS + Between SS

The Within SS represents within-group variation; the Between SS represents between-group variation.

Another way to write the Within SS is

$$\text{Within SS} = \sum_{i=1}^{k}(n_i - 1)s_i^2$$

where s_i^2 is the sample variance for the ith sample.

Another way to write the Between SS is

$$\text{Between SS} = \sum_{i=1}^{k} n_i \bar{y}_i^2 - \left(\sum_{i=1}^{k} n_i \bar{y}_i \right)^2 / \sum_{i=1}^{k} n_i$$

To test the overall hypothesis, we use the test statistic

$$F = \frac{\text{Between MS}}{\text{Within MS}} \sim F_{k-1, n-k} \text{ under } H_0$$

where MS stands for mean square and

$$\text{Between MS} = \text{Between SS}/(k - 1)$$

$$\text{Within MS} = \text{Within SS}/(n - k)$$

The values $k - 1$ and $n - k$ are referred to as the *df* pertaining to the between-group and within-group components of variation. The *p*-value $- Pr(F_{k-1, n-k} > F)$. The results are typically displayed in an analysis-of-variance (or ANOVA) table as follows:

ANOVA Table

	SS	df	MS	F-stat	p-value
Between					
Within					
Total					

Example: Determine if there are significant differences for mean SBP among the 3 alcohol-use groups. The overall mean SBP over all groups $= [110.5(357) + 109.1(440) + 114.5(23)]/820 = 109.86$. Therefore,

$$\text{Between SS} = 357(110.5 - 109.86)^2 + \ldots + 23(114.5 - 109.86)^2 = 895.55$$

$$\text{Between MS} = 895.55/2 = 447.78$$

$$\text{Within SS} = 356(13.3)^2 + \ldots + 22(14.9)^2 = 146{,}683.9$$

$$\text{Within MS} = 146{,}683.9/(820 - 3) = 179.54$$

$$F = 447.78/179.54 = 2.49 \sim F_{2,817}$$

$$p\text{-value} = Pr(F_{2,817} > 2.49) = .084$$

Thus, there is no significant difference among the means of the 3 groups.   ∎

SECTION 9.3   Comparison of Specific Groups— Least Significant Difference (LSD) Approach

Suppose we consider the data in Table 8.9 (p. 105). We wish to compare the mean blood-lead levels at 12 months among infants in the low, medium, and high cord blood-lead-level groups.

| | 12-month blood-lead level | |
Cord blood-lead group	mean ± sd	n
Low	5.8 ± 5.1	69
Medium	8.5 ± 7.6	70
High	8.8 ± 6.4	60

We first perform the overall F test for one-way ANOVA and obtain an F statistic $= 183.04/41.69 = 4.39 \sim F_{2,196}$ under H_0. The p-value $= Pr(F_{2,196} > 4.39) = .014$. Thus, we reject H_0 and conclude that some of the group means are significantly different. To compare specific groups—for example, the 1st and 2nd groups—we compute the test statistic

$$t = \frac{\bar{y}_1 - \bar{y}_2}{\sqrt{\text{Within MS}\left(\frac{1}{n_1} + \frac{1}{n_2}\right)}} \sim t_{n-k} \text{ under } H_0$$

The p-value $= 2 \times Pr(t_{n-k} > t)$ if $t \geq 0$, or $= 2 \times Pr(t_{n-k} < t)$ if $t < 0$.

In the blood-lead example, we compare each pair of groups as follows:

Groups compared	Test statistic	p-value
Low, medium	$t = \dfrac{5.8 - 8.5}{\sqrt{41.69(1/69 + 1/70)}}$ $= -2.7/1.10 = -2.47 \sim t_{196}$	.015
Low, high	$t = \dfrac{5.8 - 8.8}{\sqrt{41.69(1/69 + 1/60)}}$ $= -3.0/1.14 = -2.63 \sim t_{196}$	.009
Medium, high	$t = \dfrac{8.5 - 8.8}{\sqrt{41.69(1/70 + 1/60)}}$ $= -0.3/1.14 = -0.26 \sim t_{196}$	.79

Thus, there are significant differences between the low group and each of the medium and high groups, while there is no significant difference between the medium and high groups.

SECTION 9.4   Comparisons of Specific Groups—Bonferroni Approach

Some authors feel that first performing the overall F test for one-way ANOVA is not sufficient protection against the multiple-comparisons problem. They recommend that an additional adjustment for multiple comparisons be performed when comparing specific groups. One of the most commonly used such method is the *Bonferroni approach.* The goal is that the *experiment-wise type I error,* which is defined as the probability that at least one pair of groups will be significantly different under H_0, be set at α.

Under this approach, if we wish to achieve an experiment-wise type I error $= \alpha$, then we should use a critical value $= t_{n-k,1-\alpha*/2}$, where $\alpha* = \alpha/\binom{k}{2}$ for comparisons between specific pairs of groups where $k =$ number of groups. To conduct the significance test, we reject H_0 if $|t| > t_{n-k,1-\alpha*/2}$, and accept H_0 otherwise.

Suppose we apply this approach to the blood-lead example. If we set $\alpha = .05$, then since $k = 3$, $\alpha* = .05/3 = .0167$. Thus, the critical value $= t_{196,1-.00833} = t_{196,.99167} = 2.41$ by computer. Since $|t| = 2.47$ for low vs. medium and $|t| = 2.63$ for low vs. high, these contrasts would remain statistically significant even under the Bonferroni approach.

In general, it is harder to achieve statistical significance with the Bonferroni approach than with the LSD approach, particularly if the number of groups k is large. At the present time, it is controversial as to which approach is preferable.

SECTION 9.5   Linear Contrasts

More complex comparisons are possible in one-way ANOVA. Suppose we wish to compare the mean blood-lead level at 12 months for low cord blood-lead children with the mean blood-lead level at 12 months for the combined groups of medium and high blood-lead groups. Assume that there are an equal number of medium and high blood-lead children, and that we denote the true mean blood-lead level at 12 months for the low group by μ_1, the medium group by μ_2, and the high group by μ_3. We can form a *linear contrast* $L = \bar{y}_1 - 0.5(\bar{y}_2 + \bar{y}_3)$ to represent this comparison and test the hypothesis $H_0: \mu(L) = 0$ versus $H_1: \mu(L) \neq 0$, where $\mu(L) = \mu_1 - 0.5(\mu_2 + \mu_3)$.

In general, a *linear contrast* can be written in the form $L = \sum_{i=1}^{k} \lambda_i \bar{y}_i$, where $\sum_{i=1}^{k} \lambda_i = 0$. To test the preceding hypothesis, we compute the test statistic

$$t = \frac{L}{se(L)} = \frac{L}{\sqrt{\text{Within MS}\left(\sum_{i=1}^{k} \lambda_i^2/n_i\right)}} \sim t_{n-k}, \text{ under } H_0$$

The p-value $= 2 \times Pr(t_{n-k} > t)$ if $t > 0$ or $= 2 \times Pr(t_{n-k} < t)$ if $t \leq 0$.

In this case,

$$L = 5.8 - 0.5(8.5 + 8.8) = -2.85$$

$$se(L) = \sqrt{41.69\left(\frac{1}{69} + \frac{0.5^2}{70} + \frac{0.5^2}{60}\right)} = 0.963$$

$$t = -2.85/0.963 = -2.96 \sim t_{196} \text{ under } H_0$$

$$p\text{-value} = 2 \times Pr(t_{196} < -2.96) = .003$$

Thus, the children with low cord blood-lead levels have significantly lower mean blood-lead levels at 12 months than the children with medium or high cord blood-lead levels.

A multiple comparisons procedure also exists for testing the significance of linear contrasts. Under *Scheffe's procedure*, the experiment-wise type I error is preserved at

α, where the experiment-wise type I error is defined as the probability under H_0, that at least one linear contrast, among all possible linear contrasts, will be declared statistically significant (see section 9.4.4, text, p. 318).

SECTION 9.6   One-Way Analysis of Variance—Random-Effects Model

In the blood-lead example, we were interested in comparing specific groups, e.g., low versus high cord blood-lead groups. In some instances, we are interested in determining what proportion of the total variation is attributable to between-group versus within-group variation, but are not specifically interested in comparing individual groups.

A dietary reproducibility study was performed among 173 U.S. nurses. Each woman filled out a *diet record* on four separate weeks, approximately 3 months apart, over a 1-year period. The diet record consists of an enumeration of the type and quantity of each food eaten over a 1-week period. The question is, How reproducible are the reported intakes of specific nutrients?

To assess between- and within-person variability, we use the model

$$y_{ij} = \mu + \alpha_i + e_{ij}, \ i = 1, \dots, k = \text{person}; \ j = 1, \dots, n = \text{replicate}$$

and assume $\alpha_i \sim N(0, \sigma_A^2)$, $e_{ij} \sim N(0, \sigma^2)$. σ_A^2 and σ^2 are referred to as the between-person and within-person variance components. σ_A^2 is a measure of the true variation between different people on dietary cholesterol. σ^2 is a measure of the week-to-week variation in dietary cholesterol for an individual. To estimate the variance components, we run a one-way ANOVA and compute

$$\hat{\sigma}^2 = \text{Within MS}$$
$$\hat{\sigma}_A^2 = \max \left[(\text{Between MS} - \text{Within MS})/n, 0 \right]$$

The following data were obtained for the 1st day of intake of each of 4 weeks for dietary cholesterol:

	SS	df	MS	F-stat	p-value
Between	936.7	172	5.446	1.59	< .001
Within	1780.2	519	3.430		

Thus, as expected, there is a significant difference in mean dietary cholesterol between different individuals. The estimated within-person variance $= \hat{\sigma}^2 = 3.430$. The estimated between-person variance $= \hat{\sigma}_A^2 = (5.446 - 3.430)/4 = 0.504$. Thus, the within-person variation is 6 times as large as the between-person variation.

A useful measure of reproducibility is the *intraclass correlation coefficient,* which is defined by

$$\rho_I = \sigma_A^2/(\sigma_A^2 + \sigma^2)$$

This measure can be interpreted as the proportion of the total variance that is attributable to true between-person variation. For a very reproducible measure, ρ_I will be near 1, while for a very noisy measure, ρ_I will be near 0. For dietary cholesterol, ρ_I = 0.504/(0.504 + 3.430) = 0.504/3.934 = 0.13. This is very low and indicates a lot of within-person variation relative to between-person variation for dietary cholesterol. Only about 13% of the total variation is attributable to true between-person variation. We discuss intraclass correlation coefficients in more detail in Chapter 11.

The one-way ANOVA random-effects model can also be extended to the case of an unequal number of replicates per subject, such as might occur if some individuals did not complete all 4 weeks of diet records. See Section 9.6 (p. 325 in the main text) for a discussion of this topic.

SECTION 9.7   The Cross-Over Design

A *cross-over design* is a type of clinical trial where the same patient receives both types of treatment in different periods. Consider the following example. Erythropoietic protoporphyria (EPP) is a rare dermatologic disease where patients are very sensitive to sunlight. A study was performed comparing cysteine versus placebo for the treatment of this condition. Patients were randomized to either group 1 or group 2. Patients in group 1 received cysteine for an 8-week period, went off drug for 1 week (during a "washout period") and then received placebo for the next 8 weeks. Patients in group 2 received placebo for 8 weeks, went off drug for 1 week and then received cysteine for the next 8 weeks. There were 8 patients in group 1 and 7 patients in group 2. Patients were assigned to groups using a random-number table. Patients were phototested at the end of each active drug period (i.e., subjected to a controlled intensity of artificial light) and the time to develop erythema (excessive redness) was noted. How can we assess the relative effectiveness of cysteine versus placebo with this type of study design?

To assess the significance of the results, for each patient, we compute

$$x = x_C - x_P = \text{time to develop erythema while on cysteine}$$
$$- \text{time to develop erythema while on placebo}$$

$$\bar{d}_1 = \text{mean of } x \text{ for all subjects in group 1}$$

$$\bar{d}_2 = \text{mean of } x \text{ for all subjects in group 2}$$

$$s_{d_1}^2 = \text{variance of } x \text{ for all subjects in group 1}$$

$$s_{d_2}^2 = \text{variance of } x \text{ for all subjects in group 2}$$

We use the test statistic

$$t = \frac{\bar{d}}{\sqrt{\frac{s_d^2}{4}\left(\frac{1}{n_1} + \frac{1}{n_2}\right)}} \sim t_{n_1+n_2-2} \quad \text{under } H_0$$

where

$$\bar{d} = .5(\bar{d}_1 + \bar{d}_2)$$

$$s_d^2 = \frac{(n_1 - 1)s_{d_1}^2 + (n_2 - 1)s_{d_2}^2}{n_1 + n_2 - 2}$$

The p-value $= 2 \times Pr(t_{n_1+n_2-2} > t)$ if $t > 0$, or $= 2 \times Pr(t_{n_1+n_2-2} < t)$ if $t \leq 0$.

The following results were obtained from the EPP study:

	Group 1	Group 2
d_i [a]	6.75	21.71
s_{d_i} [a]	7.78	9.86
n_i	8	7

[a] Minutes.

Thus,

$$\bar{d} = (6.75 + 21.71)/2 = 14.23$$

$$s_d^2 = [7(7.78)^2 + 6(9.86)^2]/13 = 77.46$$

$$t = \frac{14.23}{\sqrt{\dfrac{77.46}{4}\left(\dfrac{1}{8} + \dfrac{1}{7}\right)}} = \frac{14.23}{2.28} = 6.25 \sim t_{13} \text{ under } H_0$$

$$p\text{-value} = 2 \times Pr(t_{13} > 6.25) < .001$$

Thus, cysteine significantly increases the time until symptoms develop relative to placebo.

One issue that often arises in cross-over studies is the possible presence of a carry-over effect. A *carry-over effect* is defined to be present when the effect of the treatment received in the first period is present also in the second treatment period. For example, in the preceding study, for the patients in group 1 the effect of cysteine might extend into the second period, making it hard to identify treatment effects for subjects in group 1. In this case, if a carry-over effect is present, the average time before symptoms occur over the two periods for a patient in group 1 will be higher than for patients in group 2. To test for carry-over effect, we compute

$$t = \frac{\bar{x}_1 - \bar{x}_2}{\sqrt{s^2\left(\dfrac{1}{n_1} + \dfrac{1}{n_2}\right)}} \sim t_{n_1+n_2-2} \text{ under } H_0$$

where

$x =$ average time to develop erythema for a patient over the two time periods

$\bar{x}_1 =$ mean of x over all patients in group 1

$\bar{x}_2 =$ mean of x over all patients in group 2

$s_1^2 =$ variance of x over all patients in group 1

$s_2^2 =$ variance of x over all patients in group 2

$$s^2 = \frac{(n_1 - 1)s_1^2 + (n_2 - 1)s_2^2}{n_1 + n_2 - 2}$$

The p-value $= 2 \times Pr(t_{n_1+n_2-2} > t)$ if $t > 0$, or $= 2 \times Pr(t_{n_1+n_2-2} < t)$ if $t \leq 0$.

In this case, the following results were obtained from the EPP study:

	Group 1	Group 2
$\bar{x}_i$	16.63	22.00
s_i	11.26	7.39
n_i	8	7

The pooled-variance estimate is

$$s^2 = \frac{7(11.26)^2 + 6(7.39)^2}{13} = 93.48$$

The test statistic is

$$t = \frac{16.63 - 22.00}{\sqrt{93.48\left(\frac{1}{8} + \frac{1}{7}\right)}} = \frac{-5.37}{5.00} = -1.07 \sim t_{13} \text{ under } H_0$$

The p-value $= 2 \times Pr(t_{13} < -1.07) = .30$. Thus, there is no significant carry-over effect in this study. An important assumption of the overall test for treatment efficacy given on p. 123 is that no carry-over effect is present as is the case with the EPP data.

PROBLEMS

Nutrition, Arthritis

A comparison was made of protein intake among three groups of premenopausal women: (1) women eating a standard American diet (STD), (2) women eating a lacto-ovo-vegetarian diet (LAC), and (3) women eating a strict vegetarian diet (VEG). The mean and sd for protein intake (mg) is presented in Table 9.1.

TABLE 9.1 Protein intake (mg) among three dietary groups of premenopausal women

Group	Mean	sd	n
STD	74	16	10
LAC	56	16	10
VEG	55	9	10

9.1 What parametric procedure can be used to compare the underlying means of the three groups?

9.2 Implement the procedure in Problem 9.1 using the critical-value method.

9.3 Compare the underlying means of each specific pair of groups using the t test (LSD) methodology.

9.4 Suppose that in the general population 70% of vegetarians are lacto-ovo-vegetarians, whereas 30% are strict vegetarians. Perform a statistical procedure to test if the contrast $L = 0.7\bar{y}_2 + 0.3\bar{y}_3 - \bar{y}_1$ is significantly

different from 0. What does this contrast mean?

9.5 Using the data in Table 9.1, perform a multiple-comparisons procedure to identify which specific underlying means are different.

9.6 Perform a multiple-comparisons procedure to test if the linear contrast in Problem 9.4 is significantly different from 0.

Cardiovascular Disease

Physical activity has been shown to have beneficial effects on cardiovascular-disease outcomes. As part of a study of physical-activity assessment methodology, the number of hours of sleep, light activity, moderate activity, hard activity, and very hard activity was computed for each person in a study [1]. These categories were defined in terms of levels of energy, or MET, where MET = ratio of working metabolic rate to resting metabolic rate. Using this classification, sleep = 1 MET; light activity = 1.1–2.9 MET; moderate activity = 3.0–5.0 MET; hard activity = 5.1–6.9 MET; very hard activity $\geq$ 7.0 MET. In Table 9.2, data relating the number of hours of moderate activity per week to age for males are presented.

9.7 What is the appropriate method of analysis to test for the effect of age on the number of hours of moderate activity per week?

9.8 Perform the test mentioned in Problem 9.7 and report a p-value.

TABLE 9.2 Hours per week of moderate activity for men by age

Age group											
20–34			**35–49**			**50–64**			**65–74**		
Mean	sd	n	Mean	sd	n	Mean	sd	n	Mean	sd	n
8.1	10.4	487	9.7	10.2	233	7.9	10.2	191	5.8	10.3	82

Source: Reprinted with permission from the *American Journal of Epidemiology, 121*(1), 91–106, 1985.

9.9 Comment on which specific age groups are different based on the data in Table 9.2.

9.10 Perform a test for whether or not there is a general trend for increasing or decreasing amounts of moderate activity by age based on the data in Table 9.2.

Gastroenterology

A 1985 study was performed focusing on the protein concentration of duodenal secretions from patients with cystic fibrosis [2]. Table 9.3 provides data relating protein concentration to pancreatic function as measured by trypsin secretion.

9.11 What statistical procedure can be used to compare the protein concentration among the three groups?

9.12 Perform the test mentioned in Problem 9.11 and report a *p*-value.

9.13 Compare pairs of specific groups as to protein concentration using the LSD procedure.

9.14 Provide a 95% CI for the mean difference in protein concentration between each pair of specific groups.

Pulmonary Function

In the same study referred to in Example 9.1 (p. 299, text), the authors also obtained other measures of pulmonary function on the 1050 men. In particular, the FEV data for these men are presented in Table 9.4.

9.15 Are the mean FEVs different overall in the six groups?

9.16 Analyze the data for between-group differences using the conventional *t* test procedure.

9.17 Analyze the data for between-group differences using a multiple-comparisons procedure.

Diabetes

Collagen-linked fluorescence was measured in skin-biopsy specimens from 41 subjects with longstanding type I diabetes [4]. Diabetics were subdivided by the

TABLE 9.3 Relationship between protein concentration (mg/mL) of duodenal secretions to pancreatic function as measured by trypsin secretion (u/(kg/hr))

Trypsin secretion (u/(kg/hr))					
≤ 50		**51–1000**		**>1000**	
Subject number	Protein concentration	Subject number	Protein concentration	Subject number	Protein concentration
1	1.7	1	1.4	1	2.9
2	2.0	2	2.4	2	3.8
3	2.0	3	2.4	3	4.4
4	2.2	4	3.3	4	4.7
5	4.0	5	4.4	5	5.0
6	4.0	6	4.7	6	5.6
7	5.0	7	6.7	7	7.4
8	6.7	8	7.6	8	9.4
9	7.8	9	9.5	9	10.3
		10	11.7		

Source: Reprinted with permission of the *New England Journal of Medicine, 312*(6), 329–334, 1985.

TABLE 9.4 FEV data for smoking and nonsmoking males in the White and Froeb study (3)

Group number (i)	Group name	Mean FEV (L)	sd FEV (L)	n_i
1	NS	3.72	0.65	200
2	PS	3.54	0.61	200
3	NI	3.56	0.76	50
4	LS	3.49	0.62	200
5	MS	3.08	0.61	200
6	HS	2.77	0.60	200
Overall		3.331		1050

Source: Reprinted with permission of the *New England Journal of Medicine, 302*(13), 720–723, 1980.

severity of diabetic complications. In particular, diabetics were graded according to (1) level of retinopathy (ocular abnormalities), where grade 0 = normal, grade 1 = background retinopathy, grade 2 = extensive (proliferative) retinopathy, and (2) level of nephropathy (kidney abnormalities), where grade 0 = urinary protein < 0.5 g/24 hrs, grade 1 = urinary protein ≥ 0.5 g/24 hrs, <

1 g/24 hrs, grade 2 = urinary protein ≥ 1 g/24 hrs. The results are presented in Table 9.5.

9.18 Assess whether there is any overall difference in mean fluorescence level by retinopathy grade.

9.19 Identify which specific retinopathy-grade levels are different using both ordinary t tests and the method of multiple comparisons.

9.20 Answer Problem 9.18 when patients are grouped by nephropathy grade.

9.21 Answer Problem 9.19 when patients are grouped by nephropathy grade.

Dermatology

In the dermatology study referred to in Section 9.7 (Study Guide) [5], in addition to phototesting, the patients were instructed to go out in the sun and determine the number of minutes until erythema developed. The data in Table 9.6 were obtained (not all subjects completed this part of the study).

9.22 Assess if there is a difference between the mean time of exposure to sunlight until the development of erythema for subjects while on cysteine versus placebo.

9.23 Assess if there are any carry-over effects. What does a carry-over effect mean in this context?

TABLE 9.5 Collagen-linked fluorescence in relation to the type and severity of diabetic complications

Retinopathy grade	n	Mean ± sd	Nephropathy grade	n	Mean ± sd
0	11	447 ± 17	0	28	487 ± 24
1	16	493 ± 30	1	6	481 ± 16
2	14	551 ± 35	2	7	567 ± 24

Source: Reprinted with permission of the *New England Journal of Medicine, 314*(7), 403–408, 1986.

TABLE 9.6 Mean exposure times (minutes) to symptom development: daylight exposures

Group 1			Group 2		
Patient number	Period 1	Period 2	Patient number	Period 1	Period 2
2	27.70	27.80	1	28.86	45.00
3	187.35	156.86	4	38.63	42.41
7	55.88	20.00	6	12.17	38.79
11	19.61	22.51	8	33.31	48.75
12	155.13	164.55	13	84.48	104.40
14	40.23	45.00			

SOLUTIONS

9.1 A fixed-effects one-way ANOVA can be used.

9.2 The test statistic is given by $F = $ Between MS$/$ Within MS $\sim F_{k-1,n-k}$ under H_0.

Between SS $= 10(74)^2 + 10(56)^2 + 10(55)^2$

$$- \frac{[10(74) + 10(56) + 10(55)]^2}{30}$$

$$= 54{,}760 + 31{,}360 + 30{,}250 - 114{,}083.33$$

$$= 116{,}370 - 114{,}083.33 = 2286.7$$

Between MS $= \dfrac{2286.7}{2} = 1143.3$

Within MS $= s^2$, obtained from the pooled-variance estimate,

$$s^2 = \frac{9(16)^2 + 9(16)^2 + 9(9)^2}{27}$$

$$= \frac{5337}{27} = 197.7$$

Therefore,

$$F = \frac{1143.3}{197.7} = 5.78 \sim F_{2,27} \text{ under } H_0$$

Since $F_{2,27,.95} < F_{2,20,.95} = 3.49 < F$, it follows that $p < .05$, and H_0 is rejected at the 5% level.

9.3 Use the test statistic

$$t = \frac{\bar{y}_1 - \bar{y}_2}{\sqrt{s^2(1/n_1 + 1/n_2)}} \sim t_{n-k}$$

under H_0. The results are given in the table below.

9.4 This contrast is a comparison of the general vegetarian population with the general nonvegetarian population. Compute the test statistic

$$t = \frac{L}{se(L)} \sim t_{n-k} \text{ under } H_0$$

$$= \frac{0.7(56) + 0.3(55) - 74}{\sqrt{197.7[0.7^2/10 + 0.3^2/10 + (-1)^2/10]}}$$

$$= \frac{-18.3}{\sqrt{31.231}} = \frac{-18.3}{5.589} = -3.27 \sim t_{27} \text{ under } H_0$$

Since $t_{27,.995} = 2.771$, and $t_{27,.9995} = 3.690$, it follows that $.001 < p < .01$, and we reject H_0 and conclude that among premenopausal women, the general vegetarian population has a significantly lower protein intake than the general nonvegetarian population.

9.5 We use the Bonferroni approach. For a 5% level of significance, the critical values are given by $t_{27,\alpha*/2}$, $t_{27,1-\alpha*/2}$, where $\alpha* = .05/\binom{3}{2} = .05/3 = .0167$. Thus, $t_{27,\alpha*/2} = t_{27,.0083}$ and $t_{27,1-\alpha*/2} = t_{27,.9917}$. Using a computer program (e.g., the inverse CDF program of MINITAB), we find that $t_{27,.9917} = 2.55$. From the solution to Problem 9.3, $|t(\text{STD, LAC})| = 2.86 > 2.55$, $|t(\text{STD, VEG})| = 3.02 > 2.55$ and $|t(\text{LAC, VEG})| = 0.16 < 2.55$. Thus, it follows that both the lactovegetarians and the strict vegetarians have significantly lower protein intake than the nonvegetarians, among premenopausal women.

9.6 From Problem 9.4, we have a t statistic $= -3.27$. Using the Scheffé multiple-comparisons procedure, the critical values are $c_2 = \sqrt{2F_{2,27,.95}}$, $c_1 = -c_2$. Using the computer, we obtain $F_{2,27,.95} = 3.35$. Therefore, $c_2 = \sqrt{6.70} = 2.59$, $c_1 = -2.59$. Since $t < c_1$, it follows that $p < .05$ and the linear contrast is significantly different from 0.

9.7 The fixed-effects one-way analysis of variance.

9.8 We have the test statistic $F = $ Between MS/Within MS $\sim F_{k-1,n-k}$ under H_0, where

$$\text{Between SS} = 487(8.1)^2 + \ldots + 82(5.8)^2$$

$$- [487(8.1) + \ldots + 82(5.8)]^2/993$$

$$= 68{,}553.83 - 8189.3^2/993$$

$$= 68{,}553.83 - 67{,}537.40 = 1016.43$$

Groups compared	Test statistic		p-value
STD, LAC	$t = \dfrac{74 - 56}{\sqrt{197.7(1/10 + 1/10)}}$	$= \dfrac{18}{6.288} = 2.86 \sim t_{27}$	$.001 < p < .01$
STD, VEG	$t = \dfrac{74 - 55}{6.288}$	$= \dfrac{19}{6.288} = 3.02 \sim t_{27}$	$.001 < p < .01$
LAC, VEG	$t = \dfrac{56 - 55}{6.288}$	$= \dfrac{1}{6.288} = 0.16 \sim t_{27}$	NS

Groups	Test statistic		p-value
20–34, 35–49	$t = \dfrac{8.1 - 9.7}{\sqrt{106.23\left(\frac{1}{487} + \frac{1}{233}\right)}}$	$= \dfrac{-1.6}{0.82} = -1.95 \sim t_{989}$	$.05 < p < .10$
20–34, 50–64	$t = \dfrac{8.1 - 7.9}{\sqrt{106.23\left(\frac{1}{487} + \frac{1}{191}\right)}}$	$= \dfrac{0.2}{0.88} = 0.23 \sim t_{989}$	NS
20–34, 65–74	$t = \dfrac{8.1 - 5.8}{\sqrt{106.23\left(\frac{1}{487} + \frac{1}{82}\right)}}$	$= \dfrac{2.3}{1.23} = 1.87 \sim t_{989}$	$.05 < p < .10$
35–49, 50–64	$t = \dfrac{9.7 - 7.9}{\sqrt{106.23\left(\frac{1}{233} + \frac{1}{191}\right)}}$	$= \dfrac{1.8}{1.01} = 1.79 \sim t_{989}$	$.05 < p < .10$
35–49, 65–74	$t = \dfrac{9.7 - 5.8}{\sqrt{106.23\left(\frac{1}{233} + \frac{1}{82}\right)}}$	$= \dfrac{3.9}{1.32} = 2.95 \sim t_{989}$	$.001 < p < .01$
50–64, 65–74	$t = \dfrac{7.9 - 5.8}{\sqrt{106.23\left(\frac{1}{191} + \frac{1}{82}\right)}}$	$= \dfrac{2.1}{1.36} = 1.54 \sim t_{989}$	NS

Between MS $= 1016.43/3 = 338.81$

Within MS $= s^2 = [(10.4)^2 486 + \ldots + (10.3)^2 81]/989$

$\qquad = 106.23$

Thus, $F = 338.81/106.23 = 3.19 \sim F_{3,989}$ under H_0. We note from the F table that $F > F_{3,120,.95} = 2.68 > F_{3,989,.95}$. Thus, the means are significantly different (i.e., $p < .05$).

9.9 We now perform t tests for each pair of groups to determine which specific groups are different using the LSD approach to obtain critical values. The results are given in the table above.

9.10 We note from Problem 9.9 that only two of the groups are significantly different (35–49, 65–74). However, most of the comparisons show a trend in the direction of less activity with increasing age. To test for this general trend, we form the linear contrast $L = 27.5\,\bar{y}_1 + 42.5\,\bar{y}_2 + 57.5\,\bar{y}_3 + 70\,\bar{y}_4$, where the weights are chosen to be the midpoints of the four respective age groups. Since the coefficients of the contrast must add up to 0, we subtract $(27.5 + \ldots + 70)/4 = 49.375$ from each coefficient to obtain the contrast $L = (27.5 - 49.375)\,\bar{y}_1 + (42.5 - 49.375)\,\bar{y}_2 + (57.5 - 49.375)\,\bar{y}_3 + (70 -$

$49.375)\,\bar{y}_4 = -21.875\,\bar{y}_1 - 6.875\,\bar{y}_2 + 8.125\,\bar{y}_3 + 20.625\,\bar{y}_4$. We wish to test the hypothesis $H_0: E(L) = 0$ versus $H_1: E(L) \neq 0$. We have the test statistic $t = L/se(L) \sim t_{n-k}$ under H_0, where

$$L = -21.875(8.1) + \ldots + 20.625(5.8) = -60.06$$

$$se(L) = \sqrt{106.23\left(\frac{21.875^2}{487} + \ldots + \frac{20.625^2}{82}\right)}$$

$$= \sqrt{106.23(6.719)} = \sqrt{713.75} = 26.72$$

Thus, $t = -60.06/26.72 = -2.25 \sim t_{989}$ under H_0. Since $|t| > t_{120,.975} = 1.980 > t_{989,.975}$, it follows that $p < .05$. Furthermore, since $|t| < t_{\infty,.99} = 2.326 < t_{989,.99}$, it follows that $p > .02$. Therefore, we have $.02 < p < .05$ and the linear trend is significant. Thus, there is a significant trend with the amount of overall activity declining with age. In this case, the linear contrast provides a less confusing message than did the comparison of individual groups in Problem 9.9.

9.11 The fixed-effects one-way ANOVA model is appropriate here. We use the overall F test for one-way ANOVA.

9.12 We have the following mean and sd within each group:

Group	Mean	sd	n
≤ 50	3.93	2.21	9
51–1000	5.41	3.38	10
>1000	5.94	2.55	9
Overall	5.11		28

The test statistic $= F =$ Between MS/Within MS $\sim F_{2,25}$ under H_0, where

Between SS $= 9(3.93 - 5.11)^2 + 10(5.41 - 5.11)^2$
$+ 9(5.94 - 5.11)^2$

$= 19.627$

Between MS $= 19.627/2 = 9.814$

Within SS $= 8(2.21)^2 + 9(3.38)^2 + 8(2.55)^2$

$= 193.991$

Within MS $= 193.991/25 = 7.760$

$F = 9.814/7.760 = 1.265 \sim F_{2,25}$ under H_0

p-value $= Pr(F_{2,25} > 1.265) = .30$

Thus, there is no significant difference among the mean protein concentrations for the 3 groups.

9.13 Since the F test in Problem 9.12 was not significant, we declare that each pair of means is not significantly different.

9.14 A 95% CI for the mean difference between each pair of groups is given as follows:

95% CI for mean difference in protein concentration between trypsin-secretion groups

Groups compared	95% CI
$\leq 50, 51$–1000	$3.93 - 5.41 \pm t_{25,.975}\sqrt{7.76(1/9 + 1/10)}$
	$= -1.48 \pm 2.060(1.28) = (-4.1, 1.2)$
$\leq 50, >1000$	$3.93 - 5.94 \pm 2.060\sqrt{7.76(1/9 + 1/9)}$
	$= -2.01 \pm 2.060(1.31) = (-4.7, 0.7)$
51–1000, >1000	$5.41 - 5.94 \pm 2.060\sqrt{7.76(1/10 + 1/9)}$
	$= -0.53 \pm 2.64 = (-3.2, 2.1)$

9.15 We wish to test the hypothesis H_0: $\mu_1 = ... = \mu_6$ versus H_1: at least 2 of the μ_i's are unequal. We use the F test as follows:

Between SS $= 200(3.72)^2 + ... + 200(2.77)^2$
$- 1050(3.331)^2 = 122.22$

Between MS $= 122.22/5 = 24.44$

Within MS $= s^2$, where
$$s^2 = [199(0.65)^2 + ... + 199(0.60)^2]/1050 - 6)$$
$$= 408.61/1044 = 0.391$$

Thus, we have the test statistic $F =$ Between MS/Within MS $= 24.44/0.391 = 62.46 \sim F_{5,1044}$ under H_0. Since $F_{5,1044,.999} < F_{5,120,.999} = 4.42 < F$, it follows that $p < .001$ and the 6 means are significantly different.

9.16 We use the LSD t test criterion in (**9.13**) (text, p. 308) for each pair of groups. The results are summarized as follows:

Groups	Test statistic		p-value
NS, PS	$t = \dfrac{3.72 - 3.54}{\sqrt{0.391\left(\dfrac{1}{200} + \dfrac{1}{200}\right)}}$	$= \dfrac{0.18}{0.063} = 2.88 \sim t_{1044}$	.004
NS, NI	$t = \dfrac{3.72 - 3.56}{\sqrt{0.391\left(\dfrac{1}{200} + \dfrac{1}{50}\right)}}$	$= \dfrac{0.16}{0.099} = 1.62 \sim t_{1044}$	NS
NS, LS	$t = \dfrac{3.72 - 3.49}{\sqrt{0.391\left(\dfrac{1}{200} + \dfrac{1}{200}\right)}}$	$= \dfrac{0.23}{0.063} = 3.68 \sim t_{1044}$	<.001
NS, MS	$t = \dfrac{3.72 - 3.08}{\sqrt{0.391\left(\dfrac{1}{200} + \dfrac{1}{200}\right)}}$	$= \dfrac{0.64}{0.063} = 10.23 \sim t_{1044}$	<.001

NS, HS $\quad t = \dfrac{3.72 - 2.77}{\sqrt{0.391\left(\dfrac{1}{200} + \dfrac{1}{200}\right)}} = \dfrac{0.95}{0.063} = 15.19 \sim t_{1044}$ $\qquad$ <.001

PS, NI $\quad t = \dfrac{3.54 - 3.56}{\sqrt{0.391\left(\dfrac{1}{200} + \dfrac{1}{50}\right)}} = \dfrac{-0.02}{0.099} = -0.20 \sim t_{1044}$ $\qquad$ NS

PS, LS $\quad t = \dfrac{3.54 - 3.49}{\sqrt{0.391\left(\dfrac{1}{200} + \dfrac{1}{200}\right)}} = \dfrac{0.05}{0.063} = 0.80 \sim t_{1044}$ $\qquad$ NS

PS, MS $\quad t = \dfrac{3.54 - 3.08}{\sqrt{0.391\left(\dfrac{1}{200} + \dfrac{1}{200}\right)}} = \dfrac{0.46}{0.063} = 7.35 \sim t_{1044}$ $\qquad$ <.001

PS, HS $\quad t = \dfrac{3.54 - 2.77}{\sqrt{0.391\left(\dfrac{1}{200} + \dfrac{1}{200}\right)}} = \dfrac{0.77}{0.063} = 12.31 \sim t_{1044}$ $\qquad$ <.001

NI, LS $\quad t = \dfrac{3.56 - 3.49}{\sqrt{0.391\left(\dfrac{1}{50} + \dfrac{1}{200}\right)}} = \dfrac{0.07}{0.099} = 0.71 \sim t_{1044}$ $\qquad$ NS

NI, MS $\quad t = \dfrac{3.56 - 3.08}{\sqrt{0.391\left(\dfrac{1}{50} + \dfrac{1}{200}\right)}} = \dfrac{0.48}{0.099} = 4.85 \sim t_{1044}$ $\qquad$ <.001

NI, HS $\quad t = \dfrac{3.56 - 2.77}{\sqrt{0.391\left(\dfrac{1}{50} + \dfrac{1}{200}\right)}} = \dfrac{0.79}{0.099} = 7.99 \sim t_{1044}$ $\qquad$ <.001

LS, MS $\quad t = \dfrac{3.49 - 3.08}{\sqrt{0.391\left(\dfrac{1}{200} + \dfrac{1}{200}\right)}} = \dfrac{0.41}{0.063} = 6.55 \sim t_{1044}$ $\qquad$ <.001

LS, HS $\quad t = \dfrac{3.49 - 2.77}{\sqrt{0.391\left(\dfrac{1}{200} + \dfrac{1}{200}\right)}} = \dfrac{0.72}{0.063} = 11.51 \sim t_{1044}$ $\qquad$ <.001

MS, HS $\quad t = \dfrac{3.08 - 2.77}{\sqrt{0.391\left(\dfrac{1}{200} + \dfrac{1}{200}\right)}} = \dfrac{0.31}{0.063} = 4.96 \sim t_{1044}$ $\qquad$ <.001

The p-values were all evaluated using the $N(0, 1)$ distribution that is sufficiently close to a t_{1044} distribution. The results show that (1) nonsmokers are significantly different from passive smokers and light, moderate, and heavy inhaling smokers and are not significantly different from noninhalers; (2) passive smokers are significantly different from moderate and heavy inhaling smokers, but not from light smokers or noninhalers; (3) noninhalers are significantly different from moderate and heavy smokers but not from light smokers; and (4) light, moderate, and heavy inhaling smokers are all significantly different from each other.

9.17 Using the Bonferroni approach, the critical values are $c_1 = t_{1044,.025/15} = t_{1044,.0017}$ and $c_2 = t_{1044,.9983}$. If we approximate t_{1044} by an $N(0, 1)$ distribution and refer to Table 3 (Appendix in text), we see that $c_1 = -2.94$, $c_2 = 2.94$. Thus, the results are the same as with the LSD procedure in Problem 9.16, with the exception of the (NS, PS) comparison, which is no longer significant. However, the (NS, PS) comparison is the key result in the paper, so that the LSD and Bonferroni approaches yield very different conclusions.

9.18 We use the fixed-effects one-way ANOVA as follows:

$$\text{Between SS} = 447^2(11) + 493^2(16) + 551^2(14)$$
$$- [447(11) + 493(16) + 551(14)]^2/41$$
$$= 10,337,097 - 10,269,008.80$$
$$= 68,088.20$$

$$\text{Between MS} = \text{Between SS}/2 = 34,044.10$$

$$\text{Within MS} = s^2, \text{ where}$$

$$s^2 = \frac{10(17)^2 + \dots + 13(35)^2}{38}$$

$$= \frac{32,315}{38} = 850.39$$

Thus, the test statistic $= F = 34,044.10/850.39 = 40.03 \sim F_{2,38}$ under H_0. Since $F > F_{2,30,.999} = 8.77 > F_{2,38,.999}$, it implies that $p < .001$ and there is a significant overall difference in mean fluorescence level by retinopathy grade.

9.19 We first perform ordinary t tests using critical values based on the LSD approach in the table below. Clearly, all pairs of groups are significantly different using ordinary t tests. We now use the Bonferroni method

for multiple comparisons to determine the critical values. Since there are 3 groups, the critical values are $c_1 = t_{38, \alpha*/2}, c_2 = t_{38, 1-\alpha*/2}$ where $\alpha* = .05/\binom{3}{2} = .0167$. Thus, $c_1 = t_{38,.0083}, c_2 = t_{38,.9917}$. Using the computer, we determine that $c_1 = -2.50, c_2 = 2.50$. Since the test statistics for all 3 comparisons are < -2.50, it follows that using the Bonferroni approach, there are significant differences between the mean fluorescence level for each pair of retinopathy groups.

9.20 We use the fixed-effects one-way ANOVA. We have

$$\text{Between SS} = 487^2(28) + 481^2(6) + 567^2(7)$$
$$- [487(28) + 481(6) + 567(7)]^2/41$$
$$= 10,279,321 - 10,241,001.98$$
$$= 38,319.02$$

$$\text{Between MS} = 38,319.02/2 = 19,159.51$$

$$\text{Within MS} = s^2, \text{ where}$$

$$s^2 = \frac{27(24)^2 + \dots + 6(24)^2}{38}$$

$$= \frac{20,288}{38} = 533.89$$

Thus, the F statistic $= F = 19,159.51/533.89 = 35.89 \sim F_{2,38}$ under H_0. Since $F > F_{2,30,.999} = 8.77 > F_{2,38,.999}$, it implies that $p < .001$. Thus, there is an overall difference in mean fluorescence level by nephropathy grade.

Groups compared	Test statistic		p-value
0, 1	$t = \dfrac{447 - 493}{\sqrt{850.39(1/11 + 1/16)}}$	$= \dfrac{-46}{11.422} = -4.03 \sim t_{38}$	<.001
0, 2	$t = \dfrac{447 - 551}{\sqrt{850.39(1/11 + 1/14)}}$	$= \dfrac{-104}{11.750} = -8.85 \sim t_{38}$	<.001
1, 2	$t = \dfrac{493 - 551}{\sqrt{850.39(1/16 + 1/14)}}$	$= \dfrac{-58}{10.672} = -5.43 \sim t_{38}$	<.001

Groups compared	Test statistic	p-value
0, 1	$t = \dfrac{487 - 481}{\sqrt{533.89(1/28 + 1/6)}} = \dfrac{6}{10.395} = 0.58 \sim t_{38}$	NS
0, 2	$t = \dfrac{487 - 567}{\sqrt{533.89(1/28 + 1/7)}} = \dfrac{-80}{9.764} = -8.19 \sim t_{38}$	<.001
1, 2	$t = \dfrac{481 - 567}{\sqrt{533.89(1/6 + 1/7)}} = \dfrac{-86}{12.855} = -6.69 \sim t_{38}$	<.001

9.21 We first perform ordinary t tests using critical values based on the LSD approach as shown above. Thus, there are significant differences between group 0 and group 2 and between group 1 and group 2 ($p < .001$), but no significant difference between groups 0 and 1. We now use the method of multiple comparisons based on the Bonferroni approach. Since there are 3 groups and 38 df, the critical values are the same as in Problem 9.19, namely, $c_1 = -2.50$, $c_2 = 2.50$. Since $|0.58| < 2.50$, $|-8.19| > 2.50$ and $|-6.69| > 2.50$, it follows that there are significant differences in mean fluorescence levels between nephropathy grade 2 vs. each of nephropathy grades 0 and 1, respectively, while there is no significant difference between patients in nephropathy grades 0 and 1.

9.22 The mean and sd of the difference in the time to develop erythema between the cysteine and placebo treatment periods by group are given as follows:

	Group 1	Group 2
d_i [a]	8.20	16.38
s_{d_i} [a]	19.67	8.32
n_i	6	5

[a]Minutes.

The overall mean is $\bar{d} = (8.20 + 16.38)/2 = 12.29$. We have the pooled-variance estimate

$$s_d^2 = \frac{5(19.67)^2 + 4(8.32)^2}{9}$$

$$= \frac{2210.902}{9} = 245.66$$

The test statistic is

$$t = \frac{12.29}{\sqrt{\dfrac{245.66}{4}\left(\dfrac{1}{6} + \dfrac{1}{5}\right)}}$$

$$= \frac{12.29}{4.745} = 2.59 \sim t_9 \text{ under } H_0$$

The p-value $= 2 \times Pr(t_9 > 2.59) = .029$

Thus, there is a significant difference between the time to developing erythema after exposure to sunlight while on cysteine vs. placebo as measured by patient diary.

9.23 We compute the average time to symptom development over the two periods for each patient and compute the mean and sd of the average time within each group as follows:

	Group 1	Group 2
x_i [a]	76.89	47.68
s_i [a]	69.53	26.88
n_i	6	5

[a]Minutes.

The pooled-variance estimate is

$$s^2 = \frac{5(69.53)^2 + 4(26.88)^2}{9} = 3006.73$$

The test statistic is

$$t = \frac{76.89 - 47.68}{\sqrt{3006.73(1/6 + 1/5)}}$$

$$= \frac{29.21}{33.20} = 0.88 \sim t_9$$

The p-value $= 2 \times Pr(t_9 > 0.88) = .40$. Thus, there is no significant carry-over effect in these data.

REFERENCES

[1] Sallis, J. F., Haskell, W. L., Wood, P. D., Fortmann, S. P., Rogers, T., Blair, S. N., & Paffenbarger, R. S., Jr. (1985). Physical activity assessment methodology in the five-city project. *American Journal of Epidemiology, 121*(1), 91–106.

[2] Kopelman, H., Durie, P., Gaskin, K., Weizman, Z., & Forstner, G. (1985). Pancreatic fluid secretion and protein hyperconcentration in cystic fibrosis. *New England Journal of Medicine, 312*(6), 329–334.

[3] White, J. R., & Froeb, H. F. (1980). Small-airways dysfunction in nonsmokers chronically exposed to tobacco smoke. *New England Journal of Medicine, 302*(13), 720–723.

[4] Monnier, V. M., Vishwanath, V., Frank, K. E., Elmets, C. A., Dauchot, P., & Kohn, R. R. (1986). Relation between complications of type I diabetes mellitus and collagen-linked fluorescence. *New England Journal of Medicine, 314*(7), 403–408.

[5] Mathews-Roth, M. M., Rosner, B., Benfell, K., & Roberts, J. E. (1994). A double-blind study of cysteine photoprotection in erythropoietic protoporphyria. In press.

CHAPTER TEN

HYPOTHESIS TESTING: CATEGORICAL DATA

SECTION 10.1   Comparison of Two Binomial Proportions

A case–control study was performed among 2982 cases, 5782 controls, from 10 geographic areas of the United States and Canada. The cases were newly diagnosed cases of bladder cancer in 1977–1978 obtained from cancer registries; the control group was a random sample of the population of the 10 study areas with a similar age, sex, and geographical distribution. We want to study the association between the incidence of bladder cancer and consumption of alcoholic beverages. Let

$$p_1 = \text{true proportion of drinkers among cases}$$
$$p_2 = \text{true proportion of drinkers among controls}$$

We wish to test the hypothesis $H_0: p_1 = p_2 = p$ versus $H_1: p_1 \neq p_2$.

10.1.1   Two-Sample Test for Binomial Proportions (Normal-Theory Version)

In this study, if we define a drinker as a person who consumes ≥ 1 drink/day of whiskey, then the proportion of drinkers

$$= 574/2388 = .240 \text{ for the cases} = \hat{p}_1$$
$$= 980/4660 = .210 \text{ for the controls} = \hat{p}_2$$

Not all subjects provided a drinking history, which is why the sample sizes (2388, 4660) in the two groups are less than the total sample sizes in the study (2982, 5782). We use the test statistic

$$z = \frac{\hat{p}_1 - \hat{p}_2}{\sqrt{\hat{p}\hat{q}\left(\dfrac{1}{n_1} + \dfrac{1}{n_2}\right)}} \sim N(0, 1)$$

where

$$\hat{p} = \frac{x_1 + x_2}{n_1 + n_2} = \frac{\text{total number of drinkers over both groups}}{\text{total number of subjects over both groups}}$$

The p-value $= 2 \times \Phi(z)$ if $z < 0$, or $2 \times [1 - \Phi(z)]$ if $z \geq 0$. We will only use this test if $n_1 \hat{p}\hat{q} \geq 5$ and $n_2 \hat{p}\hat{q} \geq 5$. In this case,

$$\hat{p} = \frac{574 + 980}{2388 + 4660} = \frac{1554}{7048} = .220$$

$$z = \frac{.030}{\sqrt{.220(.780)\left(\dfrac{1}{2388} + \dfrac{1}{4660}\right)}} = \frac{.0301}{.0104}$$

$$= 2.882 \sim N(0, 1) \text{ under } H_0$$

$$p\text{-value} = 2 \times [1 - \Phi(2.882)] = 2(1 - .9980) = .004$$

Thus, the cases report significantly more drinking than the controls. In this study, $n_1 \hat{p}\hat{q} = 2388(.220)(.780) = 410.4$ and $n_2 \hat{p}\hat{q} = 4660(.220)(.780) = 800.9$. Thus, it is valid to use the normal-theory test.

SECTION 10.2   The 2 × 2 Contingency-Table Approach

Another technique for the analysis of these data is the contingency-table approach. A 2×2 *contingency table* is a table where case/control status is displayed along the rows and consumption of hard liquor along the columns, as shown in the following table. A specific row and column combination is called a *cell*, and the number of people in a given cell is called the *cell count*.

		Consumption of hard liquor		
		≥1/day	<1/day	
Case/control status	Case	574	1814	2388
	Control	980	3680	4660
		1554	5494	7048

2388, 4460 are row margins; 1554, 5494 are column margins; 7048 is the grand total. In general, we use the following notation:

(1, 1) cell a	(1, 2) cell b	$a + b = R_1$
(2, 1) cell c	(2, 2) cell d	$c + d = R_2$
$a + c = C_1$	$b + d = C_2$	

(i, j) cell = ith row, jth column

The entire table is referred to as the *observed* table. To test for statistical significance, we compare the observed table with what we would expect if there were no association between being a bladder-cancer case and consuming ≥ 1 drink of hard liquor/day. We wish to test the hypothesis.

$$H_0: p_1 = p_2 = p \text{ versus } H_1: p_1 \neq p_2$$

Under H_0, the expected number of units in the ith row and jth column is

$$E_{ij} = R_i C_j / N = i\text{th row total} \times j\text{th column total}/N$$

In our case,

$$E_{11} = 2388 \times 1554/7048 = 526.5$$
$$E_{12} = 2388 \times 5494/7048 = 1861.5$$
$$E_{21} = 4660 \times 1554/7048 = 1027.5$$
$$E_{22} = 4660 \times 5494/7048 = 3632.5$$

As a check, the sum of corresponding observed and expected row and column totals should be the same, as they are in this case.

We now wish to compare the observed and expected tables. If they are reasonably close, then we will accept H_0, else we will reject H_0. The criterion used for agreement in the (i, j) cell is

$$(O_{ij} - E_{ij})^2/E_{ij}$$

For the entire table, we use the test statistic

$$X^2 \equiv \sum_{i=1}^{2} \sum_{j=1}^{2} (O_{ij} - E_{ij})^2/E_{ij} \sim \chi_1^2 \text{ under } H_0$$

The test statistic X^2 is referred to as the uncorrected chi-square statistic for 2×2 contingency tables. Under H_0, $X^2 \sim \chi_1^2$ (only 1 df, because there is one independent cell in the table; all others are determined from the row and column totals).

To better approximate the chi-square distribution, we use a continuity correction (this is controversial),

$$X_{CORR}^2 = \sum_{i=1}^{2} \sum_{j=1}^{2} (|O_{ij} - E_{ij}| - .5)^2/E_{ij} \sim \chi_1^2$$

$$= \text{Yates-corrected chi-square statistic for } 2 \times 2 \text{ contingency tables}$$

Since we only reject for large values of X^2_{CORR}, p-value $= Pr(\chi^2_1 > X^2_{CORR})$. The test procedure is referred to as the *chi-square test for 2×2 contingency tables*. We only use this test if all expected values are ≥ 5.

In this example,

<table>
<tr><td colspan="2">**Observed table**</td><td colspan="2">**Expected table**</td></tr>
<tr><td>574</td><td>1814</td><td>526.5</td><td>1861.5</td></tr>
<tr><td>980</td><td>3680</td><td>1027.5</td><td>3632.5</td></tr>
</table>

$$X^2_{CORR} = \frac{(|574 - 526.5| - .5)^2}{526.5} + \frac{(|1814 - 1861.5| - .5)^2}{1861.5}$$

$$+ \frac{(|980 - 1027.5| - .5)^2}{1027.5} + \frac{(|3680 - 3632.5| - .5)^2}{3632.5}$$

$$= \frac{47^2}{526.5} + \frac{47^2}{1861.5} + \frac{47^2}{1027.5} + \frac{47^2}{3632.5}$$

$$= 4.19 + 1.19 + 2.15 + 0.61 = 8.13 \sim \chi^2_1$$

Since $\chi^2_{1,.995} = 7.88$, $\chi^2_{1,.999} = 10.83$, and $7.88 < 8.13 < 10.83$, it follows that $1 - .999 < p < 1 - .995$ or $.001 < p < .005$.

10.2.1   Relationship Between the Chi-Square Test and the Two-Sample Test for Binomial Proportions

In general,

$$X^2 = z^2_{binomial}$$

In our case,

$$X^2 = \frac{(574 - 526.5)^2}{526.5} + \dots + \frac{(3680 - 3632.5)^2}{3632.5}$$

$$= \frac{47.5^2}{526.5} + \frac{47.5^2}{1861.5} + \frac{47.5^2}{1027.5} + \frac{47.5^2}{3632.5}$$

$$= 4.28 + 1.21 + 2.19 + 0.62$$

$$= 8.31 \sim \chi^2_1 = 2.882^2 = z^2$$

SECTION 10.3   Fisher's Exact Test

Consider a study of the relationship between early age at menarche (i.e., age at which periods begin) and breast-cancer incidence. We select 50 premenopausal breast-cancer cases and 50 premenopausal age-matched controls. We find that 5 of the cases have an age at menarche <11 yrs, and 1 control has an age at menarche <11 yrs. Is this a significant finding? We have the following observed and expected contingency tables:

	Observed table, age at menarche			Expected table, age at menarche	
	< 11	≥ 11		< 11	≥ 11
Case	5	45	50	3.0	47.0
Control	1	49	50	3.0	47.0
	6	94	100		

We can't use the chi-square test because two of the expected values are < 5. Instead, we must use a method called *Fisher's exact test*. For this test, we consider the margins of the table as fixed and ask the question, How unusual is our table among all tables with the same fixed margins?

Case	a	b	$a + b$
Control	c	d	$c + d$
	$a + c$	$b + d$	N

Let

$$p_1 = Pr(\text{age at menarche} < 11 \,|\, \text{case}) = Pr(\text{exposed} \,|\, \text{case})$$
$$p_2 = Pr(\text{age at menarche} < 11 \,|\, \text{control}) = Pr(\text{exposed} \,|\, \text{control})$$

We wish to test the hypothesis

$$H_0: p_1 = p_2 = p \text{ versus } H_1: p_1 \neq p_2$$

The exact binomial probability of observing our table given the fixed margins is given by:

$$Pr(a \text{ exposed cases, } c \text{ exposed controls} \,|\, \text{fixed margins of}$$
$$a + b, c + d, a + c, \text{ and } b + d)$$

$$= (a + b)! \,(c + d)! \,(a + c)! \,(b + d)!/(N! \,a! \,b! \,c! \,d!)$$

This is called the *hypergeometric* distribution.

Because the margins are fixed, any table is completely determined by one cell count. We usually refer to the table with cell count $= a$ in the (1, 1) cell as the "a" table. In our example, we observed the "5" table. Therefore,

$$Pr(5 \text{ table}) = \frac{50! \,50! \,6! \,94!}{100! \,5! \,45! \,1! \,49!} = \frac{50 \times 50 \times 49 \times 48 \times 47 \times 46 \times 6}{100 \times 99 \times 98 \times 97 \times 96 \times 95}$$

$$= \frac{7.628 \times 10^{10}}{8.583 \times 10^{11}}$$

$$= .089$$

How do we judge the significance of this particular table? We need to enumerate all tables that could have occurred with the same margins, and compute the probability of each such table. These are given as follows:

0	50		1	49		2	48		3	47		4	46		5	45		6	44
6	44		5	45		4	46		3	47		2	48		1	49		0	50

.013   .089   .237   .322   .237   .089   .013

10.3.1 Computation of *p*-Values with Fisher's Exact Test

There are two commonly used methods for calculation of two-tailed *p*-values, as follows:

(1)
$$p\text{-value} = 2 \times \min\left[\sum_{i=0}^{a} Pr(i), \sum_{i=a}^{a+b} Pr(i), 0.5\right]$$

(2)
$$p\text{-value} = \sum_{\{i:Pr(i)\leq Pr(a)\}} Pr(i)$$

In this case, we will use the first approach:

$$p\text{-value (2-tail)} = 2 \times \min\left[\sum_{i=0}^{5} Pr(i), \sum_{i=5}^{6} Pr(i)\right]$$

$$= 2 \times (.987, .102) = .204$$

Thus, there is no significant relationship between early menarche and breast cancer.

In general, we only need to use Fisher's exact test if at least one cell has expected value < 5. However, it is always a valid test, but is more tedious than the chi-square test.

SECTION 10.4   McNemar's Test for Correlated Proportions

A case–control study was performed to study the relationship between the source of drinking water during the prenatal period and congenital malformations. Case mothers are those with malformed infants in a registry in Australia between 1951 and 1979. Controls were individually matched by hospital, maternal age ($\pm$ 2 years), and date of birth (± 1 month). The suspected causal agent was groundwater nitrates. The following 2×2 table was obtained relating case–control status to the source of drinking water:

	Source of Drinking Water		Total	Percentage of groundwater
	Groundwater	Rainwater		
Cases	162	56	218	74.3%
Controls	123	95	218	56.4%
Total	285	151	436	

The corrected chi-square statistic $= \chi^2 = 14.63, p < .001$.

However, the assumptions of the χ^2 test are not valid, because the women in the two samples were individually matched and are not statistically independent. We instead must analyze the data in terms of *matched pairs*. The following table gives the exposure status of case–control pairs.

Case	Control	Frequency
+	+	101
+	−	61
−	+	22
−	−	34

Note: + = groundwater, − = rainwater.

We refer to the $(+, +)$ and $(−, −)$ pairs as *concordant pairs,* since the case and control members of the pair have the same exposure status. We refer to the $(+, −)$ and $(−, +)$ pairs as *discordant pairs.* For our test, we ignore the concordant pairs and only focus on the discordant pairs. Let

$$n_A = \text{the number of } (+, −) \text{ or type A discordant pairs}$$

$$n_B = \text{the number of } (−, +) \text{ or type B discordant pairs}$$

$$n_D = n_A + n_B = \text{total number of discordant pairs}$$

We wish to test the hypothesis $H_0: p = 1/2$ versus $H_1: p \neq 1/2$, where $p = \text{prob(discordant pair is of type A)}$. If $n_A + n_B \geq 20$, then we can use the normal-theory test. We use the test statistic

$$X^2 = \left(\left|n_A - \frac{n_D}{2}\right| - \frac{1}{2}\right)^2 \bigg/ \left(\frac{n_D}{4}\right) \sim \chi_1^2$$

$$p\text{-value} = Pr(\chi_1^2 > X^2)$$

In this case,

$$n_A = 61, n_B = 22, n_D = 83$$

$$X^2 = \frac{\left(\left|61 - \frac{83}{2}\right| - \frac{1}{2}\right)^2}{83/4} = 17.40 \sim \chi_1^2$$

Since

$$\chi_{1,.999}^2 = 10.83 < X^2, \text{ we obtain } p < .001$$

Thus, there is a significant association between the source of drinking water and the occurrence of congenital malformations.

The data were also analyzed separately by season of birth. The following exposure data are presented in a 2×2 table of case exposure status by control exposure status for spring births.

		Control	
		+	−
Case	+	30	14
	−	2	10

Since the number of discordant pairs $= n_A + n_B = 14 + 2 = 16 < 20$, we cannot use the normal-theory test. Instead, we must use an exact binomial test. To compute the p-value, we have

$$p = 2 \times \left[\sum_{k=0}^{n_A} \binom{n_D}{k} \left(\frac{1}{2}\right)^{n_D} \right] \quad \text{if } n_A < \frac{n_D}{2}$$

$$= 2 \times \sum_{k=n_A}^{n_D} \binom{n_D}{k} \left(\frac{1}{2}\right)^{n_D} \quad \text{if } n_A > \frac{n_D}{2}$$

$$= 1 \quad \text{if } n_A = \frac{n_D}{2}$$

In this case $n_A = 14 > \dfrac{n_D}{2} = 8$. Therefore,

$$p\text{-value} = 2 \times \sum_{k=14}^{16} \binom{16}{k} \left(\frac{1}{2}\right)^{16}$$

From Table 1 (in the Appendix of the text), under $n = 16$, $p = .50$ we have

$$p\text{-value} = 2(.0018 + .0002 + .0000) = 2 \times .002 = .004$$

Thus, there is a significant association for the subset of spring births as well.

SECTION 10.5   Sample Size for Comparing Two Binomial Proportions

To test the hypothesis, H_0: $p_1 = p_2$ versus H_1: $p_1 \neq p_2$, $|p_1 - p_2| = \Delta$ with significance level α and power $= 1 - \beta$ with an equal sample size per group, we need

$$n_1 = \left[\sqrt{2\bar{p}\bar{q}}(z_{1-\alpha/2}) + \sqrt{p_1 q_1 + p_2 q_2}(z_{1-\beta}) \right]^2 / \Delta^2 = n_2$$

subjects per group where $\bar{p} = (p_1 + p_2)/2, \bar{q} = 1 - \bar{p}$.

Example: A study is being planned among postmenopausal women to investigate the effect on breast-cancer incidence of having a family history of breast cancer. Suppose that a 5-year study is planned and it is expected that the 5-year incidence rate of breast cancer among women without a family history is 1%, while the 5-year incidence among women with a family history is 2%. If an equal number of women per group are to be studied, then how many women in each group should be enrolled to have an 80% chance of detecting a significant difference using a two-sided test with $\alpha = .05$?

In this example, $\alpha = .05$, $z_{1-.05/2} = z_{.975} = 1.96$, $1 - \beta = .8$, $z_{.8} = 0.84$, $p_1 = .01$, $q_1 = .99$, $p_2 = .02$, $q_2 = .98$, $\bar{p} = (.01 + .02)/2 = .015$, $\bar{q} = .985$, $\Delta = .01$. Therefore, we need

$$n = [\sqrt{2(.015)(.985)}(1.96) + \sqrt{.01(.99) + .02(.98)}(0.84)]^2 / .01^2$$

$$= [0.1719(1.96) + 0.1718(0.84)]^2 / .0001$$

$$= (.4812)^2 / .0001 = 2315.5$$

Therefore, we need to study 2316 subjects in each group to have an 80% chance of finding a significant difference with this number of subjects. Over 5 years we would expect about 23 breast-cancer cases among those women without a family history and 46 cases among those women with a family history. ∎

The sample-size formula can also be modified to allow for an unequal number of subjects per group—see **(10.26)**, p. 384 of the text.

Suppose the study is performed, but only 2000 postmenopausal women per group are enrolled. How much power would such a study have? The general formula is given as follows:

$$Power = \Phi\left[\frac{\Delta}{\sqrt{\frac{p_1q_1}{n_1} + \frac{p_2q_2}{n_2}}} - z_{1-\alpha/2}\frac{\sqrt{\bar{p}\bar{q}\left(\frac{1}{n_1} + \frac{1}{n_2}\right)}}{\sqrt{\frac{p_1q_1}{n_1} + \frac{p_2q_2}{n_2}}}\right]$$

where $\bar{p} = (n_1p_1 + n_2p_2)/(n_1 + n_2), \bar{q} = 1 - \bar{p}$

In this example, the power is given by

$$Power = \Phi\left[\frac{.01}{\sqrt{\frac{.01(.99)}{2000} + \frac{.02(.98)}{2000}}} - 1.96\frac{\sqrt{.015(.985)\left(\frac{1}{2000} + \frac{1}{2000}\right)}}{\sqrt{\frac{.01(.99)}{2000} + \frac{.02(.98)}{2000}}}\right]$$

$$= \Phi\left[\frac{.01}{.003841} - 1.96\left(\frac{.003844}{.003841}\right)\right]$$

$$= \Phi(2.604 - 1.962) = \Phi(0.642) = .74$$

Thus, the study would have 74% power.

SECTION 10.6   *r* × *c* **Contingency Tables**

Patients with heart failure, diabetes, cancer, and lung disease who have various infections from gram-negative organisms often receive aminoglycosides. One of the side effects of aminoglycosides is nephrotoxicity (possible damage to the kidney). A study was performed comparing the nephrotoxicity (rise in serum creatinine of at least 0.5 mg/dL) for 3 aminoglycosides. The following results were obtained:

	+*	Total	%+
Gentamicin (G)	44	121	36.4
Tobramycin (T)	21	92	22.8
Amikacin (A)	4	16	25.0

*+ = number of patients with a rise in serum creatinine of ≥ 0.5 mg/dL

Are there significant differences in nephrotoxicity among the 3 antibiotics?

We can represent the data in the form of a 2 × 3 contingency table (2 rows, 3 columns) as follows:

	Antibiotic			
	G	T	A	
Nephrotoxicity +	44	21	4	69
−	77	71	12	160
	121	92	16	229

We wish to test the hypothesis H_0: no association between row and column classifications versus H_1: some association between row and column classifications. Under H_0, the expected number of units in the ith row and jth column is E_{ij}, given by

$$E_{ij} = \frac{R_i C_j}{N}$$

where $R_i = i$th row total, $C_j = j$th column total, and $N =$ grand total. We use the test statistic

$$X^2 = \sum_{i=1}^{r} \sum_{j=1}^{c} (O_{ij} - E_{ij})^2 / E_{ij} \sim \chi^2_{(r-1)\times(c-1)} \quad \text{under } H_0$$

$$p\text{-value} = Pr(\chi^2_{(r-1)\times(c-1)} > X^2)$$

We only use this test if no more than $1/5$ of the expected values are < 5 and no expected value is < 1. We have the following expected cell counts:

$$E_{11} = 69(121)/229 = 36.5$$
$$E_{12} = 69(92)/229 = 27.7$$
$$\text{etc.}$$

The complete observed and expected tables are given as follows:

		Expected table			Observed table		
		G	T	A	G	T	A
Nephrotoxicity	**+**	36.5	27.7	4.8	44	21	4
	−	84.5	64.3	11.2	77	71	12

Only 1 of 6 cells has expected value < 5. Thus, we can use the chi-square test. We have the test statistic

$$X^2 = \frac{(44 - 36.5)^2}{36.5} + \frac{(21 - 27.7)^2}{27.7} + \frac{(4 - 4.8)^2}{4.8} + \frac{(77 - 84.5)^2}{84.5}$$
$$+ \frac{(71 - 64.3)^2}{64.3} + \frac{(12 - 11.2)^2}{11.2}$$

$$= \frac{7.5^2}{36.5} + \frac{6.7^2}{27.7} + \frac{0.8^2}{4.8} + \frac{7.5^2}{84.5} + \frac{6.7^2}{64.3} + \frac{0.8^2}{11.2}$$

$$= 1.56 + 1.63 + 0.14 + 0.67 + 0.70 + 0.06$$

$$= 4.76 \sim \chi^2_2. \text{ Since } \chi^2_{2,.95} = 5.99 > 4.76, p > .05$$

There are no significant differences in nephrotoxicity among the 3 antibiotics.

In the preceding example, the different antibiotics form a **nominal scale**; i.e., there is no specific ordering among the three antibiotics. For some exposures, there is an implicit ordering. For example, suppose we wish to relate the occurrence of bronchitis in the first year of life to the number of cigarettes per day smoked by the mother. If we focus on smoking mothers and categorize the amount smoked by

(1–4/5–14/15–24/25–44/45+) cigarettes per day, then we might construct a 2×5 contingency table as follows:

Number of cigarettes per day

		1–4	5–14	15–24	25–44	45+
Bronchitis in 1st year	+					
	–					

We could perform the chi-square test for $r \times c$ tables given above (sometimes known as the "chi-square test for heterogeneity"). However, this is equivalent to testing the hypothesis H_0: $p_1 = p_2 = \ldots = p_5$ versus H_1: at least two of the p_i's are unequal, where p_i = probability of bronchitis in the ith smoking group. However, we would expect if there is a "dose–response" relationship between bronchitis and cigarette smoking that p_i should increase as the number of cigarettes per day increases. One way to test this hypothesis is to test H_0: p_i all equal versus H_1: $p_i = \alpha + \beta S_i$, where S_i is a score variable attributable to the ith smoking group. There are different score variables that could be used. A common choice is to use $S_i = i$; i.e., $p_i = \alpha + \beta i$. In this case, β is interpreted as the increase in the probability of bronchitis for an increase of 1 cigarette-smoking group (e.g., from 1–4 to 5–14 cigarettes per day). To test this hypothesis, we use the **chi-square test for trend.** See **(10.36)**, p. 397 of the text, for details on the test procedure. This test is often more useful for establishing dose–response relationships in $2 \times k$ tables than the chi-square test for heterogeneity.

SECTION 10.7   Measures of Effect for Epidemiological Data

In some instances, we are interested in comparing binomial proportions between two groups, but the groups are heterogeneous with respect to risk and it is desirable to subdivide each group into strata that are more homogeneous. Consider the following study.

Special wards exist for the care of babies of very low birthweight. One of the leading causes of death during the 1st year of life in this group is the respiratory distress syndrome. During the period 1985–1990, surfactant was introduced to widespread clinical use. Surfactant is administered intratracheally (the trachea is more commonly called the windpipe) to infants with the respiratory distress syndrome. The data on the following page were reported regarding in-hospital mortality for the period before and after surfactant use on the wards in 14 hospitals [1]. Since mortality varied by birthweight, it was important to account for possible birthweight differences between the two time periods.

One approach is to do separate analyses within each birthweight stratum. The authors used this approach and reported their results in terms of an odds ratio for each birthweight stratum. What is an odds ratio?

The odds in favor of a disease = $p/(1 - p)$, where p = probability of disease. The *odds ratio* (OR) = odds in favor of disease for an exposed group/odds in favor of disease for an unexposed group. For example, for the stratum with birthweight

Birthweight (g)	Before surfactant			After surfactant			OR (95% CI)	p-value
	n	No. died	(%)	n	No. died	(%)		
500–749	769	479	(62)	330	177	(54)	0.7 (0.5–0.9)	.009
750–999	903	257	(28)	414	78	(19)	0.6 (0.4–0.8)	< .001
1000–1249	1027	142	(14)	415	40	(10)	0.7 (0.5–1.0)	.037
1250–1500	1223	82	(7)	548	40	(7)	1.1 (0.7–1.6)	.72

500–749 g, the odds ratio relating in-hospital mortality to surfactant use is obtained from

$$\text{Odds of in-hospital mortality after surfactant use} = 177/(330-177) = 1.157$$
$$\text{Odds of in-hospital mortality before surfactant use} = 479/(769-479) = 1.652$$
$$\text{OR relating in-hospital mortality to surfactant use} = 1.157/1.652 = 0.700$$

Thus, the odds in favor of in-hospital mortality were 30% lower after surfactant was introduced than before. An OR can also be written in the form $ad/(bc)$, where a, b, c, and d are the cells of the 2×2 table relating disease to exposure. This table is given below for the 500–749-g subgroup, where $a = 177$, $b = 153$, $c = 479$, and $d = 290$.

In-hospital mortality

		+	–	
Surfactant use	**+**	177	153	330
	–	479	290	769
		656	443	1099

To obtain a CI for OR, we use the Woolf approach. We first find a CI for $\ln(\text{OR}) = (c_1, c_2)$, where

$$c_1 = \ln(\hat{\text{OR}}) - z_{1-\alpha/2}\sqrt{1/a + 1/b + 1/c + 1/d}$$
$$c_2 = \ln(\hat{\text{OR}}) + z_{1-\alpha/2}\sqrt{1/a + 1/b + 1/c + 1/d}$$

A $100\% \times (1-\alpha)$ CI for OR is then $[\exp(c_1), \exp(c_2)]$.
  A 95% CI for $\ln(\text{OR})$ is (c_1, c_2), where

$$c_1 = \ln(0.700) - 1.96\sqrt{1/177 + 1/153 + 1/479 + 1/290}$$
$$= -0.356 - 0.261 = -0.617$$
$$c_2 = -0.356 + 0.261 = -0.095$$

The 95% CI for OR = $[\exp(-0.617), \exp(-0.095)] = (0.54, 0.91)$.

There are other comparative measures of effect for epidemiologic data, namely, the risk difference and the risk ratio defined by

$$\text{Risk difference} = p_1 - p_2$$

$$\text{Risk ratio} = p_1/p_2$$

10.7.1 Stratified Data

Since individual strata are often small, it is usually worthwhile to obtain an overall assessment of the odds ratio over all strata (in this example, over all birthweight groups). This can be accomplished with the *Mantel-Haenszel (MH) odds-ratio estimator* given by

$$\text{OR}_{\text{MH}} = \frac{\sum_{i=1}^{k} a_i d_i / n_i}{\sum_{i=1}^{k} b_i c_i / n_i}$$

where a_i, b_i, c_i, d_i, and n_i are the individual cells and grand total for the ith stratum and there are a total of k strata. In this example, the MH odds-ratio estimator can be interpreted as an odds ratio controlling for the confounding effects of birthweight. Note that this is different from simply pooling the data over all strata, where the birthweight distribution might be different before and after the use of surfactant.

In this case, the four 2×2 tables relating in-hospital mortality to surfactant use in individual birthweight strata are given as follows:

	Mortality											
	500–749			750–999			1000–1249			1250–1500		
	+	–		+	–		+	–		+	–	
Surfactant +	177	153	330	78	336	414	40	375	415	40	508	548
–	479	290	769	257	646	903	142	885	1027	82	1141	1223
	656	443	1099	335	982	1317	182	1260	1442	122	1649	1771

The MH odds-ratio estimator is given by

$$\text{OR}_{\text{MH}} = \frac{177(290)/1099 + \ldots + 40(1141)/1771}{479(153)/1099 + \ldots + 82(508)/1771} = \frac{135.29}{192.70} = 0.702$$

Thus there is an overall 30% reduction in the odds of in-hospital mortality after surfactant was introduced. It is also possible to obtain a 95% CI for the overall OR— see Equation **(10.41)**, p. 410 of the text.

A commonly used approach to assessing whether surfactant use has an effect on in-hospital mortality, while controlling for birthweight is to perform the *Mantel-Haenszel test*. This is a test of the hypothesis H_0: OR = 1 versus H_1: OR $\neq$ 1, where OR represents the underlying odds ratio relating in-hospital mortality to surfactant use within specific birthweight strata. To implement this test, use the test statistic

$$X^2_{MH} = \frac{(|O - E| - .5)^2}{V} \sim \chi^2_1 \quad \text{under } H_0$$

where

$O = \sum_{i=1}^{k} O_i =$ total observed number of in-hospital deaths after surfactant was introduced over all birthweight strata

$E = \sum_{i=1}^{k} E_i =$ total expected number of in-hospital deaths after surfactant was introduced over all birthweight strata

$V = \text{variance of } O = \sum_{i=1}^{k} V_i,$

where

$E_i = (a_i + b_i)(a_i + c_i)/n_i$

$V_i = (a_i + b_i)(c_i + d_i)(a_i + c_i)(b_i + d_i)/[n_i^2(n_i - 1)]$

The p-value $= Pr(\chi^2_1 > X^2_{MH})$.

In this example,

$$O = 177 + \dots + 40 = 335$$

$$E = 330(656)/1099 + \dots + 548(122)/1771 = 392.42$$

$$V = 330(769)(656)(443)/[1099^2(1098)]$$
$$+ \dots + 548(1223)(122)(1649)/[1771^2(1770)] = 166.39$$

$$X^2_{MH} = (|335 - 392.42| - .5)^2/166.39 = 3239.39/166.39$$
$$= 19.47 \sim \chi^2_1 \text{ under } H_0$$

Since $\chi^2_{1,.999} = 10.83 < X^2_{MH}$, it follows that $p < .001$. Thus, there was a significant reduction in in-hospital mortality after surfactant was introduced, even after controlling for possible birthweight differences between the two periods.

The Mantel-Haenszel test can also be extended to study the relationship between a dichotomous disease variable such as the prevalence of bronchitis in the first year of life and an ordered categorical variable such as the number of cigarettes smoked per day by the mother, where one wants to stratify by other potential confounding variables (such as the presence of other siblings in the household). A separate estimate of slope relating the prevalence of bronchitis to the number of cigarettes smoked can be obtained in households where there are 1+ siblings and in households where there are no siblings. A weighted average of these slope estimates can then be obtained and assessed for statistical significance. This technique is referred to as the *Mantel-Extension test*—see Equation **(10.44)**, p. 415 of the text.

SECTION 10.8 Chi-Square Goodness-of-Fit Test

Look at the distribution of serum-cholesterol changes presented in Table 2.1 (p. 2). How well does a normal distribution fit these data? A stem-and-leaf plot of the change scores is given as follows:

Stem-and-leaf plot of cholesterol change

4	981
3	6215
2	7183
1	3969932
0	828
−0	8
−1	03

The arithmetic mean $= 19.5$, sd $= 16.8$, $n = 24$. Under H_0,

$$x_i \sim N(\mu, \sigma^2)$$
$$\hat{\mu} = 19.8$$
$$\hat{\sigma}^2 = 16.8^2$$

The general approach is to divide the distribution of change scores into k groups and compute the observed and expected number of units in each group if a normal distribution holds as shown in the table.

Observed count	Expected count
O_1	E_1
$\vdots$	
O_k	E_k

We then compute the test statistic

$$X^2 = \sum_{i=1}^{k}(O_i - E_i)^2/E_i \sim \chi^2_{k-1-p}$$

where

$k =$ number of groups

$p =$ number of parameters estimated from the data

The p-value $= Pr(\chi^2_{k-1-p} > X^2)$. We will only use this test if no more than 1/5 of the expected cell counts are < 5. This test is referred to as the *chi-square goodness-of-fit test*.

For the change scores, we will use four groups ($\le 9/10$–$19/20$–$29/30+$). The observed and expected values for each group are given as follows:

	Obs	Exp
≤9	6	6.6
10–19	7	5.4
20–29	4	5.4
30+	7	6.6

To compute the expected values, we employ a continuity correction. Thus, $X \leq 9$ is actually $Y \leq 9.5$, where Y is the normal approximation. The expected probabilities within each group are given as follows:

$$Pr(X \leq 9) = \Phi\left(\frac{9.5 - 19.5}{16.8}\right) = \Phi\left(\frac{-10}{16.8}\right) = \Phi(-0.60) = .275$$

$$Pr(10 \leq X \leq 19) = \Phi\left(\frac{19.5 - 19.5}{16.8}\right) - \Phi\left(\frac{9.5 - 19.5}{16.8}\right) = .499 - .275 = .224$$

$$Pr(20 \leq X \leq 29) = \Phi\left(\frac{29.5 - 19.5}{16.8}\right) - .499 = .723 - .499 = .224$$

$$Pr(X \geq 30) = 1 - \Phi\left(\frac{29.5 - 19.5}{16.8}\right) = 1 - \Phi\left(\frac{10}{16.8}\right) = .277$$

The expected count within each group is

$$E_1 = 24 \times .275 = 6.6$$
$$E_2 = 24 \times .224 = 5.4$$
$$E_3 = 24 \times .224 = 5.4$$
$$E_4 = 24 \times .277 = 6.6$$

Thus, the test statistic is

$$X^2 = \frac{(6 - 6.6)^2}{6.6} + \frac{(7 - 5.4)^2}{5.4} + \frac{(4 - 5.4)^2}{5.4} + \frac{(7 - 6.6)^2}{6.6}$$

$$= 0.05 + 0.49 + 0.35 + 0.02 = 0.92 \sim \chi^2_{4-1-2} = \chi^2_1$$

Since $X^2 < \chi^2_{1, .95} = 3.84$, it follows that $p > .05$. Therefore, the normal distribution provides an adequate fit. The chi-square goodness-of-fit test can be used to test the goodness-of-fit of any probability model, not just the normal model.

SECTION 10.9   The Kappa Statistic

The redness of 50 eyes were graded by 2 observers using the rating scale (0/1/2/3) by comparison with reference photographs, where a higher grade corresponds to more redness. To assess the reproducibility of the grading system, the following 2 × 2 table was constructed:

		Redness rating observer B				
		0	1	2	3	Total
Redness rating observer A	0	15	2	1	0	18
	1	4	7	3	2	16
	2	1	3	5	1	10
	3	0	1	2	3	6
	Total	20	13	11	6	50

One measure of reproducibility for categorical data of this type is the *Kappa statistic,* which is defined by

$$\text{Kappa} = \kappa = (p_o - p_e)/(1 - p_e)$$

where

p_o = observed proportion of concordant responses for observer A and B

p_e = expected proportion of concordant responses for observer A and B under the assumption that the redness ratings are independent

$$= \sum_{i=1}^{c} a_i b_i ,$$

and

a_i = proportion of responses in category i for observer A

b_i = proportion of responses in category i for observer B

c = number of categories

Kappa varies between 0 and 1, with 1 indicating perfect reproducibility and 0 indicating no reproducibility at all. Kappa statistics of $> .75$ are considered excellent, between .4 and .75 good, and $< .4$ poor.

For the preceding data,

$$p_o = (15 + 7 + 5 + 3)/50 = 30/50 = .60$$

$$a_1 = 18/50 = .36, a_2 = 16/50 = .32, a_3 = 10/50 = .20,$$
$$a_4 = 6/50 = .12$$

$$b_1 = 20/50 = .40, b_2 = 13/50 = .26, b_3 = 11/50 = .22,$$
$$b_4 = 6/50 = .12$$

$$p_e = .36(.40) + \ldots + .12(.12) = .286$$

$$\text{Kappa} = (.60 - .286)/(1 - .286) = .314/.714 = .44$$

This indicates good reproducibility of the rating system.

PROBLEMS

Cardiovascular Disease

In a 1985 study of the effectiveness of streptokinase in the treatment of patients who have been hospitalized after myocardial infarction, 9 of 199 males receiving streptokinase and 13 of 97 males in the control group died within 12 months [2].

10.1 Use the normal-theory method to test for significant differences in 12-month mortality between the two groups.

10.2 Construct the observed and expected contingency tables for these data.

10.3 Perform the test in Problem 10.1 using the contingency-table method.

10.4 Compare your results in Problems 10.1 and 10.3.

10.5 Compute a 95% CI for the difference in 12-month mortality rates between the streptokinase and control groups in Problem 10.1.

10.6 Compute the odds ratio in favor of death within 12 months for streptokinase therapy versus control therapy using the data in Problem 10.1.

10.7 Provide a 95% CI for the true odds ratio corresponding to your answer to Problem 10.6.

10.8 What is the relationship between your answers to Problems 10.3 and 10.7?

Cardiovascular Disease

In the streptokinase study in Problem 10.1, 2 of 15 females receiving streptokinase and 4 of 19 females in the control group died within 12 months.

10.9 Why is Fisher's exact test the appropriate procedure to test for differences in 12-month mortality rates between these two groups?

10.10 Write down all possible tables with the same row and column margins as given in the observed data.

10.11 Calculate the probability of each of the tables enumerated in Problem 10.10.

10.12 Evaluate whether or not there is a significant difference between the mortality rates for streptokinase and control-group females using a two-sided test based on your results in Problem 10.11.

Refer to the streptokinase data presented for males in Problem 10.1 and for females in Problem 10.9.

10.13 After stratifying the data by sex, perform a significance test for association between treatment group and 12-month mortality.

10.14 After stratifying the data by sex, estimate the odds ratio in favor of 12-month mortality in the streptokinase group versus the control group.

10.15 Provide a 95% CI for the true odds ratio corresponding to your answer to Problem 10.14.

10.16 Test for the goodness of fit of the normal model for the distribution of survival times of mice given in Table 6.4 (p. 71).

Pulmonary Disease

Suppose we wish to investigate the familial aggregation of respiratory disease on a disease-specific basis. One hundred families in which the head of household or the spouse has asthma, referred to as type A families, and 200 families in which neither the head of household nor the spouse has asthma, referred to as type B families, are identified. Suppose that in 15 of the type A families the first-born child has asthma, whereas in 3 other type A families the first-born child has some nonasthmatic respiratory disease. Furthermore, in 4 of the type B households the first-born child has asthma, whereas in 2 other type B households the first-born child has some nonasthmatic respiratory disease.

10.17 Compare the prevalence rates of asthma in the two types of families. State all hypotheses being tested.

10.18 Compare the prevalence rates of nonasthmatic respiratory disease in the two types of families. State all hypotheses being tested.

Cardiovascular Disease

A 1979 study investigated the relationship between cigarette smoking and subsequent mortality in men with a prior history of coronary disease [3]. It was found that 264 out of 1731 nonsmokers and 208 out of 1058 smokers had died in the 5-year period after the study began.

10.19 Assuming that the age distributions of the two groups are comparable, compare the mortality rates in the two groups.

Obstetrics

Suppose there are 500 pairs of pregnant women who participate in a prematurity study and are paired in such a way that the body weights of the 2 women in a pair are within 5 lb of each other. One of the 2 women is given a placebo and the other drug A to see if drug A has an effect in preventing prematurity. Suppose that in 30 pairs of women, *both* women in a pair have a premature child; in 420 pairs of women, *both* women have a normal child; in 35 pairs of women, the woman taking drug A has a normal child and the woman taking the placebo has a premature child; in 15 pairs of women, the woman taking drug A has a premature child and the woman taking the placebo has a normal child.

10.20 Assess the statistical significance of these results.

Cancer

Suppose we wish to compare the following two treatments for breast cancer: simple mastectomy (S) and radical mastectomy (R). Matched pairs of women who are within the same decade of age and with the same clinical condition are formed. They receive the two treatments, and their subsequent 5-year survival is monitored. The results are given in Table 10.1. We wish to test for significant differences between the treatments.

10.21 What test should be used to analyze these data? State the hypothesis being tested.

10.22 Conduct the test mentioned in Problem 10.21.

Obstetrics

10.23 Test for the adequacy of the goodness of fit of the normal distribution when applied to the distribution of birthweights in Figure 2.8 (p. 31, text). The sample mean and standard deviation for these data are 111.26 oz and 20.95 oz, respectively.

Cardiovascular Disease

A hypothesis has been suggested that a principal benefit of physical activity is to prevent sudden death from heart attack. The following study was designed to test this hypothesis: 100 men who died from a first heart attack and 100 men who survived a first heart attack in the age group 50–59 were identified and their wives were each given a detailed questionnaire concerning their husbands' physical activity in the year preceding their heart attacks. The men were then classified as active or inactive. Suppose that 30 of the 100 who survived and 10 of the 100 who died were physically active. If we wish to test the hypothesis, then

10.24 Is a one-sample or two-sample test needed here?

10.25 Which one of the following test procedures should be used to test the hypothesis?
(a) Paired t test
(b) Two-sample t test with independent samples
(c) χ^2 test for 2×2 contingency tables
(d) Fisher's exact test
(e) McNemar's test

10.26 Carry out the test procedure(s) in Problem 10.25 and report a p-value.

10.27 Compute the odds ratio in favor of survival after an MI for physically active versus physically inactive men.

10.28 Compute a 95% CI for the odds ratio referred to in Problem 10.27.

Mental Health

An observational study is set up to assess the effects of lithium in treating manic-depressive patients. New patients in an outpatient service are matched according to age, sex, and clinical condition, with one patient receiving

TABLE 10.1 Comparison of simple and radical mastectomy in treating breast cancer

Pair	Treatment S woman	Treatment R woman	Pair	Treatment S woman	Treatment R woman
1	L[a]	L	11	D	D
2	L	D	12	L	D
3	L	L	13	L	L
4	L	L	14	L	L
5	L	L	15	L	D
6	D[b]	L	16	L	L
7	L	L	17	L	D
8	L	D	18	L	D
9	L	D	19	L	L
10	L	L	20	L	D

[a]L = lived at least 5 years.
[b]D = died within 5 years.

lithium and the other a placebo. Suppose the outcome variable is whether or not the patient has any manic-depressive episodes in the next 3 months. The results are as follows: In 20 cases both the lithium and placebo members of the pair have manic-depressive episodes; in 10 cases only the placebo member has an episode (the lithium member does not); in 2 cases only the lithium member has an episode (the placebo member does not); in 36 cases neither member has an episode.

10.29 State an appropriate hypothesis to test whether lithium has any effect in treating manic-depressive patients.

10.30 Test the hypothesis mentioned in Problem 10.29.

Cardiovascular Disease

In some studies heart disease has been associated with being overweight. Suppose this association is examined in a large-scale epidemiological study and it is found that of 2000 men in the age group 55–59, 200 have myocardial infarctions in the next 5 years. Suppose the men are grouped by body weight as given in Table 10.2.

TABLE 10.2 Association between body weight and myocardial infarction

Body weight (lb)	Number of myocardial infarctions	Total number of men
120–139	10	300
140–159	20	700
160–179	50	600
180–199	95	300
200+	25	100
Total	200	2000

10.31 Comment in detail on these data.

Cerebrovascular Disease

Atrial fibrillation (AF) is widely recognized to predispose patients to embolic stroke. Oral anticoagulant therapy has been shown to decrease the number of embolic events. However, it also increases the number of major bleeding events (i.e., bleeding events requiring hospitalization). A study is proposed in which patients with AF are randomly divided into two groups: one receives the anticoagulant warfarin, the other a placebo. The groups are then followed for the incidence of major events (i.e., embolic stroke or major bleeding events).

10.32 Suppose that 5% of treated patients and 22% of control patients are anticipated to experience a major event over 3 years. If 100 patients are to be randomized to each group, then how much power would such a study have of detecting a significant difference if a two-sided test with $\alpha = .05$ is used?

10.33 How large should such a study be to have an 80% chance of finding a significant difference given the same assumptions as in Problem 10.32?

10.34 One problem with warfarin is that about 10% of patients stop taking the medication due to persistent minor bleeding (e.g., nosebleed). If we regard the probabilities in Problem 10.32 as perfect-compliance risk estimates, then recalculate the power for the study proposed in Problem 10.32 if compliance is not perfect.

Pulmonary Disease

Each year approximately 4% of current smokers attempt to quit smoking, and about 50% of those who try to quit are successful; that is, they are able to abstain from smoking for at least 1 year from the date they quit. Investigators have attempted to identify risk factors that might influence these two probabilities. One such variable is the number of cigarettes currently smoked per day. In particular, the investigators found that among 75 current smokers who smoked ≤ 1 pack/day, 5 attempted to quit, whereas among 50 current smokers who smoked more than 1 pack/day, 1 attempted to quit.

10.35 Assess the statistical significance of these results and report a p-value.

Similarly, a different study reported that out of 311 people who had attempted to quit smoking, 16 out of 33 with less than a high school education were successful quitters; 47 out of 76 who had finished high school but had not gone to college were successful quitters; 69 out of 125 who attended college but did not finish 4 years of college were successful quitters; and 52 out of 77 who had completed college were successful quitters.

10.36 Do these data show an association between the number of years of education and the rate of successful quitting?

Infectious Disease, Hepatic Disease

Read "Foodborne Hepatitis A Infection: A Report of Two Urban Restaurant-Associated Out-Breaks" by Denes et al., in the *American Journal of Epidemiology, 105*(2) (1977), pages 156–162, and answer the following questions based on it.

10.37 The authors analyzed the results of Table 1 using a chi-square statistic. Is this method of analysis reasonable for this table? If not, suggest an alternative method.

10.38 Analyze the results in Table 1 using the method suggested in Problem 10.37. Do your results agree with the authors'?

10.39 Student's *t* test with 40 *df* was used to analyze the results in Table 2. Is this method of analysis reasonable for this table? If not, suggest an alternative method.

10.40 The authors claim that there is a significant difference ($p = .01$) between the proportion with hepatitis A among those who did and did not eat salad. Check this result using the method of analysis suggested in Problem 10.39.

Infectious Disease, Cardiology

Kawasaki's syndrome is an acute illness of unknown cause that occurs predominantly in children under the age of 5. It is characterized by persistent high fever and other clinical signs and can result in death and/or coronary-artery aneurysms. In the early 1980s, standard therapy for this condition was aspirin to prevent blood clotting. A Japanese group began experimentally treating children with intravenous gamma globulin in addition to aspirin to prevent cardiac symptoms in these patients [4].

A clinical trial is planned to compare the combined therapy of gamma globulin and aspirin vs. aspirin therapy alone. Suppose the rate of coronary-artery aneurysms is 15% in the aspirin-treated group, based on previous experience, and the investigators intend to use a two-sided significance test with $\alpha = .05$.

10.41 If the rate of coronary aneurysms in the combined therapy group is 5%, then how much statistical power will such a study have if 125 patients are to be recruited in each treatment group?

10.42 Answer Problem 10.41 if 150 patients are recruited in each group.

10.43 How many patients would have to be recruited in each group to have a 95% chance of finding a significant difference?

Obstetrics

An issue of current interest is the effect of delayed childbearing on pregnancy outcome. In a recent paper a population of first deliveries was assessed for low-birthweight deliveries (< 2500 g) according to the woman's age and prior pregnancy history [5]. The data in Table 10.3 were presented.

10.44 What test can be used to assess the effect of age on low-birthweight deliveries among women with a negative history?

TABLE 10.3 Relationship of age and pregnancy history to low-birthweight deliveries

Age	History[a]	n	Percentage low birthweight
≥ 30	No	225	3.56
≥ 30	Yes	88	6.82
< 30	No	906	3.31
< 30	Yes	153	1.31

[a]History = yes if a woman had a prior history of spontaneous abortion or infertility
= no otherwise

Source: Reprinted with permission of the *American Journal of Epidemiology, 125*(1), 101–109, 1987.

10.45 Perform the test in Problem 10.44 and report a *p*-value.

10.46 What test can be used to assess the effect of age on low-birthweight deliveries among women with a positive history?

10.47 Perform the test in Problem 10.46 and report a *p*-value.

10.48 Estimate the odds ratio relating age to the prevalence of low-birthweight deliveries, while controlling for the possible confounding effect of pregnancy history. Provide a 95% CI for the odds ratio.

10.49 Estimate the odds ratio relating pregnancy history to the prevalence of low-birthweight deliveries, while controlling for the possible confounding effect of age. Provide a 95% CI for the odds ratio.

10.50 Is there evidence of effect modification between age and pregnancy history regarding their effect on low-birthweight deliveries?

Cancer

A recent study looked at the association between breast-cancer incidence and alcohol consumption [6]. The data in Table 10.4 were presented for 50–54-year-old women.

10.51 What test procedure can be used to test if there is an association between breast-cancer incidence and alcohol consumption, where alcohol consumption is coded as (drinker/nondrinker)?

10.52 Perform the test mentioned in Problem 10.51 and report a *p*-value.

10.53 Perform a test for linear trend based on the data in Table 10.4.

TABLE 10.4 Association between alcohol consumption and breast cancer in 50–54-year-old women

	Alcohol consumption (g/day)				
Group	None	< 1.5	1.5–4.9	5.0–14.9	≥ 15.0
Breast-cancer cases	43	15	22	42	24
Total number of women	5944	2069	3449	3570	2917

Source: Reprinted with permission of the *New England Journal of Medicine, 316*(19), 1174–1180, 1987.

Cancer

A case–control study was performed looking at the association between the risk of lung cancer and the occurrence of lung cancer among first-degree relatives [7]. Lung-cancer cases were compared with controls as to the number of relatives with lung cancer. Controls were frequency matched to cases by 5-year age category, sex, vital status, and ethnicity. The following data were presented:

		Number of controls	Number of cases
Number of	0	466	393
relatives with	1	78	119
lung cancer	2+	8	20

10.54 What is a case–control study, and how does it differ from a cohort study?

10.55 What test procedure can be used to look at the association between the number of relatives with lung cancer (0/1/2+) and case–control status?

10.56 Implement the test in Problem 10.55 and report a *p*-value. Interpret the results in one or two sentences.

Cardiology

A group of patients who underwent coronary angiography between Jan. 1, 1972 and Dec. 31, 1986 in a particular hospital were identified [8]. 1493 cases with confirmed coronary-artery disease were compared with 707 controls with no plaque evidence at the time of angiography. Suppose it is found that 37% of cases and 30% of controls reported a diagnosis or treatment for hypertension at the time of angiography.

10.57 What test can be used to compare the risk of hypertension between cases and controls?

10.58 Implement the test in Problem 10.57 and report a *p*-value.

10.59 Are the proportions (37%, 30%) an example of prevalence, incidence, or neither?

Ophthalmology

A study was performed comparing the validity of different methods of reporting the ocular condition age-related macular degeneration. Information was obtained by self-report at an eye examination, surrogate (spouse) report by telephone, and by clinical determination at an eye examination [9]. The following data were reported:

TABLE 10.5 Comparison of surrogate report by telephone to self-report at eye exam for age-related macular degeneration

		Self-report at eye exam		
		No	Yes	Total
Surrogate report by telephone	No	1314	12	1326
	Yes	22	17	39
	Total	1336	29	1365[a]

TABLE 10.6 Comparison of surrogate report by telephone to clinical determination at an eye exam for age-related macular degeneration

		Clinical determination at an eye exam		
		No	Yes	Total
Surrogate report by telephone	No	1247	83	1330
	Yes	26	14	40
	Total	1273	97	1370[a]

[a]The total sample sizes in Tables 10.5 and 10.6 do not match, due to a few missing values.

10.60 What test can be performed to compare the frequency of reporting of age-related macular degeneration by self-report vs. surrogate report if neither is regarded as a gold standard?

10.61 Implement the test mentioned in Problem 10.60 and report a p-value.

10.62 Suppose the clinical determination is considered the gold standard. What measure(s) can be used to assess the validity of the surrogate report?

10.63 Provide estimates and 95% CI's for these measures.

Cancer

A prospective study was conducted among 55–59-year-old postmenopausal women looking at the association between the use of postmenopausal hormones and the incidence of breast cancer. Women were categorized as never users, current users or past users by mail questionnaire in 1976 and the incidence of breast cancer was assessed in the 10-year interval following the questionnaire (1976–1986) by follow-up mail questionnaires. The results are given in Table 10.7.

10.64 What test can be used to test the overall hypothesis that there is (or is not) a difference in risk of breast cancer among the three groups?

10.65 Implement the test in Problem 10.64 and report a p-value.

10.66 Estimate the odds ratio relating breast-cancer risk for current users versus never users. Provide a 95% CI for the odds ratio.

10.67 Estimate the odds ratio relating breast-cancer risk for past users versus never users. Provide a 95% CI for the odds ratio.

10.68 Compute risk ratios and 95% CI's corresponding to the odds ratios in Problems 10.66 and 10.67. How do the risk ratios and odds ratios compare? How do the CI's compare?

TABLE 10.7 Association between postmenopausal hormone use and the incidence of breast cancer

Status	Number of cases	Number of women
Never users	129	7175
Current users	51	2495
Past users	46	2526

SOLUTIONS

10.1 Test the hypothesis H_0: $p_1 = p_2$ versus H_1: $p_1 \neq p_2$. The test statistic is given by

$$z = \frac{\hat{p}_1 - \hat{p}_2}{\sqrt{\hat{p}\hat{q}(1/n_1 + 1/n_2)}}$$

where $\hat{p}_1 = 9/199 = .0452$, $\hat{p}_2 = 13/97 = .1340$, $\hat{p} = (9 + 13)/(199 + 97) = 22/296 = .0743$, $\hat{q} = 1 - \hat{p} = .9257$,

$$z = \frac{.0452 - .1340}{\sqrt{.0743(.9257)(1/199 + 1/97)}}$$

$$= \frac{-.0888}{.0325} = -2.734 \sim N(0, 1) \text{ under } H_0$$

Since $z < -1.96$, reject H_0 at the 5% level.

10.2 The observed table is given by

12-month mortality status—observed table

	Dead	Alive	
Streptokinase	9	190	199
Control	13	84	97
	22	274	296

The expected cell counts are obtained from the row and column margins as follows:

$$E_{11} = \frac{199 \times 22}{296} = 14.79$$

$$E_{12} = \frac{199 \times 274}{296} = 184.21$$

$$E_{21} = \frac{97 \times 22}{296} = 7.21$$

$$E_{22} = \frac{97 \times 274}{296} = 89.79$$

These values are displayed as follows:

12-month mortality status—expected table

	Dead	Alive	
Streptokinase	14.79	184.21	199
Control	7.21	89.79	97
	22	274	296

10.3 Compute the Yates-corrected chi-square statistic as follows:

$$X^2 = \frac{(|9 - 14.79| - .5)^2}{14.79} + \dots + \frac{(|84 - 89.79| - .5)^2}{89.79}$$

$$= \frac{5.29^2}{14.79} + \dots + \frac{5.29^2}{89.79}$$

$$= 1.892 + 0.152 + 3.882 + 0.312$$

$$= 6.24 \sim \chi_1^2 \text{ under } H_0$$

Since $\chi_{1,.95}^2 = 3.84 < X^2$, reject H_0 at the 5% level.

10.4 The decisions reached in Problems 10.1 and 10.3 were the same (reject H_0 at the 5% level). If a chi-square test had been used without continuity correction, the results would have been identical, since $X^2 = 5.79^2/14.79 + \dots + 5.79^2/89.79 = 7.47 = 2.734^2 = z^2$ and the same p-value is obtained whether z is compared to an $N(0, 1)$ distribution or $X^2 = z^2$ to a χ_1^2 distribution.

10.5 A 95% CI for $p_1 - p_2$ is given by

$$\hat{p}_1 - \hat{p}_2 \pm 1.96\sqrt{\hat{p}_1\hat{q}_1/n_1 + \hat{p}_2\hat{q}_2/n_2}$$

$$= .0452 - .1340$$

$$\pm 1.96\sqrt{.0452(.9548)/199 + .1340(.8660)/97}$$

$$= -.0888 \pm 1.96(.0376) = -.0888 \pm .0737$$

$$= (-.162, -.015)$$

10.6 The odds ratio is given by OR $= (9/190)/(13/84) = .0474/.1548 = 0.306$.

10.7 A 95% CI for $\ln$(OR) is given by (c_1, c_2), where

$$c_1 = \ln(0.306) - 1.96\sqrt{1/9 + 1/190 + 1/13 + 1/84}$$

$$= -1.184 - 0.888 = -2.072$$

$$c_2 = -1.184 + 0.888 = -0.296$$

The corresponding 95% CI for OR $= (e^{-2.072}, e^{-0.296}) = (0.13, 0.74)$.

10.8 The results in Problem 10.3 are statistically significant and the 95% CI in Problem 10.7 excludes 1. This is usually, but not always, the case for these hypothesis-testing and confidence-interval procedures.

10.9 Form the following observed 2 × 2 table:

12-month mortality status

	Dead	**Alive**	
Streptokinase	2	13	15
Control	4	15	19
	6	28	34

The smallest expected value $= E_{11} = (15 \times 6)/34 = 2.65 < 5$. Thus, Fisher's exact test must be used.

10.10

| 0 | 15 |
| 6 | 13 |

| 1 | 14 |
| 5 | 14 |

| 2 | 13 |
| 4 | 15 |

| 3 | 12 |
| 3 | 16 |

| 4 | 11 |
| 2 | 17 |

| 5 | 10 |
| 1 | 18 |

| 6 | 9 |
| 0 | 19 |

10.11 Use the recursion rule as follows:

$$Pr(0) = 1 \times Pr(0)$$

$$Pr(1) = \frac{6 \times 15}{1 \times 14}Pr(0) = 6.429\, Pr(0)$$

$$Pr(2) = \frac{5 \times 14}{2 \times 15}Pr(1) = 15.000\, Pr(0)$$

$$Pr(3) = \frac{4 \times 13}{3 \times 16}Pr(2) = 16.250\, Pr(0)$$

$$Pr(4) = \frac{3 \times 12}{4 \times 17}Pr(3) = 8.603\, Pr(0)$$

$$Pr(5) = \frac{2 \times 11}{5 \times 18}Pr(4) = 2.103\, Pr(0)$$

$$Pr(6) = \frac{1 \times 10}{6 \times 19}Pr(5) = 0.184\, Pr(0)$$

Thus,

$$Pr(0) \times (1 + 6.429 + \dots + 0.184) = 1$$

or

$$Pr(0) \times 49.569 = 1$$

or

$$Pr(0) = \frac{1}{49.569} = .0202$$

Furthermore,

$$Pr(1) = .130, \quad Pr(2) = .303, \quad Pr(3) = .328,$$
$$Pr(4) = .174, \quad Pr(5) = .042, \quad Pr(6) = .004$$

10.12 Our table is the "2" table. Therefore, the two-tailed p-value is given by

$$p = 2 \times \min[Pr(0) + Pr(1) + Pr(2),$$
$$Pr(2) + Pr(3) + \dots + Pr(7)]$$
$$= 2 \times \min(.452, .850)$$
$$= 2 \times .452 = .905$$

Clearly, there is no significant difference in 12-month mortality status between the two treatment groups for females.

10.13 We have the following observed and expected values for the 2 × 2 tables for males and females respectively (with expected values in parentheses):

Males—12-month mortality status

	Dead	**Alive**	
Streptokinase	9	190	199
	(14.79)	(184.21)	
Control	13	84	97
	(7.21)	(89.79)	
	22	274	296

Females—12-month mortality status

	Dead	**Alive**	
Streptokinase	2	13	15
	(2.65)	(12.35)	
Control	4	15	19
	(3.35)	(15.65)	
	6	28	34

We compute the test statistic

$$X^2_{MH} = \frac{(|O - E| - 0.5)^2}{V}$$

where

$$O = 9 + 2 = 11$$
$$E = 14.79 + 2.65 = 17.44$$
$$V = \frac{199 \times 97 \times 22 \times 274}{296^2 \times 295} + \frac{15 \times 19 \times 6 \times 28}{34^2 \times 33}$$
$$= 4.502 + 1.255 = 5.757$$

Thus,

$$X^2_{MH} = \frac{(|11 - 17.44| - .5)^2}{5.757}$$
$$= \frac{5.94^2}{5.757} = 6.12 \sim \chi^2_1 \text{ under } H_0$$

Since $\chi^2_{1,.975} = 5.02$, $\chi^2_{1,.99} = 6.63$ and $5.02 < 6.12 < 6.63$, it follows that $1 - .99 < p < 1 - .975$ or $.01 < p < .025$. Thus, there is a significant association between 12-month mortality status and treatment group. This confirms the significant association that was found for males alone in Problem 10.3.

10.14 We have

$$OR_{MH} = \frac{\sum_{i=1}^{k} a_i d_i / n_i}{\sum_{i=1}^{k} b_i c_i / n_i}$$

$$= \frac{(9 \times 84)/296 + (2 \times 15)134}{(13 \times 190)/296 + (4 \times 13)/34}$$

$$= \frac{2.554 + 0.882}{8.345 + 1.529}$$

$$= \frac{3.436}{9.874} = 0.35$$

10.15 We use the approach in Equation (10.41) (text, p. 410) as follows:

$$Var(\ln\widehat{OR}) = A + B + C, \text{ where}$$

$$A = \frac{93(9)(84)/296^2 + 17(2)(15)/34^2}{2[9(84)/296 + 2(15)/34]^2}$$

$$= \frac{1.2436}{23.6178} = 0.0527$$

$$B =$$

$$\frac{[93(13)(190)/296^2 + 17(4)(13)/34^2 + 203(9)(84)/296^2 + 17(2)(15)/34^2]}{2[9(84)/296 + 2(15)/34][13(190)/296 + 4(13)/34]}$$

$$= 5.5793/67.8622 = 0.0822$$

$$C = \frac{203(13)(190)/296^2 + 17(4)(13)/34^2}{2[13(190)/296 + 4(13)/34]^2}$$

$$= \frac{6.4875}{194.9920} = 0.0333$$

Thus, $Var(\ln\widehat{OR}) = 0.1681$ and a 95% CI for $\ln(\widehat{OR}) = \ln\widehat{OR} \pm 1.96 se[\ln(\widehat{OR})] = \ln(0.35) \pm 1.96\sqrt{0.1681} = -1.0555 \pm 0.8037 = (-1.8592, -0.2518)$. The corresponding 95% CI for OR $= (e^{-1.8592}, e^{-0.2518}) = (0.16, 0.78)$.

10.16 We first compute the mean and standard deviation for the sample of survival times. We have $\bar{x} = 16.13$, $s = 2.66$, $n = 429$. We compute the probabilities under a normal model for the groups 10–12, 13–15, 16–18, 19–21, 22–24 as follows:

Group	Probability
10–12	$\Phi[(12.5 - 16.13)/2.66] = \Phi(-1.36) = .086$
13–15	$\Phi[(15.5 - 16.13)/2.66] - .086$
	$= \Phi(-0.24) - .087 = .407 - .087 = .320$
16–18	$\Phi[(18.5 - 16.13)/2.66] - .407$
	$= \Phi(0.89) - .407 = .814 - .407 = .407$
19–21	$\Phi[(21.5 - 16.13)/2.66] - .814$
	$= \Phi(2.02) - .814 = .978 - .814 = .164$
22–24	$1 - .978 = .022$

We now compute the observed and expected number of units in each group:

Group	Observed number of units	Expected number of units
10–12	45	$429 \times .086 = 36.9$
13–15	121	$429 \times .320 = 137.5$
16–18	184	$429 \times .407 = 174.7$
19–21	69	$429 \times .164 = 70.6$
22–24	10	$429 \times .022 = 9.3$

We compute the chi-square goodness-of-fit statistic:

$$X^2 = \sum_{i=1}^{k} \frac{(O_i - E_i)^2}{E_i}$$

$$= \frac{(45 - 36.9)^2}{36.9} + \dots + \frac{(10 - 9.3)^2}{9.3}$$

$$= 1.754 + 1.974 + 0.492 + 0.034 + 0.055$$

$$= 4.31 \sim \chi^2_{g-1-p} = \chi^2_{5-1-2} = \chi^2_2 \quad \text{under } H_0$$

Since $\chi^2_{2,.95} = 5.99 > X^2$, it follows that $p > .05$, and we accept the null hypothesis that the normal model fits the data adequately.

10.17 Test the hypothesis $H_0: p_A = p_B$ versus $H_1: p_A \neq p_B$, where

$$p_A = Pr(\text{first-born child has asthma in a type A family})$$

$$p_B = Pr(\text{first-born child has asthma in a type B family})$$

The observed and expected 2×2 tables are shown as follows.

Observed table

Asthma

		+	−	
Type of family	A	15	85	100
	B	4	196	200
		19	281	300

Expected table

Asthma

		+	−	
Type of family	A	6.3	93.7	100
	B	12.7	187.3	200
		19.0	281.0	300

Note: + = asthma, − = no asthma.

The χ^2 test for 2×2 tables will be used, since the expected table has no expected value < 5. We have the following Yates-corrected chi-square statistic

$$X^2 = \frac{(|15 - 6.3| - .5)^2}{6.3} + \dots + \frac{(|196 - 187.3| - .5)^2}{187.3}$$

$$= 16.86 \sim \chi^2_1$$

The p-value for this result is $< .001$, since

$$\chi^2_{1,.999} = 10.83 < 16.86$$

Thus, there is a highly significant association between the type of family and the asthma status of the child.

10.18 The 2×2 table is shown as follows:

Nonasthmatic respiratory disease status

		+	−	
Type of family	A	3	97	100
	B	2	198	200
		5	295	300

Note: + = nonasthmatic respiratory disease,
− = no nonasthmatic respiratory disease.

There are two expected values < 5; in particular,

$$E_{11} = \frac{5(100)}{300} = 1.7 \quad E_{21} = \frac{5(200)}{300} = 3.3$$

Thus, Fisher's exact test must be used to analyze this table. We write all possible tables with the same margins as the observed table, as follows:

0	100
5	195

1	99
4	196

2	98
3	197

3	97
2	198

4	96
1	199

5	95
0	200

Use the recursion rule to compute the probability of each table.

$$Pr(0) = 1 \times Pr(0)$$

$$Pr(1) = Pr(0) \times \frac{5 \times 100}{1 \times 196} = 2.551 \, Pr(0)$$

$$Pr(2) = Pr(1) \times \frac{4 \times 99}{2 \times 197} = 2.564 \, Pr(0)$$

$$Pr(3) = P4(2) \times \frac{3 \times 98}{3 \times 198} = 1.269 \, Pr(0)$$

$$Pr(4) = Pr(3) \times \frac{2 \times 97}{4 \times 199} = 0.309 \, Pr(0)$$

$$Pr(5) = Pr(4) \times \frac{1 \times 96}{5 \times 200} = 0.030 \, Pr(0)$$

Thus,

$$Pr(0)(1 + 2.551 + 2.564 + 1.269 + 0.309 + 0.030) = 1$$

$$Pr(0) = \frac{1}{7.723} = .1295$$

$$Pr(1) = .330$$
$$Pr(2) = .332$$
$$Pr(3) = .164$$
$$Pr(4) = .040$$
$$Pr(5) = .004$$

Thus, since the observed table is the "3" table, the two-tailed p-value is given by

$$2 \times \min(.164 + .040 + .004, .164 + .332 + .330 + .1295)$$
$$= 2 \times .208 = .416$$

The results are not statistically significant and indicate that there is no significant difference in the prevalence of nonasthmatic respiratory disease among households in which the parents do or do not have asthma.

10.19 We have the following observed table:

5-year mortality incidence

		Dead	Alive	
Smoking status	Nonsmokers	264	1467	1731
	Smokers	208	850	1058
		472	2317	2789

The smallest expected value $= (472 \times 1058)/2789 = 179.1 \geq 5$. Thus, we can use the chi-square test for 2×2 contingency tables. We have the following test statistic:

$$X^2 = \frac{n(|ad - bc| - n/2)^2}{(a + b)(c + d)(a + c)(b + d)}$$

$$= \frac{2789(|264(850) - 208(1467)| - 2789/2)^2}{1731 \times 1058 \times 472 \times 2317}$$

$$= \frac{2789(79{,}341.5)^2}{1731 \times 1058 \times 472 \times 2317} = 8.77 \sim \chi_1^2 \text{ under } H_0$$

Since $\chi_{1,.995}^2 = 7.88 < X^2 < \chi_{1,.999}^2 = 10.83$, it follows that $.001 < p < .005$. Thus, cigarette smokers with a prior history of coronary disease have a significantly higher mortality incidence in the subsequent 5 years $(208/1058 = .197)$ than do nonsmokers with a prior history of coronary disease $(264/1731 = .153)$.

10.20 We can use McNemar's test in this situation. We have the following table based on matched pairs:

Placebo

		Premature	Normal	
Drug A	Premature	30	15	45
	Normal	35	420	455
		65	435	500

We can ignore the $30 + 420$ concordant pairs and focus on the remaining 50 discordant pairs. We have the test statistic

$$X^2 = \frac{\left(\left|35 - \frac{50}{2}\right| - \frac{1}{2}\right)^2}{50/4} = 7.22 \sim \chi_1^2 \text{ under } H_0$$

Since $\chi_{1,.99}^2 = 6.63$, $\chi_{1,.995}^2 = 7.88$, it follows that the two-sided p-value is given by $.005 < p < .01$. Thus, there is a significant difference between the two treatments, with drug A women having lower prematurity rates than placebo women.

10.21 We wish to test whether or not treatment S differs from treatment R. We will use McNemar's test for correlated proportions since we have matched pairs. The hypotheses being tested in this case are H_0: $p = 1/2$ versus H_1: $p \neq 1/2$, where $p =$ probability that a discordant pair is of type A; i.e., where the treatment S woman lives for ≥ 5 years and the treatment R woman dies within 5 years.

10.22 We have the following 2×2 table of matched pairs:

Treatment R woman

		L	D
Treatment	**L**	10	8
S woman	**D**	1	1

We must compute the p-value using an exact binomial test, because the number of discordant pairs (9) is too small to use the normal approximation. We refer to the exact binomial tables (Table 1, Appendix, text) and obtain

$$p\text{-value} = 2 \times \sum_{k=8}^{9} {}_9C_k(.5)^9$$

$$= 2 \times (.0176 + .0020) = .039$$

Thus, we reject H_0 and conclude that treatment S is better than treatment R.

10.23 We divide the distribution of birthweights (oz) into the groups ≤ 79, 80–89, 90–99, 100–109, 110–119, 120–129, 130–139, 140+. We will assume that each birthweight is rounded to the nearest ounce and thus 75 ounces actually represents the interval 74.5–75.5. Thus, we can compute the expected number of infants in each group under a normal model as follows:

$$E_1 = 100Pr(X \leq 79)$$

$$= 100\Phi\left(\frac{79.5 - 111.26}{20.95}\right)$$

$$= 100\Phi\left(\frac{-31.76}{20.95}\right)$$

$$= 100\Phi(-1.52) = 100(1 - .9352) = 6.5$$

$$E_2 = 100Pr(80 \leq X \leq 89)$$

$$= 100\left[\Phi\left(\frac{89.5 - 111.26}{20.95}\right) - \Phi\left(\frac{79.5 - 111.26}{20.95}\right)\right]$$

$$= 100\left[\Phi(-1.04) - \Phi(-1.52)\right]$$

$$= 100(1 - .8505) - 100(1 - .9352) = 8.5$$

$$\vdots$$

$$E_8 = 100Pr(X \geq 140)$$

$$= 100\left[1 - \Phi\left(\frac{139.5 - 111.26}{20.95}\right)\right]$$

$$= 100[1 - \Phi(1.35)] = 100(1 - .9112) = 8.9$$

We have the following table of observed and expected cell counts:

Birthweight	Observed	Expected
≤ 79	5	6.5
80–89	10	8.5
90–99	11	13.8
100–109	19	17.9
110–119	17	18.6
120–129	20	15.5
130–139	12	10.3
≥ 140	6	8.9

We can use the chi-square goodness-of-fit test because all expected values are ≥ 5. We have the following test statistic

$$X^2 = \frac{(5 - 6.5)^2}{6.5} + \dots + \frac{(6 - 8.9)^2}{8.9}$$

$$= 3.90 \sim \chi^2_{g-p-1} = \chi^2_5 \text{ under } H_0$$

since there are 8 groups and 2 parameters estimated from the data. Because $\chi^2_{5,.95} = 11.07 > X^2$, it follows that $p > .05$. Thus, the results are not statistically significant and the goodness of fit of the normal distribution is adequate.

10.24 A two-sample test is needed here, because samples of men who survived and died, respectively, from a first heart attack are being compared.

10.25 The observed table is shown as follows:

Observed table relating sudden death from a first heart attack and previous physical activity

Physical activity

		Active	Inactive	
Mortality status	Survived	30	70	100
	Died	10	90	100
		40	160	200

The smallest expected value is

$$\frac{40 \times 100}{200} = 20 \geq 5$$

Thus, the χ^2 test for 2×2 contingency tables can be used here.

10.26 The test statistic is given by

$$X^2 = \frac{n\left[\left|ad - bc\right| - \frac{n}{2}\right]^2}{(a + b)(c + d)(a + c)(b + d)}$$

$$= \frac{200(\left|30 \times 90 - 10 \times 70\right| - 100)^2}{100(100)(40)(160)}$$

$$= \frac{200(1900)^2}{100(100)(40)(160)} = 11.28 \sim \chi_1^2 \text{ under } H_0$$

Since $\chi_{1,.999}^2 = 10.83 < X^2$, it follows that $p < .001$, and we can conclude that there is a significant association between physical activity and survival after an MI.

10.27 The odds ratio in favor of survival after an MI for physically active versus physically inactive men is given by

$$\widehat{OR} = \frac{30 \times 90}{70 \times 10} = \frac{2700}{700} = 3.86$$

10.28 We use the Woolf method. We first obtain a 95% CI for ln(OR) given by

$$\ln\widehat{OR} \pm 1.96\sqrt{\frac{1}{a} + \frac{1}{b} + \frac{1}{c} + \frac{1}{d}}$$

$$= \ln(3.86) \pm 1.96\sqrt{\frac{1}{30} + \frac{1}{70} + \frac{1}{10} + \frac{1}{90}}$$

$$= 1.350 \pm 0.781 = (0.569, 2.131)$$

The corresponding 95% CI for OR $= (e^{0.569}, e^{2.131}) = (1.77, 8.42)$.

10.29 This is a classic example illustrating the use of McNemar's test for correlated proportions. There are two groups of patients, one receiving lithium and one receiving placebo, but the two groups are matched on age, sex, and clinical condition and thus represent dependent samples. Let a type A discordant pair be a pair of people such that the lithium member of the pair has a manic-depressive episode and the placebo member does not. Let a type B discordant pair be a pair of people such that the placebo member of the pair has a manic-depressive episode and the lithium member does not. Let $p =$ probability that a discordant pair is of type A. Then test the hypothesis

$$H_0: p = \frac{1}{2} \text{ versus } H_1: p \neq \frac{1}{2}$$

10.30 There are 2 type A discordant pairs and 10 type B discordant pairs, giving a total of 12 discordant pairs. Since the number of discordant pairs is < 20, an exact binomial test must be used. Under H_0,

$$Pr(k \text{ type A discordant pairs}) = \binom{12}{k}\left(\frac{1}{2}\right)^{12}$$

In particular, from Table 1 (Appendix, text)

$$Pr(k \leq 2) = \left(\frac{1}{2}\right)^{12}\left[\binom{12}{0} + \binom{12}{1} + \binom{12}{2}\right]$$

$$= .0002 + .0029 + .0161 = .0192$$

Since a two-sided test is being performed,

$$p = 2 \times .0192 = .038$$

Thus, H_0 is rejected and we conclude that the placebo patients are more likely to have manic-depressive episodes when the results differ in the two members of a pair.

10.31 We form the following 2×5 contingency table:

Body weight

		120-139	140-159	160-179	180-199	200+	
MI	Yes	10	20	50	95	25	200
	No	290	680	550	205	75	1800
		300	700	600	300	100	2000
	% yes	3.3	2.9	8.3	31.7	25.0	

We perform the chi-square test for trend using the score statistic 1, 2, 3, 4, 5 for the five columns in the table. We have the test statistic $X_1^2 = A^2/B$, where

$$A = \sum_{i=1}^{k} x_i S_i - \bar{x}S$$

$$= 10(1) + 20(2) + \ldots + 25(5)$$

$$-200\left[\frac{300(1) + \ldots + 100(5)}{2000}\right]$$

$$= 705 - \frac{200(5200)}{2000} = 705 - 520 = 185$$

$$B = \bar{p}\bar{q}\left[\sum_{i=1}^{k} n_i S_i^2 - \left(\sum_{i=1}^{k} n_i S_i\right)^2/N\right]$$

$$= \frac{200}{2000} \times \frac{1800}{2000}$$

$$\times \left[300(1^2) + \ldots + 100(5^2) - \frac{5200^2}{2000}\right]$$

$$= .09(15,800 - 13,520) = .09(2280) = 205.2$$

Thus, we have

$$X_1^2 = \frac{185^2}{205.2} = 166.8 \sim \chi_1^2 \text{ under } H_0$$

Since $X_1^2 > \chi_{1,.999}^2 = 10.83$, we have $p < .001$ and there is a significant linear trend relating body weight and the incidence of MI.

10.32 We use the power formula in **(10.27)** (text, p. 385) as follows:

$$\text{Power} = \Phi \left[\frac{\Delta}{\sqrt{p_1 q_1/n_1 + p_2 q_2/n_2}} \right.$$
$$\left. - z_{1-\alpha/2} \frac{\sqrt{\bar{p}\bar{q}(1/n_1 + 1/n_2)}}{\sqrt{p_1 q_1/n_1 + p_2 q_2/n_2}} \right]$$

where $p_1 = .05$, $p_2 = .22$, $n_1 = n_2 = 100$, $\alpha = .05$, $\bar{p} = (.05 + .22)/2 = .135$, $\bar{q} = .865$. We have

$$\text{Power} = \Phi \left\{ \frac{.22 - .05}{\sqrt{[.05(.95) + .22(.78)]/100}} \right.$$
$$\left. - z_{.975} \frac{\sqrt{.135(.865)(1/100 + 1/100)}}{\sqrt{[.05(.95) + .22(.78)]/100}} \right\}$$

$$= \Phi \left(\frac{.17}{.0468} - 1.96 \times \frac{.0483}{.0468} \right)$$

$$= \Phi(3.632 - 2.024) = \Phi(1.608) = .946$$

Thus, such a study should have a 95% chance of detecting a significant difference.

10.33 We use the sample-size formula in **(10.26)** (text, p. 384) as follows:

$$n = \frac{(\sqrt{2\bar{p}\bar{q}} \, z_{1-\alpha/2} + \sqrt{p_1 q_1 + p_2 q_2} \, z_{1-\beta})^2}{\Delta^2}$$

$$= \frac{[\sqrt{2(.135)(.865)} z_{.975} + \sqrt{.05(.95) + .22(.78)} z_{.80}]^2}{(.22 - .05)^2}$$

$$= \frac{[.4833(1.96) + .4681(0.84)]^2}{(.17)^2}$$

$$= \frac{1.3404^2}{(.17)^2} = 62.2$$

Thus, we need 63 patients in each group to have an 80% probability of finding a significant difference.

10.34 We obtain an estimate of power adjusted for noncompliance as presented in Section 10.7.3 (text, p. 392). We have that $\lambda_1 = .10$, $\lambda_2 = 0$. Therefore,

$$p_1^* = .05(.9) + .22(.1) = .067, q_1^* = .933$$
$$p_2^* = p_2 = .22, q_2^* = .78$$
$$\bar{p}^* = (.067 + .22)/2 = .1435, \bar{q}^* = .8565$$
$$\Delta^* = |p_1^* - p_2^*| = .153$$

We use equation **(10.27)** (text, p. 385) with p_1, p_2, q_1, q_2, $\bar{p}, \bar{q}$, and Δ replaced by $p_1^*, p_2^*, q_1^*, q_2^*, \bar{p}^*, \bar{q}^*$, and Δ^* as follows:

$$\text{Power} = \Phi \left[\frac{.153}{\sqrt{[.067(.933) + .22(.78)]/100}} \right.$$
$$\left. - z_{.975} \frac{\sqrt{.1435(.8565)(2/100)}}{\sqrt{[.067(.933) + .22(.78)]/100}} \right]$$

$$= \Phi \left[\frac{.153}{.0484} - 1.96 \left(\frac{.0496}{.0484} \right) \right]$$

$$= \Phi(3.162 - 2.008) = \Phi(1.154) = .876$$

Therefore, the power is reduced from 95 to 88% if lack of compliance is taken into account.

10.35 We have the following 2×2 table:

		Attempt to quit		
		Yes	No	
Packs/day	>1	1	49	50
	≤1	5	70	75
		6	119	125

The expected number of units in the (1, 1) cell $= 6 \times 50/125 = 2.4 < 5$. Thus, we must use Fisher's exact test to assess the significance of this table. We construct all possible tables with the same row and column margins as the observed table as follows:

0	50		1	49		2	48		3	47		4	46
6	69		5	70		4	71		3	72		2	73

5	45		6	44
1	74		0	75

We have from the recursion rule that

$$Pr(0) = 1 Pr(0)$$

$$Pr(1) = \frac{6 \times 50}{1 \times 70} Pr(0) = 4.286 \, Pr(0)$$

$$Pr(2) = \frac{5 \times 49}{2 \times 71} Pr(1) = 7.394 \, Pr(0)$$

$$Pr(3) = \frac{4 \times 48}{3 \times 72} Pr(2) = 6.573 \, Pr(0)$$

$$Pr(4) = \frac{3 \times 47}{4 \times 73} Pr(3) = 3.174 \, Pr(0)$$

$$Pr(5) = \frac{2 \times 46}{5 \times 74} Pr(4) = 0.789 \, Pr(0)$$

$$Pr(6) = \frac{1 \times 45}{6 \times 75} Pr(5) = 0.079 \, Pr(0)$$

Thus, $Pr(0) (1 + 4.286 + \ldots + 0.079) = 1$

or $Pr(0) = \dfrac{1}{23.295} = .0429$

We now derive the other probabilities as follows: $Pr(1) = .184$, $Pr(2) = .317$, $Pr(3) = .282$, $Pr(4) = .136$, $Pr(5) = .034$, $Pr(6) = .003$. Since we observed the "1" table, the two-tailed p-value is given by $p = 2 \times (.043 + .184) = .454$. Thus, there is no significant relationship between amount smoked and propensity to quit.

10.36 We have the following 2×4 table relating success in quitting smoking to level of education:

Years of education

		<12	12	>12,<16	16+	
Successful quitter	Yes	16	47	69	52	184
	No	17	29	56	25	127
Percentage of successful quitters		33	76	125	77	311
		(48)	(62)	(55)	(68)	

We will perform the chi-square test for linear trend to detect if there is a significant association between the proportion of successful quitters and the number of years of education. We assign scores of 1, 2, 3, and 4 to the four education groups. We have the test statistic $X_1^2 = A^2/B$ where

$$A = \sum_{i=1}^{k} x_i S_i - \bar{x}S$$
$$= 16(1) + \ldots + 52(4) - 184 \times \frac{33(1) + \ldots + 77(4)}{311}$$
$$= 525 - 184 \times \frac{868}{311}$$
$$= 525 - 513.54 = 11.46$$
$$B = \overline{pq} \left[\sum_{i=1}^{k} n_i S_i^2 - \left(\sum_{i=1}^{k} n_i S_i \right)^2 / N \right]$$
$$= \frac{184}{311} \times \frac{127}{311} \times \left[33(1^2) + \ldots + 77(4^2) - \frac{868^2}{311} \right]$$
$$= .2416(2694 - 2422.59)$$
$$= .2416(271.41) = 65.57$$

Therefore, $X_1^2 = 11.46^2/65.57 = 2.00 \sim \chi_1^2$ under H_0. Since $\chi_{1,.75}^2 = 1.32$, $\chi_{1,.90}^2 = 2.71$ and $1.32 < 2.00 < 2.71$, it follows that $1 - .90 < p < 1 - .75$ or $.10 < p < .25$. Thus, there is no significant association between success in quitting smoking and number of years of education.

10.37 The data are in the form of a 2×2 table, so the chi-square test may be an appropriate method of analysis if the expected cell counts are large enough. The smallest expected value is given by $(12 \times 22)/50 = 5.28 > 5$. Thus, this is a reasonable method of analysis.

10.38 The observed table is given as follows:

Association between working status and health status

	Ill	Well	
Worked	10	12	22
Did not work	2	26	28
	12	38	50

Compute the following chi-square statistic:

$$X^2 = \frac{n \left(|ad - bc| - \dfrac{n}{2} \right)^2}{(a + b)(c + d)(a + c)(b + d)}$$
$$= \frac{50(|10 \times 26 - 2 \times 12| - 25)^2}{22(28)(12)(38)}$$
$$= 50(211)^2/[22(28)(12)(38)] = 7.92 \sim \chi_1^2$$

Referring to the χ^2 tables, we find that

$$\chi_{1,.995}^2 = 7.88, \quad \chi_{1,.999}^2 = 10.83$$

Thus, $.001 < p < .005$. The authors found a chi-square of 7.8, $p = .01$, and thus our results are somewhat more significant than those claimed in the article.

10.39 The t test is *not* a reasonable test to use in comparing binomial proportions from two independent samples. Instead, either the chi-square test for 2×2 tables with large expected values or Fisher's exact test for tables with small expected values should be used.

10.40 The 2×2 table is given as follows:

Association between salad consumption and health status

		Ill	Well		Percentage ill
Ate salad	Yes	25	8	33	(76)
	No	3	6	9	(33)
		28	14	42	

The smallest expected value $= (14 \times 9)/42 = 3 < 5$, which implies that Fisher's exact test must be used. First rearrange the table so that the smaller row total is in row 1 and the smaller column total is in column 1, as follows:

		Well	Ill	
Ate salad	No	6	3	9
	Yes	8	25	33
		14	28	42

Now enumerate all tables with the same row and column margins as follows:

0	9
14	19

1	8
13	20

2	7
12	21

3	6
11	22

4	5
10	23

5	4
9	24

6	3
8	25

7	2
7	26

8	1
6	27

9	0
5	28

Now use the recursion rule to compute the exact probability of each table.

$$Pr(0) = 1 \ Pr(0)$$
$$Pr(1) = \frac{14 \times 9}{1 \times 20} \ Pr(0) = 6.300 \ Pr(0)$$
$$Pr(2) = \frac{13 \times 8}{2 \times 21} \ Pr(1) = 15.600 \ Pr(0)$$
$$Pr(3) = \frac{12 \times 7}{3 \times 22} \ Pr(2) = 19.855 \ Pr(0)$$
$$Pr(4) = \frac{11 \times 6}{4 \times 23} \ Pr(3) = 14.243 \ Pr(0)$$
$$Pr(5) = \frac{10 \times 5}{5 \times 24} \ Pr(4) = 5.935 \ Pr(0)$$
$$Pr(6) = \frac{9 \times 4}{6 \times 25} \ Pr(5) = 1.424 \ Pr(0)$$

$$Pr(7) = \frac{8 \times 3}{7 \times 26} \ Pr(6) = 0.188 \ Pr(0)$$
$$Pr(8) = \frac{7 \times 2}{8 \times 27} \ Pr(7) = 0.012 \ Pr(0)$$
$$Pr(9) = \frac{6 \times 1}{9 \times 28} \ Pr(8) = 0.00029 \ Pr(0)$$

Thus, $Pr(0)(1 + 6.300 + 15.600 + 19.855$
$$+ \ 14.243 + 5.935 + 1.424$$
$$+ \ 0.188 + 0.012 + 0.00029) = 1$$

or $Pr(0) = 1/64.557 = .0155$
$$Pr(1) = .098$$
$$Pr(2) = .242$$
$$Pr(3) = .308$$
$$Pr(4) = .221$$
$$Pr(5) = .092$$
$$Pr(6) = .022$$
$$Pr(7) = .003$$
$$Pr(8) = .0002$$
$$Pr(9) = 4.49 \times 10^{-6}$$

Since our observed table is the "6" table, the two-sided p-value is given by

$$p = 2 \times \min\left[\sum_{i=0}^{6} Pr(i), \ \sum_{i=6}^{9} Pr(i)\right]$$
$$= 2 \times \min(.997, .0252) = .050$$

Thus, the results are on the margin of being statistically significant ($p = .05$) as opposed to the p-value of .01 given in the paper.

10.41 We use the power formula in (**10.27**) (text, p. 385) as follows:

Power =

$$\Phi\left[\frac{\Delta}{\sqrt{(p_1 q_1 + p_2 q_2)/n}} - z_{1-\alpha/2}\frac{\sqrt{2\bar{p}\,\bar{q}/n}}{\sqrt{(p_1 q_1 + p_2 q_2)/n}}\right]$$

where $p_1 = .15$, $p_2 = .05$, $\Delta = |.15 - .05| = .10$, $\alpha = .05$, $n = 125$, $\bar{p} = (.15 + .05)/2 = .10$. We have

$$\text{Power} = \Phi\left\{\frac{.10}{\sqrt{[.15(.85) + .05(.95)]/125}}\right.$$
$$\left. - \ z_{.975}\frac{\sqrt{2(.10)(.90)/125}}{\sqrt{[.15(.85) + .05(.95)]/125}}\right\}$$
$$= \Phi\left(\frac{.10}{.0374} - 1.96 \times \frac{.0379}{.0374}\right)$$
$$= \Phi(2.673 - 1.988) = \Phi(0.685) = .75$$

Thus, such a study would have 75% power.

10.42 We let $n = 150$ and keep all other parameters the same. We have

$$\text{Power} = \Phi\left\{\frac{.10}{\sqrt{[.15(.85) + .05(.95)]/150}} - 1.988\right\}$$

$$= \Phi\left(\frac{.10}{.0342} - 1.988\right)$$

$$= \Phi(2.928 - 1.988) = \Phi(0.940) = .83$$

Thus, the power increases to 83% if the sample size is increased to 150 patients in each group.

10.43 We use the sample-size formula in **(10.26)** (text, p. 384) as follows:

$$n = \frac{(\sqrt{2\bar{p}\bar{q}}z_{1-\alpha/2} + \sqrt{p_1q_1 + p_2q_2}z_{1-\beta})^2}{\Delta^2}$$

$$= \frac{[\sqrt{2(.10)(.90)}z_{.975} + \sqrt{.15(.85) + .05(.95)}z_{.95}]^2}{(.10)^2}$$

$$= \frac{[.4243(1.96) + .4183(1.645)]^2}{.01}$$

$$= \frac{1.5197^2}{.01} = 231.0$$

Thus, we would need to recruit 231 patients in each group in order to achieve a 95% power.

10.44 Form the following 2 × 2 table to assess age effects among women with a negative history:

Women with a negative history

Low birthweight

Age		Yes	No	
	≥ 30	8	217	225
	< 30	30	876	906
		38	1093	1131

The smallest expected cell count $= E_{11} = (38 \times 225)/1131 = 7.56 \geq 5$. Therefore, the Yates-corrected chi-square test can be used.

10.45 The test statistic is given by

$$X^2 = \frac{1131(|8 \times 876 - 30 \times 217| - 1131/2)^2}{38 \times 1093 \times 906 \times 225}$$

$$= \frac{1131(498 - 565.5)^2}{8.467 \times 10^9}$$

$$= \frac{5.153 \times 10^6}{8.467 \times 10^9} = 0.00061 \sim \chi_1^2 \text{ under } H_0$$

Clearly, since $\chi_{1,.50}^2 = 0.45$ and $X^2 < 0.45$, it follows that $p > .50$, and there is no significant effect of age on low-birthweight deliveries in this strata.

10.46 Form the following 2 × 2 contingency table among women with a positive history:

Women with a positive history

Low birthweight

Age		Yes	No	
	≥ 30	6	82	88
	< 30	2	151	153
		8	233	241

The smallest expected value $= (8 \times 88)/241 = 2.92 < 5$. Therefore, Fisher's exact test must be used to perform the test.

10.47 First form all possible tables with the same row and column margins, as follows:

0	88	1	87	2	86	3	85	4	84
8	145	7	146	6	147	5	148	4	149

5	83	6	82	7	81	8	80
3	150	2	151	1	152	0	153

Now use the recursion rule to compute the exact probabilities of these tables.

$$Pr(0) = 1\, Pr(0)$$

$$Pr(1) = \frac{8 \times 88}{1 \times 146}\, Pr(0) = 4.822\, Pr(0)$$

$$Pr(2) = \frac{7 \times 87}{2 \times 147}\, Pr(1) = 9.988\, Pr(0)$$

$$Pr(3) = \frac{6 \times 86}{3 \times 148}\, Pr(2) = 11.608\, Pr(0)$$

$$Pr(4) = \frac{5 \times 85}{4 \times 149}\, Pr(3) = 8.277\, Pr(0)$$

$$Pr(5) = \frac{4 \times 84}{5 \times 150}\, Pr(4) = 3.708\, Pr(0)$$

$$Pr(6) = \frac{3 \times 83}{6 \times 151}\, Pr(5) = 1.019\, Pr(0)$$

$$Pr(7) = \frac{2 \times 82}{7 \times 152}\, Pr(6) = 0.157\, Pr(0)$$

$$Pr(8) = \frac{1 \times 81}{8 \times 153}\, Pr(7) = 0.010\, Pr(0)$$

Thus, $Pr(0)[1 + 4.822 + \ldots + 0.010] = 1$

or $Pr(0) = 1/40.591 = .0246$

Thus, it follows that

$Pr(1) = .119$, $Pr(2) = .246$, $Pr(3) = .286$, $Pr(4) = .204$, $Pr(5) = .091$, $Pr(6) = .025$, $Pr(7) = .004$, $Pr(8) = .0003$

Since our table is the "6" table, a two-sided p-value is computed as follows:

$$p = 2 \times \min\left[\sum_{k=0}^{6} Pr(k), \sum_{k=6}^{8} Pr(k)\right]$$

$= 2 \times \min(.025 + \ldots + .025, .025 + .004 + .0003)$

$= 2 \times \min(.996, .0292) = .058$

Thus, for women with a positive history, there is a trend toward significance, with older women having a higher incidence of low-birthweight deliveries.

10.48 We use the Mantel-Haenszel odds-ratio estimator as follows:

$$\hat{OR}_{MH} = \frac{\sum\limits_{i=1}^{2} a_i d_i / n_i}{\sum\limits_{i=1}^{2} b_i c_i / n_i}$$

$$= \frac{8(876)/1131 + 6(151)/241}{30(217)/1131 + 2(82)/241}$$

$$= \frac{9.956}{6.436} = 1.55$$

To obtain a 95% CI, we use the Robins method given in Equation (**10.41**) (text, p. 410). We have the following summary statistics:

History	i	P_i	Q_i	R_i	S_i	P_iR_i	P_iS_i	Q_iR_i	Q_iS_i
Negative	1	.7816	.2184	6.1963	5.7560	4.8431	4.4989	1.3532	1.2571
Positive	2	.6515	.3485	3.7593	0.6805	2.4490	0.4433	1.3103	0.2372
Total				9.9556	6.4365	7.2921	4.9422	2.6635	1.4942

Therefore,

$\text{Var}(\ln \hat{OR}_{MH}) =$

$$\frac{\sum\limits_{i=1}^{2} P_iR_i}{2\left(\sum\limits_{i=1}^{2} R_i\right)^2} + \frac{\sum\limits_{i=1}^{2}\left(P_iS_i + Q_iR_i\right)}{2\left(\sum\limits_{i=1}^{2} R_i\right)\left(\sum\limits_{i=1}^{2} S_i\right)} + \frac{\sum\limits_{i=1}^{2} Q_iS_i}{2\left(\sum\limits_{i=1}^{2} S_i\right)^2}$$

$$= \frac{7.2921}{2(9.9556)^2} + \frac{4.9422 + 2.6635}{2(9.9556)(6.4365)} + \frac{1.4942}{2(6.4365)^2}$$

$= 0.0368 + 0.0593 + 0.0180 = 0.1142$

Therefore, a 95% CI for $\ln(OR) = \ln(1.55) \pm 1.96\sqrt{0.1142} = 0.4362 \pm 0.6623 = (-0.2261, 1.0984)$. A 95% CI for $OR = (e^{-0.2261}, e^{1.0984}) = (0.80, 3.00)$.

10.49 We construct 2×2 tables relating low birthweight to pregnancy history within specific age groups as follows:

Age < 30

Low birthweight

		+	**−**	
Pregnancy history	**+**	2	151	153
	−	30	876	906
		32	1027	1059

Age ≥ 30

Low birthweight

		+	**−**	
Pregnancy history	**+**	6	82	88
	−	8	217	225
		14	299	313

We use the Mantel-Haenszel odds-ratio estimator as follows:

$$\widehat{OR}_{MH} = \frac{2(876)/1059 + 6(217)/313}{30(151)/1059 + 8(82)/313}$$

$$= \frac{5.8141}{6.3735} = 0.91$$

To obtain a 95% CI, we use the Robins method as follows:

Age	i	P_i	Q_i	R_i	S_i	P_iR_i	P_iS_i	Q_iR_i	Q_iS_i
< 30	1	.8291	.1709	1.6544	4.2776	1.3716	3.5465	0.2828	0.7311
≥ 30	2	.7125	.2875	4.1597	2.0958	2.9637	1.4932	1.1961	0.6026
Total				5.8141	6.3735	4.3353	5.0397	1.4789	1.3338

Therefore,

$$Var(\ln \widehat{OR}_{MH}) = \frac{4.3353}{2(5.8141)^2}$$

$$+ \frac{5.0397 + 1.4789}{2(5.8141)(6.3735)} + \frac{1.3338}{2(6.3735)^2}$$

$$= 0.0641 + 0.0880 + 0.0164$$

$$= 0.1685$$

A 95% CI for ln (OR) $= \ln(0.91) \pm 1.96\sqrt{0.1685} = -0.0919 \pm 0.8045 = (-0.8964, 0.7127)$. The corresponding 95% CI for OR $= (e^{-0.8964}, e^{0.7127}) = (0.41, 2.04)$.

10.50 To test for effect modification, we use the Woolf test for homogeneity of odds ratios over different strata given in Equation (**10.42**) (text, p. 412). We have the test statistic

$$X^2_{HOM} = \sum_{i=1}^{2} w_i[\ln(\widehat{OR}_i)]^2 - \left(\sum_{i=1}^{2} w_i\ln\widehat{OR}_i\right)^2 / \sum_{i=1}^{2} w_i \sim \chi^2_1$$

$$\text{under } H_0$$

where we test H_0: $OR_1 = OR_2$ versus H_1: $OR_1 \neq OR_2$ where OR_1 = odds ratio relating low birthweight to age for women with a negative pregnancy history and OR_2 = the corresponding odds ratio for women with a positive pregnancy history. We have the following summary statistics:

Thus, we have

$$X^2_{HOM} = 4.2947 - (2.9426)^2/7.5533$$

$$= 3.15 \sim \chi^2_1 \text{ under } H_0$$

Since $3.15 < 3.84 = \chi^2_{1,.95}$ it follows that $p > .05$ and there is no significant heterogeneity among the two odds ratios. Thus, there is no significant effect modification.

10.51 We first combine together the data from all drinking women and form the following 2 × 2 contingency table:

	Drinking status		
	Nondrinker	Drinker	
Case	43	103	146
Control	5901	11,902	17,803
	5944	12,005	17,949

The smallest expected value in this table $= E_{11} = 146 \times 5944/17,949 = 48.3 \geq 5$. Thus, we can use the Yates-corrected chi-square test for 2 × 2 contingency tables to test this hypothesis.

History	i	w_i	$\widehat{OR}_i$	$\ln(\widehat{OR}_i)$	$w_i\ln(\widehat{OR}_i)$	$w_i(\ln(\widehat{OR}_i))^2$
Negative	1	6.0945	1.0765	0.0737	0.4492	0.0331
Positive	2	1.4588	5.5244	1.7092	2.4934	4.2616
Total		7.5533			2.9426	4.2947

10.52 We have the test statistic

$$X^2 = \frac{n(|ad - bc| - n/2)^2}{(a + b)(c + d)(a + c)(b + d)}$$

$$= \frac{17{,}949[|43(11{,}902) - 103(5901)| - 17{,}949/2]^2}{146(17{,}803)(5944)(12{,}005)}$$

$$= \frac{1.360 \times 10^{14}}{1.855 \times 10^{14}} = 0.73 \sim \chi_1^2 \text{ under } H_0$$

Since $\chi_{1,.50}^2 = 0.45$, $\chi_{1,.75}^2 = 1.32$, and $0.45 < 0.73 < 1.32$, it follows that $1 - .75 < p < 1 - .50$ or $.25 < p < .50$. Thus, there is no significant difference in breast-cancer incidence between drinkers and nondrinkers.

10.53 We use the chi-square test for linear trend using scores of 1, 2, 3, 4, 5 for the 5 alcohol-consumption groups. Compute the test statistic $X_1^2 = A^2/B$, where

$A = 1(43) + 2(15) + 3(22) + 4(42) + 5(24)$

$\quad - 146[1(5944) + 2(2069) + \dots + 5(2917)]/17{,}949$

$\quad = 427 - 146(49{,}294)/17{,}949$

$\quad = 427 - 400.97 = 26.03$

$B = [146(17{,}803)/17{,}949^2]$

$\quad \times [1(5944) + 4(2069) + \dots + 25(2917)$

$\quad - 49294^2/17{,}949]$

$\quad = .00807(175{,}306 - 135{,}377.93)$

$\quad = .00807(39{,}928.07) = 322.14$

Thus, $X_1^2 = 26.03^2/322.14 = 2.10 \sim \chi_1^2$ under H_0. Since $X_1^2 < \chi_{1,.95}^2 = 3.84$, it follows that $p > .05$, and there is no significant association between alcohol consumption and breast cancer in this age group.

10.54 A *case–control study* is a study where a group of subjects who already have a disease (the cases) are identified and compared with another group of patients who do not have the disease (the controls). The health habits of cases and controls are then compared. A *cohort study* is a study where a group of disease-free individuals (the cohort) are identified and then followed over time. The members of the cohort are then subdivided according to risk-factor status at baseline and a comparison is made between the proportion of subjects who develop disease in different risk-factor groups.

10.55 The chi-square test for trend.

10.56 We have the test statistic $X_1^2 = A^2/B \sim \chi_1^2$ under H_0. We will use scores of 0, 1, and 2 corresponding to the number of relatives with lung cancer $= 0$, 1, and 2+, respectively. We have the following 2×3 table:

Number of relatives with lung cancer

	0	1	2+	Total
Cases	393	119	20	532
Controls	466	78	8	552
	859	197	28	1084

For this data set,

$A = 0(393) + 1(119) + 2(20)$

$\quad - (532/1084)[0(859) + 1(197) + 2(28)]$

$\quad = 159 - 124.166 = 34.834$

$B = [532(552)/1084^2]\{0^2(859) + 1^2(197) + 2^2(28)$

$\quad - [0(859) + 1(197) + 2(28)]^2/1084\}$

$\quad = .2499(309 - 59.049) = 62.467$

Thus,

$\quad X_1^2 = 34.834^2/62.467 = 19.42 \sim \chi_1^2$ under H_0

Since $19.42 > 10.83 = \chi_{1,.999}^2$, it follows that $p < .001$. Since $A > 0$, we conclude that the cases have a significantly greater number of relatives with lung cancer than the controls.

10.57 We wish to test the hypothesis $H_0: p_1 = p_2$ versus $H_1: p_1 \neq p_2$, where

$\quad p_1 =$ true prevalence of hypertension among cases

$\quad p_2 =$ true prevalence of hypertension among controls

Under H_0, the best estimate of the common proportion p is

$$\hat{p} = \frac{n_1 \hat{p}_1 + n_2 \hat{p}_2}{n_1 + n_2} = \frac{1493(.37) + 707(.30)}{1493 + 707}$$

$$= \frac{552 + 212}{2200} = \frac{764}{2200} = .347$$

Since $n_1 \hat{p}\hat{q} = 1493(.347)(.653) = 338.43$ and $n_2 \hat{p}\hat{q} = 707(.347)(.653) = 160.26$, it follows that we can use the two-sample test for binomial proportions.

10.58 The test statistic is

$$z = \frac{\hat{p}_1 - \hat{p}_2}{\sqrt{\hat{p}\hat{q}(1/n_1 + 1/n_2)}}$$

$$= \frac{.37 - .30}{\sqrt{.347(.653)(1/1493 + 1/707)}}$$

$$= \frac{.07}{.0217} = 3.221 \sim N(0, 1) \text{ under } H_0$$

The p-value $= 2 \times [1 - \Phi(3.221)] = 2 \times (1 - .9994) = .0013$. Thus, we can reject H_0 and conclude that the two underlying prevalence rates are not the same.

10.59 The proportions are an example of prevalence, because the subjects were asked whether they have hypertension at one point in time, viz. at the time of coronary angiography.

10.60 We have that each person is used as his or her own control. Thus, these are paired samples and we must use McNemar's test for correlated proportions to analyze the data.

10.61 We test the hypothesis H_0: $p = 1/2$ versus H_1: $p \neq 1/2$, where $p =$ proportion of discordant pairs that are of type A. We have the test statistic

$$X^2 = \frac{\left(\left| 12 - \frac{12 + 22}{2} \right| - .5 \right)^2}{(12 + 22)/4}$$

$$= \frac{4.5^2}{8.5} = 2.38 \sim \chi_1^2 \text{ under } H_0$$

The p-value $= Pr(\chi_1^2 > 2.38) = 2 \times Pr[N(0, 1) > \sqrt{2.38}] = 2 \times Pr[N(0, 1) > 1.543] = 2 \times .061 = .123$. Thus, there is no significant difference between the surrogate report by telephone and the self-report at the eye exam.

10.62 The sensitivity and specificity are the appropriate measures.

10.63 The sensitivity of the surrogate report $= Pr(\text{test} + | \text{true} +) = Pr(\text{surrogate report} + | \text{clinical determination} +) = 14/97 = .144$ (very poor!).

A 95% CI for the sensitivity $= .144 \pm 1.96\sqrt{.144(.856)/97} = .144 \pm .070 = (.074, .214)$.

The specificity $= Pr(\text{test} - | \text{true} -) = 1247/1273 = .980$ (good). A 95% CI for the specificity $= .980 \pm 1.96\sqrt{.980(.020)/1273} = .980 \pm .008 = (.972, .987)$.

10.64 We can set up the data in the form of a 3×2 contingency table as follows:

PMH status	Breast cancer +	−	Total	Percentage of breast cancer
Never use	129	7046	7175	(1.8)
Current use	51	2444	2495	(2.0)
Past use	46	2480	2526	(1.8)
	226	11,970	12,196	

The smallest expected value $= 226 \times 2495/12,196 = 46.2 \geq 5$. Thus, we can use the chi-square test for $r \times c$ tables.

10.65 We compute the expected table corresponding to the observed table in Problem 10.64. We have

$$E_{11} = \frac{226 \times 7175}{12,196} = 133.0$$

$$E_{12} = 7175 - 133.0 = 7042.0$$

$$E_{21} = \frac{226 \times 2495}{12,196} = 46.2$$

$$E_{22} = 2495 - 46.2 = 2448.8$$

$$E_{31} = \frac{226 \times 2526}{12,196} = 46.8$$

$$E_{32} = 2526 - 46.8 = 2479.2$$

The chi-square statistic is given by

$$X^2 = \frac{(129 - 133.0)^2}{133} + \frac{(7046 - 7042.0)^2}{7042.0}$$

$$+ \frac{(51 - 46.2)^2}{46.2} + \frac{(2444 - 2448.8)^2}{2448.8}$$

$$+ \frac{(46 - 46.8)^2}{46.8} + \frac{(2480 - 2479.2)^2}{2479.2}$$

$$= 0.118 + 0.002 + 0.491 + 0.009$$

$$+ 0.014 + 0.000$$

$$= 0.635 \sim \chi_2^2 \text{ under } H_0$$

Since $\chi_{2,.50}^2 = 1.39 > 0.635$, and $\chi_{2,.25}^2 = 0.58 < 0.635$, it follows that $1 - .5 < p < 1 - .25$ or $.5 < p < .75$, and we accept H_0 that the incidence of breast cancer is the same in the three groups.

10.66 The estimated odds ratio $= (51/2444)/(129/7046) = 1.14$. A 95% CI for $\ln(\text{OR})$ is given by $\ln(1.14) \pm 1.96 \times \sqrt{1/51 + 1/2444 + 1/129 + 1/7046} = 0.1308 \pm 0.3274 = (-0.1966, 0.4583)$. The corresponding 95% CI for OR $= [\exp(-0.1966), \exp(0.4583)] = (0.82, 1.58)$.

10.67 The estimated odds ratio $= (46/2480)/(129/7046) = 1.01$. A 95% CI for $\ln(\text{OR})$ is given by $\ln(1.01) \pm 1.96 \times \sqrt{1/46 + 1/2480 + 1/129 + 1/7046} = 0.0130 \pm 0.3397 = (-0.3267, 0.3527)$. The corresponding 95% CI for OR $= [\exp(-0.3267), \exp(0.3527)] = (0.72, 1.42)$.

10.68 For current users versus never users, RR $= (51/2495)/(129/7175) = 1.14$. To obtain confidence limits, we use the formula in Equation **(10.12)** (text, p. 364). A 95% CI for $\ln(\text{RR}) = \ln(1.14) \pm 1.96\sqrt{2444/[51(2495)] + 7046/[129(7175)]} = 0.1283 \pm 0.3210 = (-0.1927, 0.4493)$. The corresponding 95% CI for RR $= [\exp(-0.1927), \exp(0.4493)] = (0.82, 1.57)$.

For past users versus never users, RR $= (46/2526)/(129/7175) = 1.01$. A 95% CI for $\ln(\text{RR}) = \ln(1.01) \pm$

$1.96\sqrt{2480/[46(2526)] + 7046/[129(7175)]} = 0.0128 \pm 0.3335 = (-0.3207, 0.3463)$. The corresponding 95% CI for RR = [exp(-0.3207), exp(0.3463)] = (0.73, 1.41). Both the point estimates and the 95% CI's for OR and RR are very similar for both current and past users, respectively, versus never users. Bigger differences would be expected if either OR or RR were further away from 1.0.

REFERENCES

[1] Schwartz, R. M., Luby, A. M., Scanlon, J. W., & Kellogg, R. J. (1991). Effect of surfactant on morbidity, mortality and resource use in newborn infants weighing 500 to 1500 g. *New England Journal of Medicine, 330*(21), 1476–1480.

[2] Kennedy, J. W., Ritchie, J. L., Davis, K. B., Stadius, M. L., Maynard, C., & Fritz, J. K. (1985). The western Washington randomized trial of intracoronary streptokinase in acute myocardial infarction: A 12-month follow-up report. *New England Journal of Medicine, 312*(17), 1073–1078.

[3] The Coronary Drug Project Research Group. (1979). Cigarette smoking as a risk factor in men with a history of myocardial infarction. *Journal of Chronic Diseases, 32*(6), 415–425.

[4] Furusko, K., Sato, K., Socda, T., et al. (1983, December 10). High dose intravenous gamma globulin for Kawasaki's syndrome [Letter]. *Lancet,* 1359.

[5] Barkan, S. E., & Bracken, M. (1987). Delayed childbearing: No evidence for increased risk of low birth weight and preterm delivery. *American Journal of Epidemiology, 125*(1), 101–109.

[6] Willett, W., Stampfer, M. J., Colditz, G. A., Rosner, B. A., Hennekens, C. H., & Speizer, F. E. (1987). Moderate alcohol consumption and the risk of breast cancer. *New England Journal of Medicine, 316*(19), 1174–1180.

[7] Shaw, G. L., Falk, R. T., Pickle, L. W., Mason, T. J., & Buffler, P. A. (1991). Lung cancer risk associated with cancer in relatives. *Journal of Clinical Epidemiology, 44*(4/5), 429–437.

[8] Applegate, W. B., Hughes, J. P., & Vanderzwaag, R. (1991). Case–control study of coronary heart disease risk factors in the elderly. *Journal of Clinical Epidemiology, 44*(4/5), 409–415.

[9] Linton, K. L. P., Klein, B. E. K., & Klein, R. (1991). The validity of self-reported and surrogate-reported cataract and age-related macular degeneration in the Beaver Dam Eye Study. *American Journal of Epidemiology, 134*(12), 1438–1446.

CHAPTER ELEVEN

REGRESSION AND CORRELATION ANALYSIS

REVIEW OF KEY CONCEPTS

SECTION 11.1　The Linear-Regression Model

A study was performed relating vegetable and fruit consumption to systolic blood pressure (SBP) in a group of 60 lactovegetarians (vegetarians who eat dairy products but not meat, poultry, or fish). The descriptive statistics for these variables are given as follows:

	Mean	sd	n	Range	
SBP	109.3	11.9	60	82–144	mm Hg
Vegetables	3.13	2.06	60	0–7.95	Number of servings/day
Fruit	3.41	2.44	60	0–10.63	Number of servings/day

We focus on fruit consumption, subdividing fruit consumption into "one serving per day" groups and computing mean SBP in each fruit-consumption group. These are presented numerically in Table 11.1 and graphically in Figure 11.1.

TABLE 11.1 Mean systolic blood pressure (SBP) by number of servings per day of fruit

Fruit consumption (servings per day)	n	Mean SBP	Predicted SBP
0.0	8	115.9	114.5
0.01–0.99	4	114.4	113 8
1.00–1.99	6	109.4	112.2
2.00–2.99	8	110.7	110.7
3.00–3.99	12	109.1	109.2
4.00–4.99	6	112.8	107.7
5.00–5.99	7	104.2	106.1
6.0 +	9	101.9	104.6[a]

[a]It is assumed that the average fruit consumption in this group is 6.5 servings per day.

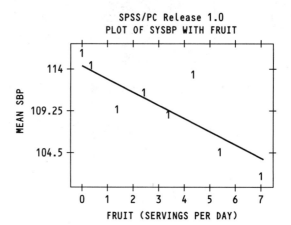

Figure 11.1
Plot of systolic blood
pressure versus fruit
consumption
(servings per day)
in a group of 60
lactovegetarians

It appears that mean SBP decreases as fruit consumption increases, with the decline being roughly linear. This suggests the following *linear-regression model:*

General Linear-Regression Model

$$y_i = \alpha + \beta x_i + e_i \quad y_i = \text{SBP } i\text{th person}, \quad x_i = \text{fruit consumption } i\text{th person}$$

$$e_i \sim N(0, \sigma^2) \text{ error term}$$

The parameter σ^2 is a measure of the variability of the points about the line. If all points fall exactly on a line, then $\sigma^2 = 0$. In general, $\sigma^2 > 0$. The parameters α and β are the intercept and slope of the regression line. The slope is likely to be negative ($\beta < 0$) for SBP versus fruit consumption, positive ($\beta > 0$) for SBP versus body weight and zero ($\beta = 0$) for SBP versus birthday as depicted in Figure 11.2.

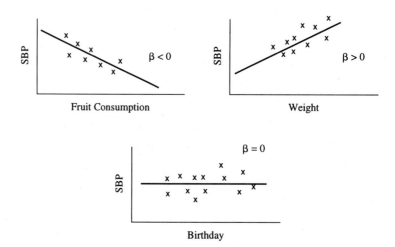

Figure 11.2
Relationship of SBP to
fruit consumption,
body weight, and
birthday, respectively

SECTION 11.2   Fitting Regression Lines—The Method of Least Squares

We use the *method of least squares* to estimate the parameters of a regression line. The coefficients a and b are estimated parameters, corresponding to the true parameters α and β. We find a and b so as to minimize S, where

$$S = \sum_{i=1}^{n} \left[y_i - (a + bx_i) \right]^2 = \sum_{i=1}^{n} d_i^2$$

The least-squares estimates are

$$b = \frac{L_{xy}}{L_{xx}}, \quad a = \bar{y} - b\bar{x} = \frac{\Sigma y_i - b\Sigma x_i}{n}$$

where L_{xy} = corrected sum of cross-products

$$= \sum_{i=1}^{n} (x_i - \bar{x})(y_i - \bar{y})$$

L_{xx} = corrected sum of squares for x

$$= \sum_{i=1}^{n} (x_i - \bar{x})^2$$

L_{yy} = corrected sum of squares for y

$$= \sum_{i=1}^{n} (y_i - \bar{y})^2$$

The line $y = a + bx$ is called the *least-squares or regression line*. In this example,

$$\Sigma y_i = 6559 \qquad \Sigma y_i^2 = 725{,}343 \qquad \Sigma x_i = 204.83$$

$$\Sigma x_i^2 = 1051.31 \qquad \Sigma x_i y_i = 21{,}853.86$$

$$L_{yy} = 725{,}343 - \frac{6559^2}{60} = 8334.98$$

$$L_{xx} = 1051.31 - \frac{204.83^2}{60} = 352.05$$

$$L_{xy} = 21{,}853.86 - \frac{204.83(6559)}{60} = -537.47$$

$$b = \frac{-537.47}{352.05} = -1.53 \qquad a = \frac{6559 + 1.53 \times 204.83}{60} = 114.53$$

$$y = 114.53 - 1.53x$$

This is the least-squares or regression line.

SECTION 11.3   Testing for the Statistical Significance of a Regression Line

We wish to test the hypothesis H_0: $\beta = 0$ versus H_1: $\beta \neq 0$. We can subdivide the Total sum of squares (SS) $= \sum_{i=1}^{n} (y_i - \bar{y})^2$ into two components called the Regression SS and the Residual SS as shown in the equation:

$$\underset{\text{Total SS}}{\sum_{i=1}^{n}(y_i - \overline{y})^2} = \underset{\text{Residual SS}}{\sum_{i=1}^{n}(y_i - \hat{y}_i)^2} + \underset{\text{Regression SS}}{\sum_{i=1}^{n}(\hat{y}_i - \overline{y})^2}$$

where $\hat{y}_i$ is the predicted value of y for a specific value of x, given by $a + bx_i$. The Residual SS is a measure of how close the individual sample points are from the regression line. The Regression SS is a measure of how far the slope of the regression line is from zero (in absolute value). For a good-fitting regression line, we want a high Regression SS and a low Residual SS. We use $\dfrac{\text{Regression SS}}{\text{Residual SS}}$ as a test criterion to test the hypothesis and reject H_0 for large values of Regression SS/Residual SS. For convenience, the test statistic is usually formulated in terms of the Regression mean square (MS) and the Residual mean square (MS) defined by

$$\text{Regression MS} = \text{Regression SS}/1$$
$$\text{Residual MS} = \text{Residual SS}/(n - 2)$$

The test statistic is given by $F = \text{Regression MS}/\text{Residual MS} \sim F_{1, n-2}$ under H_0. The p-value $= Pr(F_{1, n-2} > F)$. This is called the *F test for simple linear regression*. We can arrange the results in the form of an ANOVA table as follows:

	SS	df	MS	F-stat	p-value
Regression					
Residual					
Total					

11.3.1 Short Computational Form for the *F* Test

The Regression SS and Residual SS and test statistic F can also be written in the more computationally convenient forms

$$\text{Regression SS} = b^2 L_{xx} = b L_{xy} = L_{xy}^2/L_{xx}$$
$$\text{Residual SS} = \text{Total SS} - \text{Regression SS} = L_{yy} - L_{xy}^2/L_{xx}$$
$$F = \frac{L_{xy}^2/L_{xx}}{(L_{yy} - L_{xy}^2/L_{xx})/(n - 2)} \sim F_{1, n-2} \text{ under } H_0$$

In our example, $L_{xy} = -537.47$, $L_{xx} = 352.05$, $L_{yy} = 8334.98$. Thus,

$$\text{Regression SS} = (-537.47)^2/352.05 = 820.55$$
$$\text{Residual SS} = 8334.98 - 820.55 = 7514.44$$
$$\text{Residual MS} = 7514.44/58 = 129.56$$
$$F = \frac{\text{Regression MS}}{\text{Residual MS}} = \frac{820.55}{129.56} = 6.33 \sim F_{1, 58}$$

Because $F_{1, 40, .975} = 5.42$, $F_{1, 40, .99} = 7.31$, $5.42 < 6.33 < 7.31$

and $F_{1, 60, .975} = 5.29$, $F_{1, 60, .99} = 7.08$, $5.29 < 6.33 < 7.08$,

it follows that $.01 < p < .025$, which implies that there is a significant relationship between fruit consumption and systolic blood pressure. The results can be displayed in an ANOVA table as follows:

	SS	df	MS	F-stat	p-value
Regression	820.55	1	820.55	6.33	$.01 < p < .025$
Residual	7514.44	58	129.56		
Total	8334.98	59			

SECTION 11.4 The *t* Test Approach to Significance Testing for Linear Regression

Another approach to hypothesis testing is based on the estimated slope b. It can be shown that

$$se(b) = \sqrt{\frac{\text{Residual MS}}{L_{xx}}}$$

We also denote Residual MS by $s_{y \cdot x}^2 =$ variance of y for a given x as opposed to the unconditional variance of $y = s_y^2$. We use the test statistic

$$t = b/se(b) \sim t_{n-2} \quad \text{under } H_0$$

The p-value $= 2 \times Pr(t_{n-2} > t)$ if $t > 0$, or $2 \times Pr(t_{n-2} < t)$ if $t \le 0$. This test is called the *t test for simple linear regression*. In our case,

$$b = -1.53, \, se(b) = \sqrt{129.56/352.05} = 0.61$$
$$t = b/se(b) = -1.53/0.61 = -2.517 \sim t_{58} \quad \text{under } H_0$$

The p-value $= 2 \times Pr(t_{58} < -2.517) = .015$.

The F test and the t test for simple linear regression are equivalent in the sense that $t^2 = F$. In this case, $t^2 = (-2.517)^2 = 6.33 = F$. Also, the p-values obtained from either test are the same.

SECTION 11.5 Interval Estimation for Linear Regression

We can obtain interval estimates for the parameters of a regression line. A 95% CI for β is given by

$$b \pm t_{n-2,.975} \, se(b) = -1.53 \pm t_{58,.975}(0.61)$$
$$= -1.53 \pm 2.001(0.61)$$
$$= -1.53 \pm 1.21 = (-2.74, -0.31)$$

A 95% CI for α is given by $a \pm t_{n-2,.975} \, se(a)$, where

$$se(a) = \sqrt{s_{y \cdot x}^2 \left(\frac{1}{n} + \frac{\bar{x}^2}{L_{xx}} \right)}$$

The intercept is usually more of interest in the comparison of two regression lines.

We can also obtain interval estimates for predictions made from regression lines.

Suppose $\quad\quad\quad x_1 = x$-value for a new individual not used in constructing the regression line

$\bar{x} =$ mean value for the subjects used in the regression

Then $\quad\quad\quad \hat{y} =$ expected value of y for that individual $= a + bx_1$

There are two types of se's used in predictions from regression lines:

$$se_1(\hat{y}) = \sqrt{s_{y \cdot x}^2 \left[1 + \frac{1}{n} + \frac{(x_1 - \bar{x})^2}{L_{xx}} \right]}$$

$= $ standard error of $\hat{y}$ for a given value of $x(x_1)$ for one individual

$$se_2(\hat{y}) = \sqrt{s_{y \cdot x}^2 \left[\frac{1}{n} + \frac{(x_2 - \bar{x})^2}{L_{xx}} \right]}$$

$se_2(\hat{y}) = $ standard error of $\hat{y}$ for a given value of $x(x_2)$ over many individuals with the value x_2

$se_1(\hat{y})$ is used in making predictions for one specific individual; $se_2(\hat{y})$ is used in making predictions for the average value of y over many individuals with the same value of x.

SECTION 11.6 R^2

A commonly used measure of goodness of fit for a regression line is R^2, which is defined by

$$R^2 = \frac{\text{Regression SS}}{\text{Total SS}}$$ and is interpreted as the proportion of the variance of y explained by x

To see this, note that it can be shown algebraically that

$$R^2 \approx 1 - \frac{s_{y \cdot x}^2}{s_y^2}$$

or $R^2 \approx 1 -$ proportion of variance of y that is not explained by $x =$ proportion of variance of y that is explained by x.

If the fit is perfect, then $R^2 = 1$. If x has no relationship to y, then $R^2 = 0$. In our case, $R^2 = \frac{820.55}{8334.58} = .10$. Only 10% of the variance in y is explained by x.

Caution: We can have a highly significant regression with a small R^2 due to a large n. Conversely, we can also have a marginally significant regression with a large R^2 if n is small. Thus, a high R^2 and a low p-value are not synonymous.

Assessing Goodness of Fit of Regression Lines

Assumptions Made in Linear Regression

(1) Linearity Variables must be in the proper scale so that the average value of y for a given value of x is a linear function of x.

(2) Homoscedasticity Residual variance is the same for all data points, and in particular, is independent of x.

(3) Independence All residuals are assumed to be independent; e.g., usually, you can't enter multiple data points on the same person.

To check the validity of the first two assumptions, we can use *residual analysis.* The *residual* for an individual point (x_i, y_i) is $y_i - \hat{y}_i$, where $\hat{y}_i = a + bx_i$ = fitted value of y for the value $x = x_i$. With computer packages such as SAS or MINITAB, we can plot the residuals versus either x_i or $\hat{y}_i$. If assumptions 1 and 2 hold, then no pattern should be evident in the residual plot. For example, violations of assumption 1 will often result in clumps of positive and/or negative residuals for specific ranges of x. Violations of assumption 2 will often result in greater spread of the residuals for either large or small values of x. Residual analysis is also helpful in identifying *outliers* (points that are very far from the regression line), and *influential points* (points that have an extreme influence on the regression parameter estimates). A point is influential if the regression slope and/or intercept estimates change a lot in magnitude when that point is removed. It is almost always worthwhile to perform some residual analyses to check for the validity of regression models.

In Figure 11.1, we have plotted the mean SBP versus fruit consumption, where the latter is grouped in one-serving-per-day groups, and have superimposed the fitted regression line. The observed mean values appear to be randomly scattered about the regression line, indicating no particular pattern in the residuals. The actual observed mean and predicted SBP for each fruit-consumption group are also provided in Table 11.1. To compute the predicted value for a group, the midpoint of the group interval was used. For example, for the 1.00–1.99-servings-per-day group, the midpoint = 1.5 and the predicted SBP was $114.53 - 1.53(1.5) = 112.2$ mm Hg. It appears that the linearity assumption is reasonable for these data.

Multiple Regression

The purpose of multiple regression is to relate a single normally distributed outcome variable y to several $(k > 1)$ predictor variables $x_1, \ldots, x_k$.

11.8.1 ### Multiple-Regression Model

$$y_i = \alpha + \sum_{j=1}^{k} \beta_j x_{ij} + e_i \quad \text{where } x_{ij} = j\text{th variable for the } i\text{th person, } e_i \sim N(0, \sigma^2)$$

We use the method of least squares to fit the model.

11.8.2 **Interpretation of Regression Coefficients**

The regression coefficients β_j in a multiple-regression model are referred to as *partial-regression coefficients*. β_j can be interpreted as the expected increase in y per unit increase in x_j, *holding all other variables constant*. This is different from the regression coefficient β in a simple linear-regression model, where no other variables are controlled for.

11.8.3 **Hypothesis Testing**

There are two types of hypotheses that can be tested with multiple regression. The first type of hypothesis is

$$H_0: \beta_1 = \beta_2 = \ldots = \beta_k = 0 \quad \text{versus} \quad H_1: \text{at least one } \beta_j \neq 0$$

This is a test of the overall hypothesis that at least some of the β_j are different from zero without specifying which one. An F test can be employed to test this hypothesis. The second type of hypothesis is $H_0: \beta_j = 0$, other $\beta_l \neq 0$ versus $H_1:$ all $\beta_j \neq 0$. This is a test of the hypothesis that the jth variable has a significant effect on y after controlling for the other covariates in the model. For example, suppose

$$y_i = \text{SBP for the } i\text{th subject}$$
$$x_{i1} = \text{age} = \text{age (yrs) for the } i\text{th subject}$$
$$x_{i2} = \text{sex for the } i\text{th subject} = 1 \text{ if male}$$
$$\qquad\qquad\qquad\qquad\qquad\qquad\quad 2 \text{ if female}$$
$$x_{i3} = \text{weight (lb) for the } i\text{th subject}$$
$$x_{i4} = \text{fruit} = \text{number of servings of fruit/day}$$

We fit the multiple-regression model

$$y_i = \alpha + \beta_1 x_{i1} + \beta_2 x_{i2} + \beta_3 x_{i3} + \beta_4 x_{i4} + e_i$$

The results are as follows, using SAS PROC REG:

STATISTICAL ANALYSIS SYSTEM

MODEL:	MODEL02		SSE	5096.259	F RATIO	8.74	
			DFE	55	PROB>F	0.0001	
DEP VAR: MSYS			MSE	92.659246	R-SQUARE	0.3885	

VARIABLE	DF	PARAMETER ESTIMATE	STANDARD ERROR	T RATIO	PROB>|T|	VARIABLE LABEL
INTERCEPT	1	96.541534	12.464154	7.7455	0.0001	
AGE	1	0.013926	0.176773	0.0788	0.9375	
SEX	1	-6.738063	2.782668	-2.4214	0.0188	
WEIGHT	1	0.196530	0.065980	2.9786	0.0043	
FRUIT	1	-1.137050	0.538329	-2.1122	0.0392	

We can test the hypothesis H_0: $\beta_4 = 0$, β_1, β_2, $\beta_3 \neq 0$ versus H_1: all $\beta_j \neq 0$; this is a test of the hypothesis that fruit consumption has an effect on BP after controlling for age, sex, and weight. We have

$$b_4 = -1.14 \pm 0.54, p = .039$$

This indicates that fruit consumption has a significant effect on SBP even after controlling for age, sex, and weight. However, the magnitude of the effect is much smaller than in the simple linear-regression model, where $b = -1.53$. Roughly $1/4$ of the effect of fruit consumption can be explained by the other variables in the model.

We also see that sex and weight have significant effects on SBP, with males and heavier people having higher blood pressure. At the top of the SAS output is a test of the first hypothesis H_0: all $\beta_j = 0$ versus H_1: at least one $\beta_j \neq 0$. The p-value (listed under PROB>F) is 0.0001, indicating that at least some of the variables are having a significant effect on SBP. Finally, the $R^2 = 0.39$, indicating that 39% of the variation in SBP can be explained by age, sex, weight, and fruit consumption.

11.8.4   **Special Types of Multiple-Regression Models**

(1) One-Way ANOVA Model A continuous normally distributed outcome variable y (e.g., SBP) and a single categorical predictor variable indicating group membership—called a CLASS variable; (e.g., ethnic group) represented by $k-1$ "dummy variables" for a variable with k categories.

(2) Two-Way ANOVA Model A continuous normally distributed outcome variable y (e.g., SBP) and two categorical variables indicating group membership on two distinct characteristics (e.g., ethnic group and sex). Each can be represented by sets of dummy variables, possibly together with another set of dummy variables representing the interaction effect between the two variables.

(3) Analysis of Covariance Model A continuous normally distributed outcome variable y (e.g., SBP), one or more CLASS variables (e.g., ethnic group, sex) of primary interest and one or more other variables (usually continuous) that are important predictors of the outcome variable and need to be controlled for (e.g., weight) before looking at the relationship between y and the key outcome variable(s).

All three of these models are special cases of multiple-regression models and can be fit using multiple-regression software in statistical packages.

SECTION 11.9   **The Correlation Coefficient**

The correlation coefficient r is defined as $L_{xy}/\sqrt{L_{xx}L_{yy}} = b s_x/s_y$. It has the advantage over the regression coefficient of being a dimensionless quantity $(-1 \leq r \leq 1)$, and is not affected by the units of measurement. If $r > 0$, then y tends to increase as x increases [e.g., SBP versus weight in Figure 11.3(a)]; if $r < 0$, then y tends to decrease as x increases [e.g., SBP versus fruit consumption in Figure 11.3(b)]; if $r = 0$, then y is unrelated to x [e.g., SBP versus birthday in Figure 11.3(c)]. In general, r is a more appropriate measure than b if one is interested in whether there is an association between two variables, rather than in predicting one variable from another. Conversely, if prediction is the main interest, then b is the more appropriate measure.

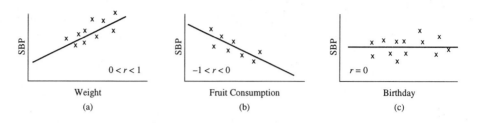

Figure 11.3
Scatter plot of SBP
versus weight, fruit
consumption, and
birthday, respectively

11.9.1   **One-Sample Inference**

We may wish to test the hypothesis H_0: $\rho = 0$ versus H_1: $\rho \neq 0$, where ρ is the population correlation coefficient. For this purpose, we use the test statistic

$$t = r\sqrt{n-2}/\sqrt{1 - r^2} \sim t_{n-2} \quad \text{under } H_0$$

The p-value $= 2 \times Pr(t_{n-2} > t)$ if $t > 0$, or $= 2 \times Pr(t_{n-2} < t)$ if $t \leq 0$. It is mathematically equivalent to either the F test or the t test for simple linear regression discussed in Sections 11.3 and 11.4 (it always yields the same p-value).

In some instances, we expect a nonzero correlation and are interested in the magnitude (or strength) of the correlation. For example, suppose we know from large studies that the familial correlation of SBP between two sibs in the same family is .3. We measure the correlation between SBPs of twin pairs and find a correlation of .5 based on 50 twin pairs. Is this a significant difference?

We wish to test the hypothesis H_0: $\rho = \rho_0 \neq 0$, versus H_1: $\rho \neq \rho_0$. The appropriate test is called *Fisher's one-sample z-test for correlation*. The test statistic for this test is

$$\lambda = (z - z_0)\sqrt{n - 3} \sim N(0, 1) \quad \text{under } H_0$$

where $z =$ Fisher's z transformation of r

$$= .5 \ln[(1 + r)/(1 - r)] \quad \text{(given in Table 11, in the}$$
$$\text{Appendix of the text)}$$

$$z_0 = \text{Fisher's } z \text{ transformation of } \rho_0$$

The p-value $= 2 \times \Phi(\lambda)$ if $\lambda < 0$, or $= 2 \times [1 - \Phi(\lambda)]$ if $\lambda \geq 0$.

In this case, a one-sided alternative is more appropriate, since twins will only plausibly have a stronger correlation than sib pairs. We test the hypothesis H_0: $\rho = .3$ versus H_1: $\rho > .3$.

We have

$$z_0 = \frac{1}{2}\ln(1.3/.7) = .310$$

$$z = \frac{1}{2}\ln(1.5/.5) = .549$$

$$\lambda = \sqrt{47}(.549 - .310) = 1.644 \sim N(0, 1)$$

One-tailed $p = .050$

11.9.2 **Two-Sample Inference**

The tracking correlation is an index of the relationship between the same physiologic measure at different times. Suppose we measure systolic blood pressure at two points in time three years apart. We find that the correlation between the two repeated measures (the tracking correlation) = .7 for 50 people age 30–49 at baseline and .5 for 40 people age 50–69 at baseline. Suppose we wish to compare the two tracking correlations. We wish to test the hypothesis

$$H_0: \rho_1 = \rho_2 \text{ versus } H_1: \rho_1 \neq \rho_2$$

We use the test statistic

$$\lambda = \frac{z_1 - z_2}{\sqrt{\dfrac{1}{n_1 - 3} + \dfrac{1}{n_2 - 3}}} \sim N(0, 1)$$

where z_1, z_2 are Fisher's z transformation of r_1 and r_2, respectively. The p-value $= 2 \times \Phi(\lambda)$ if $\lambda < 0$, or $2 \times [1 - \Phi(\lambda)]$ if $\lambda \geq 0$.
In this case,

$$z_1 = .867 \quad z_2 = .549$$

$$\lambda = \frac{.318}{\sqrt{\dfrac{1}{47} + \dfrac{1}{37}}} = \frac{.318}{.220} = 1.447 \sim N(0, 1)$$

p-value $= 2 \times [1 - \Phi(1.447)] = 2(1 - .926) = .148$

Thus, there is no significant difference between the correlation coefficients for the two age groups.

11.9.3 **Special Types of Correlation**

(1) **Partial Correlation** The correlation between a normally distributed outcome variable y (e.g., SBP) and a single predictor variable (e.g., fruit consumption) after controlling both the outcome and predictor variable for one or more other variables in a multiple-regression model (e.g., age, sex, weight).

(2) **Multiple Correlation** The correlation between a normally distributed outcome variable y (e.g., SBP) and the linear predictor $\beta_1 x_1 + \ldots + \beta_k x_k$, where $\beta_1, \ldots, \beta_k$ are the coefficients in a multiple-regression model. The multiple correlation represents the highest possible correlation between y and a linear function of the x's. For example, in the SAS multiple-regression model shown in Section 11.8, the multiple correlation $= \sqrt{0.3885} = .62$.

(3) **Intraclass Correlation** The correlation between replicate measures obtained from the same subject, where the replicates are indistinguishable from each other. It is generally used as an index of reproducibility and is estimated using the one-way ANOVA random-effects model (see Section 9.6). It also is the same as the square of the correlation between a single replicate and the average of an infinite number of replicates obtained from a subject. In this context, it is also referred to as the *reliability coefficient* because it measures how close one replicate measure is to the "true value" for a subject, where the true value is conceived as the average of an infinite number of replicate measures. For example, we can think of the reliability of a single cholesterol measurement as how related that measurement is to the "true" cholesterol of a subject, which is the average of many cholesterol measurements obtained from the same subject.

SECTION 11.10   Multiple Logistic Regression

In Section 10.7, we described a study of the effect of surfactant on in-hospital mortality among babies with very low birthweights. Since birthweight itself is a powerful predictor of in-hospital mortality, the data were stratified by birthweight and the Mantel-Haenszel test was used to compare in-hospital mortality before and after the use of surfactant, while controlling for birthweight. The authors were also concerned with other potential risk factors for in-hospital mortality, including gender and race. How can we look at the effect of surfactant while controlling for several potential confounding variables? The authors used *multiple logistic regression* to accomplish this task.

11.10.1   Multiple Logistic-Regression Model

$$\text{logit}(p) = \ln[p/(1-p)] = \alpha + \beta_1 x_1 + \ldots + \beta_4 x_4$$

where

p = probability of in-hospital mortality

x_1 = 1 if male sex, 0 otherwise

x_2 = 1 if white race, 0 otherwise

x_3 = birthweight (g)

x_4 = 1 after surfactant became available
  = 0 before surfactant became available

11.10.2   Interpretation of Parameters

For dichotomous predictor variables, such as x_4 = surfactant availability, the regression parameter has the following interpretation: $\exp(\beta_4)$ = odds ratio relating in-hospital mortality to surfactant availability among infants that are the same for all other risk factors (the same sex, race, and birthweight). For continuous predictor variables, such as x_3 = birthweight, the regression parameter has the following interpretation: $\exp(\beta_3)$ = odds ratio relating in-hospital mortality to a 1-g difference in birthweight (i.e., the odds in favor of in-hospital mortality for the heavier infant versus the lighter infant), among babies that are the same for all other risk factors (sex, race, and surfactant availability). Since 1 g is a small increment, the authors chose a different increment for the purpose of reporting results (viz., 100 g). In this case, the odds ratio relating in-hospital mortality to a 100-g difference in birthweight = $\exp(100\beta_4)$. Thus, for both dichotomous and continuous predictor variables, the regression coefficients play the same role as do partial-regression coefficients in multiple linear regression.

11.10.3 **Hypothesis Testing**

We wish to test the hypothesis H_0: $\beta_j = 0$, all other $\beta_l \neq 0$, versus H_1: all $\beta_j \neq 0$. This is a test of the specific effect of the jth independent variable after controlling for the effect of all other variables in the model. We use the test statistic

$$z = \hat{\beta}_j / se(\hat{\beta}_j) \sim N(0, 1) \text{ under } H_0$$

The p-value $= 2 \times \Phi(z)$ if $z < 0$, or $= 2 \times [1 - \Phi(z)]$, if $z \geq 0$. The standard error of $\hat{\beta}_j$ is difficult to compute and is always obtained by computer.

The results of the study were as follows:

Variable	Odds ratio	95% CI	p-value
Male sex	1.6	(1.4–1.8)	< .001
White race	1.0	(0.9–1.2)	> .05
Surfactant availability	0.7	(0.6–0.8)	< .001
Birthweight	0.6[a]	(0.6–0.7)	< .001

[a]For a 100-g increase in birthweight.

The results indicate that the odds in favor of in-hospital mortality were 30% lower after surfactant became available than before, after controlling for the effects of sex, race, and birthweight ($p < .001$). This is similar to the results in Section 10.7 using the Mantel-Haenszel test, where only birthweight was controlled for. Also, the odds in favor of in-hospital mortality are 60% higher for boys than girls ($p < .001$). Finally, the odds in favor of in-hospital mortality are 40% lower for the heavier infant than the lighter infant of the same race, sex, and time period, where the heavier infant is 100 g heavier than the lighter infant. No significant effect of race was seen.

Similar goodness-of-fit procedures can be performed with multiple logistic regression as with linear regression, including residual analyses that can help identify outlying and influential observations.

PROBLEMS

Cardiology

The data in Table 11.2 are given for 27 patients with acute dilated cardiomyopathy [1].

11.1 Fit a regression line relating age (x) to LVEF (y).

11.2 What is the expected LVEF for a 45-year-old patient with this condition?

11.3 Test for the significance of the regression line in Problem 11.1 using the F test.

11.4 What is the R^2 for the regression line in Problem 11.1?

11.5 What does R^2 mean in Problem 11.4?

11.6 Test for the significance of the regression line in Problem 11.1 using the t test.

11.7 What are the standard errors of the slope and intercept for the regression line in Problem 11.1?

11.8 Provide 95% CI's for the slope and intercept.

11.9 What is the z transformation of .435?

TABLE 11.2 Data for patients with acute dilated cardiomyopathy

Patient number	Age, x	Left ventricular ejection fraction (LVEF), y
1	35	0.19
2	28	0.24
3	25	0.17
4	75	0.40
5	42	0.40
6	19	0.23
7	54	0.20
8	35	0.20
9	30	0.30
10	65	0.19
11	26	0.24
12	56	0.32
13	60	0.32
14	47	0.28
15	50	0.24
16	43	0.18
17	30	0.22
18	56	0.23
19	23	0.14
20	26	0.14
21	58	0.30
22	65	0.07
23	34	0.12
24	63	0.13
25	23	0.17
26	23	0.24
27	46	0.19

Note: $\Sigma x_i = 1137$ $\Sigma x_i^2 = 54{,}749$ $\Sigma y_i = 6.05$
$\Sigma y_i^2 = 1.5217$ $\Sigma x_i y_i = 262.93$

Source: Reprinted with permission of the *New England Journal of Medicine, 312*(14), 885–890, 1985.

Hypertension

Many hypertension studies involve more than one observer, and it is important to verify that the different observers are in fact comparable. A new observer and an experienced observer simultaneously take blood pressure readings on 50 people (x_i, y_i) $i = 1, \ldots, 50$, and the sample correlation coefficient is .75. Suppose the minimum acceptable population correlation is .9.

11.10 Are the sample data compatible with this population correlation?

Pathology

11.11 Assuming normality, use the data in Table 6.1 (p. 69 of this Study Guide) to test if there is any relation between total heart weight and body weight within each of the two groups.

TABLE 11.3 SBP measurements from 20 married couples

Couple	Male	Female	Couple	Male	Female
1	136	110	11	156	135
2	121	112	12	98	115
3	128	128	13	132	125
4	100	106	14	142	130
5	110	127	15	138	132
6	116	100	16	126	146
7	127	98	17	124	127
8	150	142	18	137	128
9	180	143	19	160	135
10	172	150	20	125	110

TABLE 11.4 The relationship of kallikrein to blood pressure

Observation	ln[a] (kallikrein), x	SBP z score[b] y
1	2.773	1.929
2	5.545	−1.372
3	3.434	−0.620
4	3.434	1.738
5	2.639	0.302
6	3.091	0.679
7	4.836	0.999
8	3.611	0.656
9	4.554	0.027
10	3.807	−0.057
11	4.500	1.083
12	2.639	−2.265
13	3.555	0.963
14	3.258	−1.062
15	4.605	2.771
16	3.296	−0.160
17	4.787	−0.217
18	3.401	−1.290

[a]The ln transformation was used to better normalize the underlying distribution.

[b]Instead of raw blood pressure a z score was used, which is defined as the raw blood pressure standardized for body weight and expressed in standard deviation units, with positive scores indicating high blood pressure and negative scores indicating low blood pressure.

Hypertension

Suppose that 20 married couples, each in the age group 25–34, have their systolic blood pressures taken, with the data listed in Table 11.3.

11.12 Test the hypothesis that there is no correlation between the male and female scores.

11.13 Suppose a study in the literature has found $\rho = .1$ based on a sample of size 100. Are the data consistent with this finding?

Hypertension

The level of a vasoactive material called *kallikrein* in the urine is a variable that has been associated in some studies with level of blood pressure in adults. Generally, people with low levels of kallikrein tend to have high levels of blood pressure. A study was undertaken in a group of 18 infants to see if this relationship persisted. The data in Table 11.4 were obtained.

Suppose we are given the following statistics:

$$\sum_{i=1}^{18} x_i = 67.765 \quad \sum_{i=1}^{18} x_i^2 = 267.217 \quad \sum_{i=1}^{18} y_i = 4.104$$

$$\sum_{i=1}^{18} y_i^2 = 28.767 \quad \sum_{i=1}^{18} x_i y_i = 17.249$$

11.14 Assuming that a linear relationship exists between SBP z score and ln kallikrein, derive the best-fitting linear relationship between these two variables.

11.15 Is there a significant relationship between these two variables based on your answer to Problem 11.14? Report a p-value.

11.16 In words, what does a p-value mean in the context of Problem 11.15?

11.17 Assess the goodness of fit of the regression line fitted in Problem 11.14.

Pediatrics

It is often mentioned anecdotally that very young children have a higher metabolism that gives them more energy than older children and adults. We decide to test this hypothesis by measuring the pulse rates on a selected group of children. The children are randomly chosen from a community census so that two male children are selected from each 2-year age group starting with age 0 and ending with age 21 (i.e., 0–1/2–3/4–5/.../20–21). The data are given in Table 11.5.

We are given the following basic statistics:

TABLE 11.5 Pulse rate and age in children aged 0–21

Age group	Age, x_i	Pulse rate, y_i
0–1	$\begin{cases}1\\0\end{cases}$	$\begin{cases}103\\125\end{cases}$
2–3	$\begin{cases}3\\3\end{cases}$	$\begin{cases}102\\86\end{cases}$
4–5	$\begin{cases}5\\5\end{cases}$	$\begin{cases}88\\78\end{cases}$
6–7	$\begin{cases}6\\6\end{cases}$	$\begin{cases}77\\68\end{cases}$
8–9	$\begin{cases}9\\8\end{cases}$	$\begin{cases}90\\75\end{cases}$
10–11	$\begin{cases}11\\11\end{cases}$	$\begin{cases}78\\66\end{cases}$
12–13	$\begin{cases}12\\13\end{cases}$	$\begin{cases}76\\82\end{cases}$
14–15	$\begin{cases}14\\14\end{cases}$	$\begin{cases}58\\56\end{cases}$
16–17	$\begin{cases}16\\17\end{cases}$	$\begin{cases}72\\70\end{cases}$
18–19	$\begin{cases}19\\18\end{cases}$	$\begin{cases}56\\64\end{cases}$
20–21	$\begin{cases}21\\21\end{cases}$	$\begin{cases}81\\74\end{cases}$

$$n = 22 \qquad \sum_{i=1}^{22} y_i = 1725 \qquad \sum_{i=1}^{22} y_i^2 = 140{,}933$$

$$\sum_{i=1}^{22} x_i = 233 \qquad \sum_{i=1}^{22} x_i^2 = 3345 \qquad \sum_{i=1}^{22} x_i y_i = 16{,}748$$

11.18 Suppose we hypothesize that there is a linear-regression model relating pulse rate and age. What are the assumptions for such a model?

11.19 Fit the parameters for the model in Problem 11.18.

11.20 Test the model fitted in Problem 11.19 for statistical significance.

11.21 What is the predicted pulse for an average 12-year-old child?

11.22 What is the standard error of the estimate in Problem 11.21?

11.23 Suppose John Smith is 12 years old. What is his estimated pulse rate from the regression line?

11.24 What is the standard error of the estimate in Problem 11.23?

11.25 How does John compare with other children in his age group if his actual pulse rate is 75?

11.26 What is the difference between the two predictions and standard errors in Problems 11.21–11.24?

11.27 Plot the data and assess whether the assumptions of the model are satisfied. In particular, does the linearity assumption appear to hold?

Hypertension, Genetics

Much research has been devoted to the etiology of hypertension. One general problem is to determine to what extent hypertension is a genetic phenomenon. This issue can be examined in 20 families by measuring the SBP of the mother, father, and first-born child in the family. The data are given in Table 11.6.

11.28 Test the assumption that the mother's SBP and father's SBP are uncorrelated. (We would expect a lack of correlation, since the mother and father are genetically unrelated.)

11.29 We would expect from genetic principles that the correlation between the mother's SBP and the child's SBP is .5. Can this expectation be tested?

11.30 Suppose a nonzero correlation has been found in Problem 11.29. Is there some explanation other than a genetic one for the existence of this correlation?

We would like to be able to predict the child's SBP on the basis of the parents' SBP.

11.31 Find the best-fitting linear relationship between the child's SBP and the mother's SBP.

11.32 Test for the significance of this relationship.

11.33 What would be the expected average child's SBP if the mother's SBP is 130 mm Hg?

11.34 What would be the expected average child's SBP if the mother's SBP is 150 mm Hg?

11.35 What would be the expected average child's SBP if the mother's SBP is 170 mm Hg?

11.36 Find the standard errors for the estimates in Problems 11.33–11.35.

11.37 Why are the three standard errors calculated in Problem 11.36 not the same?

The two-variable regression model with the father's SBP and the mother's SBP as independent variables and the child's SBP as the dependent variable is fitted in Table 11.7. Suppose we wish to assess the independent effects of the mother's SBP and father's SBP.

11.38 What hypotheses should be tested for this assessment?

11.39 Test the hypotheses proposed in Problem 11.38.

TABLE 11.6 Familial systolic blood-pressure (SBP) relationships

Family	SBP mother (mm Hg), y	SBP father (mm Hg), x	SBP child (mm Hg), t	
1	130	140	90	$\sum_{i=1}^{20} x_i = 2980$
2	125	120	85	
3	140	180	120	$\sum_{i=1}^{20} x_i^2 = 451{,}350$
4	110	150	100	
5	145	175	105	
6	160	120	100	$\sum_{i=1}^{20} y_i = 2620$
7	120	145	110	
8	180	160	140	$\sum_{i=1}^{20} y_i^2 = 351{,}350$
9	120	190	115	
10	130	135	105	$\sum_{i=1}^{20} x_i y_i = 390{,}825$
11	125	150	100	
12	110	125	80	$\sum_{i=1}^{20} t_i = 2030$
13	90	140	70	
14	120	170	115	$\sum_{i=1}^{20} t_i^2 = 210{,}850$
15	150	150	90	
16	145	155	90	$\sum_{i=1}^{20} x_i t_i = 305{,}700$
17	130	160	115	
18	155	125	110	$\sum_{i=1}^{20} y_i t_i = 269{,}550$
19	110	140	90	
20	125	150	100	

11.40 What are the standardized regression coefficients in Table 11.7 and what do they mean?

11.41 Assess the goodness of fit of the multiple-regression model in Table 11.7.

Obstetrics

Interest has increased in possible relationships between the health habits of the mother during pregnancy and adverse delivery outcomes. An issue that is receiving attention is the effect of marijuana usage on the occurrence of congenital malformations. One potential problem is that many other maternal factors are associated with adverse pregnancy outcomes, including age, race, socioeconomic status (SES), smoking, and so forth. Since these factors may also be related to marijuana usage, they need to be simultaneously controlled for in the analysis. Therefore, data from the Delivery Interview Program (DIP) were used, and a multiple logistic regression was performed relating the occurrence of major congenital malformations to several maternal risk factors [2]. The results are given in Table 11.8.

TABLE 11.7 Two-variable regression model predicting child's blood pressure as a function of the parental blood pressures

```
                SPSSX/PC Release 1.0

* * * *   M U L T I P L E   R E G R E S S I O N   * * * *

Equation Number 1    Dependent Variable..  CBP  CHILD BP

Variable(s) Entered on Step Number

1..    FBP      FATHER BP
2..    MBP      MOTHER BP

Analysis of Variance

                 DF    Sum of Squares   Mean Square
Regression        2       2870.08458    1435.04229
Residual         17       1934.91542     113.81855

* * * *   M U L T I P L E   R E G R E S S I O N   * * * *

Equation Number 1    Dependent Variable..  CBP  CHILD BP

Variable              B          SE B

FBP               0.41500       0.12482
MBP               0.42255       0.11852
(Constant)      -15.68925      23.65025
```

TABLE 11.8 Multiple logistic-regression analysis relating the occurrence of major congenital malformations to several risk factors, using 12,424 women in the DIP study

Variable[a]	Regression coefficient	Standard error
Marijuana usage (any frequency)	0.307	0.173
Any previous miscarriage	0.239	0.141
White race	0.191	0.161
Alcohol use in pregnancy	0.174	0.145
Age 35+	0.178	0.186
Any previous stillbirth	0.049	0.328
On welfare	0.030	0.182
Smoking 3+ cigarettes per day at delivery	−0.174	0.144
Any previous induced abortion	−0.198	0.168
Parity > 1[b]	−0.301	0.113

[a]All variables are coded as 1 if yes and 0 if no.
[b]Women with at least one previous pregnancy.

Source: Reprinted with permission of the *American Journal of Public Health,* 73(10), 1161–1164, 1983.

11.42 Perform a test to assess the significance of each of the risk factors after controlling for the other risk factors.

11.43 Compute an odds ratio and an associated 95% CI relating the presence or absence of each independent variable to the risk of malformations.

11.44 How do you interpret the effect of marijuana usage on congenital malformations, given your findings in Problems 11.42 and 11.43?

Nutrition

A study was performed looking at the reproducibility of dietary intake among a group of 181 children ages 45–60 months at baseline [3]. The children provided dietary recall data in year 1 and also in year 3. Suppose the correlation coefficient was computed for polyunsaturated fat between these two assessments and was 0.27.

11.45 What test can be used to assess if the true correlation coefficient is different from zero?

11.46 Implement the test in Problem 11.45 and report a *p*-value.

11.47 Provide a 95% CI for the true correlation (ρ).

11.48 Suppose in the same study that the correlation

coefficient for saturated fat is 0.43, and we wish to compare the correlation coefficients for saturated and polyunsaturated fat. Can the two-sample test for correlations [Equation **(11.44)**, p. 514, text] be employed to compare these correlations? Why or why not? (Do not perform the test.)

Hypertension, Nutrition

A study was performed to investigate the relationship between alcohol intake and systolic blood pressure (SBP).

Alcohol intake (g/day)	Mean SBP	sd	n
0	130.8	18.2	473
≥ 30	128.5	20.9	185

11.49 What test can be performed to compare mean SBP between the high-intake, ≥ 30 (g/day), and no-intake group, if we assume that the underlying variances of the two groups are the same? (Do not implement this test.)

11.50 It is well known that age is related to SBP, and it is suspected that age may be related to alcohol intake as well. Can an analysis be performed to look at differences in SBP between the high-intake and no-intake group, while controlling for age? Write down the type of model and the hypotheses to be tested under H_0 and H_1. (Do not implement this analysis.)

11.51 The authors also studied a low alcohol intake group (alcohol intake > 0, but < 30 g/day). Suppose we want to consider all three groups and wish to test the hypothesis that the mean SBP is the same in all three groups versus the alternative hypothesis that mean SBP is different between at least two of the groups while controlling for age. What type of analysis could be performed to answer this question (answer in 1 or 2 sentences by specifying the type of model to be used and the hypotheses to be tested). (Do not implement this analysis.)

Pulmonary Disease

In Table 6.3 (p. 70 of this Study Guide) we have provided replicate measures for each of two different methods of measuring flow rates at 50% of forced vital capacity.

11.52 Estimate the intraclass correlation coefficient between replicate measures, using the manual method, and provide a 95% CI about this estimate.

11.53 Answer the same questions posed in Problem 11.52 based on the digitizer method.

SOLUTIONS
_____

11.1

$$L_{xx} = \Sigma x_i^2 - (\Sigma x_i)^2/n$$

$$= 54{,}749 - \frac{1137^2}{27}$$

$$= 54{,}749 - 47{,}880.33 = 6868.67$$

$$L_{xy} = \Sigma x_i y_i - \frac{(\Sigma x_i)(\Sigma y_i)}{n}$$

$$= 262.93 - \frac{1137(6.05)}{27}$$

$$= 262.93 - 254.77 = 8.16$$

$$b = \frac{L_{xy}}{L_{xx}} = \frac{8.16}{6868.67} = 0.0012$$

$$a = \frac{\Sigma y_i - b\Sigma x_i}{n}$$

$$= \frac{6.05 - 0.0012(1137)}{27} = \frac{6.05 - 1.350}{27}$$

$$= \frac{4.700}{27} = 0.174$$

Thus, the regression line is given by $y = 0.174 + 0.0012x$.

11.2 The expected LVEF $= 0.174 + 0.0012(45) = 0.174 + 0.053 = 0.228$.

11.3 First compute L_{yy}.

$$L_{yy} = \Sigma y_i^2 - \frac{(\Sigma y_i)^2}{n}$$

$$= 1.5217 - \frac{6.05^2}{27} = 1.522 - 1.356$$

$$= 0.166$$

Now compute the regression and residual sum of squares and mean square:

$$\text{Reg SS} = \frac{L_{xy}^2}{L_{xx}} = \frac{8.16^2}{6868.67} = 0.0097 = \text{Reg MS}$$

$$\text{Res SS} = 0.166 - 0.0097 = 0.156$$

$$\text{Res MS} = \frac{0.156}{25} = 0.0063$$

Finally, the F statistic is given by $F = \text{Reg MS}/\text{Res MS} = 0.0097/0.0063 = 1.55 \sim F_{1,25}$ under H_0. Since $F_{1,25,.95} > F_{1,30,.95} = 4.17 > 1.55$, it follows that $p > .05$. Therefore, H_0 is accepted and we conclude that there is no significant relationship between LVEF and age.

11.4 $R^2 = \text{Reg SS}/\text{Total SS} = 0.0097/0.166 = 0.058$

11.5 It means that only 5.8% of the variance in LVEF is explained by age.

11.6 Compute the standard error of the estimated slope as follows:

$$se(b) = \sqrt{\frac{\text{Res MS}}{L_{xx}}} = \sqrt{\frac{0.0063}{6868.67}}$$

$$= 0.00095$$

Thus, the t statistic is given by

$$t = \frac{b}{se(b)} = \frac{0.0012}{0.00095}$$

$$= 1.24 \sim t_{25} \text{ under } H_0$$

Since $t_{25,.975} = 2.060 > t$, it follows that $p > .05$, and H_0 is accepted at the 5% level.

11.7 $se(b) = 0.00095$ as given in Problem 11.6. The standard error of the intercept is given by

$$se(a) = \sqrt{\text{Res MS}\left(\frac{1}{n} + \frac{\bar{x}^2}{L_{xx}}\right)}$$

$$= \sqrt{0.0063\left[\frac{1}{27} + \frac{(1137/27)^2}{6868.67}\right]}$$

$$= \sqrt{0.0063(0.0370 + 0.2582)}$$

$$= \sqrt{0.0063(0.2952)} = 0.043$$

11.8 A 95% CI for the slope is given by

$$b \pm t_{25,.975}\, se(b) = 0.0012 \pm 2.060(0.00095)$$

$$= 0.0012 \pm 0.0020$$

$$= (-0.0008, 0.0032)$$

A 95% CI for the intercept is given by

$$a \pm t_{25,.975}\, se(a) = 0.174 \pm 2.060(0.043)$$

$$= 0.174 \pm 0.089 = (0.086, 0.263)$$

11.9 We have that

$$z = .5 \times \ln\left(\frac{1 + .435}{1 - .435}\right) = .5 \times \ln\left(\frac{1.435}{0.565}\right)$$

$$= .5 \times \ln(2.540) = .5(0.932) = 0.466$$

11.10 We wish to test the hypothesis H_0: $\rho = .9$ versus H_1: $\rho < .9$. We use the one-sample z test with test statistic

$$\lambda = \sqrt{n-3}(z - z_0) \sim N(0, 1) \quad \text{under } H_0$$

where from Table 11 (in the Appendix of the text)

$$z = .5 \times \ln\left(\frac{1 + .75}{1 - .75}\right) = 0.973$$

$$z_0 = .5 \times \ln\left(\frac{1 + .9}{1 - .9}\right) = 1.472$$

We have $\lambda = \sqrt{47}(0.973 - 1.472) = 6.856(-0.499) = -3.421 \sim N(0, 1)$ under H_0. The p-value $= \Phi(-3.421) < .001$. Thus, the observed correlation is significantly lower than .9, indicating that the new observer does *not* meet the standard.

11.11   Compute the correlation coefficient between THW and BW in each of the two groups. We first have the following summary statistics ($x = $ THW, $y = $ BW):

Left-heart disease

$$\sum_{i=1}^{11} x_i = 4950$$

$$\sum_{i=1}^{11} x_i^2 = 2,421,650$$

$$\sum_{i=1}^{11} y_i = 611.7$$

$$\sum_{i=1}^{11} y_i^2 = 35,350.49$$

$$\sum_{i=1}^{11} x_i y_i = 280,031.5$$

$L_{xx} = 194,150$

$L_{yy} = 1334.41$

$L_{xy} = 4766.5$

$r = L_{xy}/(L_{xx}L_{yy})^{1/2}$

$\quad = .296$

$t = r\sqrt{n-2}/\sqrt{1-r^2}$

$\quad = .296(3)/\sqrt{1-.296^2}$

$\quad = 0.93 \sim t_9$, NS

Normal

$$\sum_{i=1}^{10} x_i = 3170$$

$$\sum_{i=1}^{10} x_i^2 = 1,024,850$$

$$\sum_{i=1}^{10} y_i = 562.3$$

$$\sum_{i=1}^{10} y_i^2 = 32,816.33$$

$$\sum_{i=1}^{10} x_i y_i = 181,462$$

$L_{xx} = 19,960$

$L_{yy} = 1198.20$

$L_{xy} = 3212.9$

$r = L_{xy}/(L_{xx}L_{yy})^{1/2}$

$\quad = .657$

$t = r\sqrt{n-2}/\sqrt{1-r^2}$

$\quad = .657\sqrt{8}/\sqrt{1-.657^2}$

$\quad = 2.465 \sim t_8,$

$\quad .02 < p < .05$

Thus, there is a significant association between total heart weight and body weight in the normal group, but not in the left-heart disease group.

11.12   If $x = $ male, $y = $ female, then we have that $r = L_{xy}/\sqrt{L_{xx} \times L_{yy}}$, where $L_{xx} = 8923.8$, $L_{yy} = 4412.95$, $L_{xy} = 4271.9$. Therefore,

$$r = \frac{4271.9}{\sqrt{8923.8 \times 4412.95}} = .681$$

We use the t test with the following test statistic:

$$t = \frac{r\sqrt{n-2}}{\sqrt{1-r^2}} = \frac{.681\sqrt{18}}{\sqrt{1-0.681^2}}$$

$$= \frac{2.888}{0.733} = 3.94 \sim t_{18} \quad \text{under } H_0$$

Since $t_{18, .9995} = 3.922 < 3.94$, we have that $p < .001$. Thus, there is a significant correlation between spouse blood pressures.

11.13   Let $\rho_1 = $ our population correlation coefficient, $\rho_2 = $ literature population correlation coefficient. We wish to test the hypothesis H_0: $\rho_1 = \rho_2$ versus H_1: $\rho_1 \neq \rho_2$. We have the test statistic

$$\lambda = \frac{z_1 - z_2}{\sqrt{\dfrac{1}{n_1 - 3} + \dfrac{1}{n_2 - 3}}} \sim N(0, 1) \text{ under } H_0$$

We have

$$z_1 = .5 \times \ln\left(\frac{1 + .681}{1 - .681}\right) = 0.830$$

$$z_2 = .5 \times \ln\left(\frac{1 + .10}{1 - .10}\right) = 0.100$$

Thus,

$$\lambda = \frac{0.830 - 0.100}{\sqrt{\dfrac{1}{17} + \dfrac{1}{97}}}$$

$$= \frac{0.730}{0.263} = 2.777 \sim N(0, 1) \text{ under } H_0$$

The p-value $= 2 \times [1 - \Phi(\lambda)] = 2 \times [1 - \Phi(2.777)] = 2 \times (1 - .9973) = .005$. Thus, we can say that our correlation is significantly different from the correlation in the literature.

11.14   We fit a least-squares line of the form $y = a + bx$ where $b = L_{xy}/L_{xx}$, $a = \bar{y} - b\bar{x}$. We have that

$$L_{xx} = \Sigma x_i^2 - (\Sigma x_i)^2/18$$
$$= 267.217 - 67.765^2/18 = 12.101$$

$$L_{yy} = \Sigma y_i^2 - (\Sigma y_i)^2/18$$
$$= 28.767 - 4.104^2/18 = 27.831$$

$$L_{xy} = \Sigma x_i y_i - (\Sigma x_i)(\Sigma y_i)/18$$
$$= 17.249 - 67.765(4.104)/18 = 1.799$$

Thus,

$$b = 1.799/12.101 = 0.149$$

$$a = \frac{\Sigma y_i - b\Sigma x_i}{18}$$

$$= \frac{4.104 - 0.149 \times 67.765}{18} = -0.332$$

The best-fitting linear relationship is $y = -0.332 + 0.149x$.

11.15   We wish to test the hypothesis H_0: $\beta = 0$ versus H_1: $\beta \neq 0$. We use the test statistic $F = $ Reg MS/Res MS $\sim F_{1, n-2}$ under H_0. We have

$$\text{Reg SS} = \frac{L_{xy}{}^2}{L_{xx}}$$

$$= \frac{1.799^2}{12.101} = 0.267 = \text{Reg MS}$$

Total SS $= L_{yy} = 27.831$

Res SS $= $ Tot SS $-$ Reg SS $= 27.831 - 0.267$

$$= 27.564$$

Res MS $= \dfrac{\text{Res SS}}{n-2}$

$$= \frac{27.564}{16} = 1.723$$

Thus, $F = $ Reg MS/Res MS $= 0.267/1.723 = 0.155 \sim$ $F_{1,16}$ under H_0. We see from the F table that $F_{1,16,.90} = 3.05 > F$. Thus, $p > .10$ and there is no significant relationship between these two variables in this data set.

11.16 A p-value in this context is the probability of obtaining an estimated slope at least as large (in absolute value) as the one obtained between blood pressure and kallikrein, when in fact no relationship exists.

11.17 We use MINITAB to run the regression of SBP z score on ln(kallikrein). A scatter plot of the raw data (labeled A) and the fitted regression line (labeled B) are given as follows. They do not reveal any outliers, influential points, or systematic departures from the linear-regression assumptions. This is reinforced by the plot of the Studentized residuals (SRES1) versus ln(kallikrein).

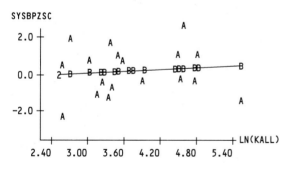

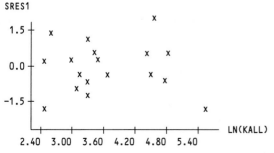

11.18 The underlying model here is $y_i = \alpha + \beta x_i + e_i$ where $e_i \sim N(0, \sigma^2)$, the e_i are independent random variables, $y_i = $ pulse rate for the ith child, and $x_i = $ age for the ith child.

11.19 We first compute

$$L_{xy} = 16{,}748 - \frac{233(1725)}{22} = -1521.32$$

$$L_{xx} = 3345 - \frac{233^2}{22} = 877.32$$

$$L_{yy} = 140{,}933 - \frac{1725^2}{22} = 5677.32$$

We then have

$$b = \frac{L_{xy}}{L_{xx}} = \frac{-1521.32}{877.32} = -1.73$$

$$a = \frac{\Sigma y_i - b\Sigma x_i}{22} = \frac{1725 + 1.73(233)}{22} = 96.8$$

The least-squares line is thus $y = 96.8 - 1.73x$.

11.20 We test the hypothesis $H_0: \beta = 0$ versus $H_1: \beta \neq 0$. We will use the F test for statistical significance. We have that

$$\text{Reg SS} = \frac{L_{xy}{}^2}{L_{xx}} = \frac{(-1521.32)^2}{877.32} = 2638.05$$

Total SS $= L_{yy} = 5677.32$

Res SS $= 5677.32 - 2638.05 = 3039.27$

We thus have the following ANOVA table:

	SS	df	MS	F-stat
Regression	2638.05	1	2638.05	17.36
Residual	3039.27	20	151.96	
Total	5677.32	21		

We compute $F = $ Reg MS/Res MS $= 17.36 \sim F_{1,20}$ under H_0. Since $F > F_{1,20,.999} = 14.82$, we have that $p < .001$ and we conclude that there is a highly significant correlation between pulse rate and age.

11.21 From the regression equation, we have that $\hat{y} = 96.8 - 1.73(12) = 76.0$ beats per minute.

11.22 The standard error is given by

$$\sqrt{s_{y \cdot x}^2 \left[\frac{1}{n} + \frac{(x - \bar{x})^2}{L_{xx}}\right]}$$

$$= \sqrt{151.96\left[\frac{1}{22} + \frac{(12 - 233/22)^2}{877.32}\right]}$$

$$= \sqrt{151.96(0.048)} = 2.69$$

11.23 The estimated pulse is the same as for an average 12-year-old = 76.0 beats per minute (as given in the solution to Problem 11.21).

11.24 The standard error is given by

$$se = \sqrt{s_{y \cdot x}^2 \left[1 + \frac{1}{n} + \frac{(x - \bar{x})^2}{L_{xx}}\right]}$$

$$= \sqrt{151.96\left[1 + \frac{1}{22} + \frac{(12 - 233/22)^2}{877.32}\right]}$$

$$= \sqrt{151.96(1.048)} = 12.62$$

11.25 John is about average for children in his age group since his pulse is 75 while the expected pulse is 76.0, with a standard error of 12.62.

11.26 The predicted pulse is the same for a particular 12-year-old and for an average 12-year-old (76.0). However, the standard error for a particular 12-year-old (12.62) is much larger than for an average 12-year-old (2.69).

11.27 A plot of the data in Table 11.5 is provided as follows. The fit of the regression line appears adequate. There is one mild outlier corresponding to a pulse of 125 at age 0.

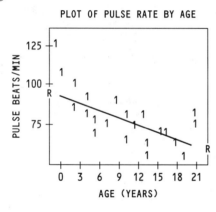

PLOT OF PULSE RATE BY AGE

11.28 Test the hypothesis H_0: $\rho = 0$ versus H_1: $\rho \neq 0$. Use the test statistic

$$t = \frac{r\sqrt{n - 2}}{\sqrt{1 - r^2}} \sim t_{n-2} \text{ under } H_0$$

In this case

$$r = \frac{L_{xy}}{\sqrt{L_{xx}L_{yy}}}$$

We have

$$L_{xx} = 451{,}350 - \frac{2980^2}{20} = 7330$$

$$L_{yy} = 351{,}350 - \frac{2620^2}{20} = 8130$$

$$L_{xy} = 390{,}825 - \frac{2980(2620)}{20} = 445$$

Thus, $\qquad r = \frac{445}{\sqrt{7330 \times 8130}} = \frac{445}{7719.64} = .058$

The test statistic

$$t = \frac{.058\sqrt{18}}{\sqrt{1 - .058^2}} = \frac{.245}{.998} = .245 \sim t_{18}$$

Since $t_{18, .975} = 2.101$, this is clearly not statistically significant, and there is no significant correlation between the mother's and father's SBP, which is what would be expected on purely genetic grounds.

11.29 Compute

$$r_{yt} = \frac{L_{yt}}{\sqrt{L_{yy} \times L_{tt}}}$$

We have

$$L_{tt} = 210{,}850 - \frac{2030^2}{20} = 4805$$

$$L_{yt} = 269{,}550 - \frac{2620(2030)}{20} = 3620$$

and

$$r = \frac{3620}{\sqrt{8130 \times 4805}} = \frac{3620}{6250.17} = .579$$

Test the null hypothesis H_0: $\rho = .5$ versus the alternative H_1: $\rho \neq .5$. Use the Fisher z statistic, which yields

$$z = \frac{1}{2}\left[\ln(1 + r) - \ln(1 - r)\right]$$

$$= \frac{1}{2}\left[\ln(1.579) - \ln(0.421)\right]$$

$$= \frac{1}{2}\left(0.4569 + 0.8656\right) = 0.661$$

Also, from Table 11 (Appendix, text), $z_0 = 0.549$. Use the test statistic

$$\lambda = (z - z_0)\sqrt{n - 3} \sim N(0, 1) \text{ under } H_0$$

Thus,

$$\lambda = \sqrt{17}(0.661 - 0.549) = 0.461 \sim N(0, 1)$$

The p-value $= 2 \times [1 - \Phi(0.461)] = .64$. Thus, we accept H_0 that $\rho = .5$.

11.30 The mother and first-born child were both living in the same environment, which might explain all or part of the observed correlation of blood pressure.

11.31 Use the model $t_i = \alpha + \beta y_i + e_i$. Test the null hypothesis that $\beta = 0$ versus the alternative hypothesis that $\beta \neq 0$.

$$b = \frac{L_{ty}}{L_{yy}}$$

where

$$L_{ty} = 3620$$

$$L_{yy} = 8130 \quad \text{from Problems 11.28 and 11.29}$$

Thus, $b = \frac{3620}{8130} = 0.445$

Furthermore,

$$a = \frac{\sum\limits_{i=1}^{20} t_i - b\sum\limits_{i=1}^{20} y_i}{20} = \frac{2030 - 0.445(2620)}{20}$$

$$= 43.2$$

We thus have the linear relation $t = 43.2 + 0.445y$.

11.32 Test the hypothesis $H_0: \beta = 0$ versus $H_1: \beta \neq 0$. Use the test statistic

$$F = \frac{\text{Reg MS}}{\text{Res MS}} \sim F_{1,n-2} \text{ under } H_0$$

We have

$$\text{Reg MS} = \frac{L_{ty}^2}{L_{yy}} = \frac{3620^2}{8130} = 1611.86 = \text{Reg SS}$$

$$\text{Res MS} = \frac{L_{tt} - \text{Reg SS}}{18} = \frac{L_{tt} - 1611.86}{18}$$

Thus, $\text{Res MS} = \frac{4805 - 1611.86}{18} = 177.40 = s_{t \cdot y}^2$

and $F = \frac{1611.86}{177.40} = 9.09 \sim F_{1, 18}$ under H_0

Since $F_{1,18,.99} = 8.29$, $F_{1,18,.995} = 10.22$, $.005 < p < .01$, and there is a significant relationship between the mother's and child's SBP.

11.33 The child's SBP can be predicted from the mother's SBP from the relationship in Problem 11.31; $E(t) = 43.2 + 0.445 \times 130 = 101.1$ mm Hg.

11.34 In this case $E(t) = 43.2 + 0.445 \times 150 = 110.0$ mm Hg.

11.35 In this case $E(t) = 43.2 + 0.445 \times 170 = 118.9$ mm Hg.

11.36 The standard errors are given by the formula

$$se = \sqrt{s_{t \cdot y}^2 \left[\frac{1}{n} + \frac{(y - \bar{y})^2}{L_{yy}} \right]}$$

$$= \sqrt{177.40 \left[\frac{1}{20} + \frac{(y - \bar{y})^2}{8130} \right]}$$

We have

$$\bar{y} = \frac{2620}{20} = 131 \text{ mm Hg}$$

Thus, for $y = 130$, 150, and 170 mm Hg, we have, respectively,

$$se(130) = \sqrt{177.40 \left[\frac{1}{20} + \frac{(130 - 131)^2}{8130} \right]} = 2.98$$

$$se(150) = \sqrt{177.40 \left[\frac{1}{20} + \frac{(150 - 131)^2}{8130} \right]} = 4.09$$

$$se(170) = \sqrt{177.40 \left[\frac{1}{20} + \frac{(170 - 131)^2}{8130} \right]} = 6.49$$

11.37 The three standard errors are not the same because the standard error increases the further the value of the mother's SBP is from the mean SBP for all mothers (131 mm Hg). Thus, the se corresponding to 130 mm Hg is smallest, whereas the se corresponding to 170 mm Hg is largest.

11.38 To test for the independent effect of the father's SBP on the child's SBP after controlling for the effect of the mother's SBP, test the hypothesis $H_0: \beta_1 = 0$, $\beta_2 \neq 0$ versus $H_1: \beta_1 \neq 0$, $\beta_2 \neq 0$. Similarly, to test for the independent effect of the mother's SBP on the child's SBP after controlling for the effect of the father's SBP, test the hypothesis $H_0: \beta_2 = 0$, $\beta_1 \neq 0$ versus $H_1: \beta_2 \neq 0$, $\beta_1 \neq 0$.

11.39 Compute the following test statistics:

Father: $t_1 = b_1/se(b_1) = 0.4150/0.1248$

$\qquad = 3.32 \sim t_{17}, \ .001 < p < .01$

Mother: $t_2 = b_2/se(b_2) = 0.4226/0.1185$

$\qquad = 3.57 \sim t_{17}, \ .001 < p < .01$

Thus, each parent's SBP is significantly associated with the child's SBP after controlling for the other spouse's SBP.

11.40 From Problems 11.28 and 11.29,

$$s_t^2 = \frac{L_{tt}}{n-1} = \frac{4805}{19} = 252.89$$

$$s_x^2 = \frac{L_{xx}}{n-1} = \frac{7330}{19} = 385.79$$

$$s_y^2 = \frac{L_{yy}}{n-1} = \frac{8130}{19} = 427.89$$

The standardized regression coefficients are given by

Father: $b_{s_1} = b_1 \times s_x/s_t$

$\qquad = 0.4150 \times \sqrt{\dfrac{385.79}{252.89}} = 0.513$

Mother: $b_{s_2} = b_2 \times s_y/s_t$

$\qquad = 0.4226 \times \sqrt{\dfrac{427.89}{252.89}} = 0.550$

The standardized coefficients represent the increase in the number of standard deviations in the child's SBP that would be expected per standard-deviation increase in the parent's SBP after controlling for the other spouse's SBP.

11.41 We have rerun the multiple-regression model in Table 11.7 using MINITAB, with results given as follows:

The regression equation is

$$SBP_{child} = -15.7 + 0.423\, SBP_{mother} + 0.415\, SBP_{father}$$

```
--------------------------------------------------------------
       PREDICTOR        COEF       STDEV     T-RATIO        P
       CONSTANT       -15.69       23.65       -0.66    0.516
       SBP-MOTH       0.4225      0.1185        3.57    0.002
       SBP-FATH       0.4150      0.1248        3.32    0.004

       S = 10.67      R-SQ = 59.7%      R-SQ(ADJ) = 55.0%

       ANALYSIS OF VARIANCE

       SOURCE        DF          SS         MS          F        P
       REGRESSION     2      2870.1     1435.0      12.61    0.000
       ERROR         17      1934.9      113.8
       TOTAL         19      4805.0
--------------------------------------------------------------
```

To assess goodness of fit, we first plot the Studentized residuals from the multiple-regression model (SRES1) versus the mother's and father's SBP, respectively (see the following figures). The plots do not reveal any violation of regression assumptions.

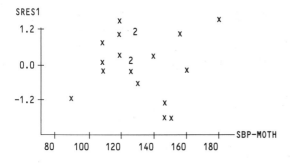

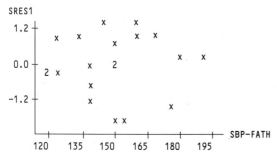

We also obtain partial-residual plots to assess the relationship between the child's SBP and one parent's SBP, after controlling each for the effects of the other parent's SBP. Since the Student version of MINITAB does not support partial-residual plots, we construct them ourselves. Specifically, to obtain the partial-residual plot of child's SBP on mother's SBP, we (1) run a regression of child's SBP on father's SBP and obtain residuals (labeled RESC_F), (2) run a regression of mother's SBP on father's SBP and obtain residuals (labeled RESM_F), and (3) generate a scatter plot of RESC_F on RESM_F. In a similar manner, we obtain a partial-residual plot of child's SBP on father's SBP after controlling for mother's SBP. Both plots appear approximately linear with no outliers or violation of regression assumptions.

Partial-residual plot of child's SBP on mother's SBP after controlling for father's SBP

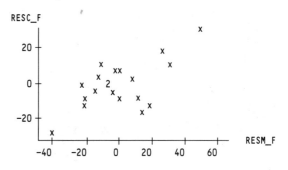

Partial-residual plot of child's SBP on father's SBP after controlling for mother's SBP

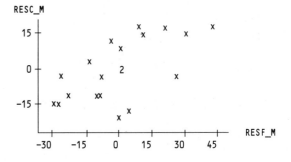

11.42 We wish to test $H_0: \beta_i = 0$ versus $H_1: \beta_i \neq 0$ for each of the risk factors. We will use the test statistic $z_i = \hat{\beta}_i/se(\hat{\beta}_i) \sim N(0,1)$ under H_0 to test for the effect of each variable after simultaneously controlling for the effects of all other variables. We have the following table:

Variable	$\hat{\beta}_i / se(\hat{\beta}_i)$	Two-tail p-value
Marijuana usage	1.775	.076
Any previous miscarriage	1.695	.090
White race	1.186	.24
Alcohol use in pregnancy	1.200	.23
Age 35+	0.957	.34
Any previous stillbirth	0.149	.88
On welfare	0.165	.87
Smoking 3+ cigarettes per day at delivery	−1.208	.23
Any previous induced abortion	−1.179	.24
Parity > 1	−2.664	.008

The only risk factor significantly affecting the risk of congenital malformations is parity, with children born from women with no previous pregnancies at higher risk than children born from women who have had at least 1 previous pregnancy. In addition, marijuana usage at the current pregnancy and previous miscarriages showed a trend toward increasing the risk of congenital malformations, but were not statistically significant after controlling for the other risk factors.

11.43 The estimated odds ratios are given by $\exp(\hat{\beta}_i)$ and the associated 95% CI's by $\exp[\hat{\beta}_i \pm 1.96\, se(\hat{\beta}_i)]$. These are given in Table 11.9.

TABLE 11.9

Variable	Odds ratio	95% CI
Marijuana usage	$\exp(0.307)$ = 1.36	$\exp[0.307 \pm 1.96(0.173)]$ = (0.97, 1.91)
Any previous miscarriage	$\exp(0.239)$ = 1.27	$\exp[0.239 \pm 1.96(0.141)]$ = (0.96, 1.67)
White race	$\exp(0.191)$ = 1.21	$\exp[0.191 \pm 1.96(0.161)]$ = (0.88, 1.66)
Alcohol use in pregnancy	$\exp(0.174)$ = 1.19	$\exp[0.174 \pm 1.96(0.145)]$ = (0.90, 1.58)
Age 35+	$\exp(0.178)$ = 1.19	$\exp[0.178 \pm 1.96(0.186)]$ = (0.83, 1.72)
Any previous stillbirth	$\exp(0.049)$ = 1.05	$\exp[0.049 \pm 1.96(0.328)]$ = (0.55, 2.00)
On welfare	$\exp(0.030)$ = 1.03	$\exp[0.030 \pm 1.96(0.182)]$ = (0.72, 1.47)
Smoking 3+ cigarettes per day at delivery	$\exp(-0.174)$ = 0.84	$\exp[-0.174 \pm 1.96(0.144)]$ = (0.63, 1.11)
Any previous induced abortion	$\exp(-0.198)$ = 0.82	$\exp[-0.198 \pm 1.96(0.168)]$ = (0.59, 1.14)
Parity > 1	$\exp(-0.301)$ = 0.74	$\exp[-0.301 \pm 1.96(0.113)]$ = (0.59, 0.92)

11.44 Marijuana usage in the current pregnancy shows a trend toward statistical significance with odds ratio = 1.36 after controlling for all other risk factors. This means that the odds in favor of a congenital malformation are 36% higher for offspring of users of marijuana compared with offspring of nonusers of marijuana, all other factors held constant. However, we should be cautious in interpreting these findings, due to the borderline nature of the results.

11.45 The one-sample t test for correlation.

11.46 We test the hypothesis H_0: $\rho = 0$ versus H_1: $\rho \neq 0$. We have the test statistic

$$t = r\sqrt{n - 2}/\sqrt{1 - r^2}$$
$$= 0.27\sqrt{179}/\sqrt{1 - .27^2}$$
$$= 3.612/0.963 = 3.75 \sim t_{179}$$

Since $t > t_{120,.9995} = 3.373 > t_{179,.9995}$, it follows that $p < .001$. Therefore, we reject H_0 and conclude that the correlation coefficient between polyunsaturated fat intake at years 1 and 3 is significantly greater than zero.

11.47 A 95% CI for $\rho = (\rho_1, \rho_2)$, where

$$\rho_1 = [\exp(2z_1) - 1]/[\exp(2z_1) + 1]$$
$$\rho_2 = [\exp(2z_2) - 1]/[\exp(2z_2) + 1]$$
$$z_1 = z - 1.96\sqrt{1/(n_1 - 3)}$$
$$z_2 = z + 1.96\sqrt{1/(n_2 - 3)}$$

From Table 11 (Appendix, text), the z transform corresponding to $r = .27$ is 0.277. Thus,

$$z_1 = 0.277 - 1.96\sqrt{1/178}$$
$$= 0.277 - 0.147 = 0.130$$
$$z_2 = 0.277 + 0.147 = 0.424$$

Thus, $\quad \rho_1 = [\exp(0.260) - 1]/[\exp(0.260) + 1]$
$$= 0.297/2.297 = .129$$
$$\rho_2 = [\exp(0.848) - 1]/[\exp(0.848) + 1]$$
$$= 1.335/3.335 = .400$$

Thus, the 95% CI for $\rho = (.129, .400)$

11.48 No. The two-sample z test can only be used for independent samples. Since in this case the samples are derived from the same subjects, they are not independent samples.

11.49 The two-sample t test with equal variances.

11.50 We can use a multiple-regression model of the form

$$y_i = \alpha + \beta_1 x_{1i} + \beta_2 x_{2i} + e_i$$

where $\quad y_i$ = SBP for the ith subject

$\quad\quad x_{1i}$ = age for the ith subject

$\quad\quad x_{2i}$ = alcohol-intake group for the ith subject

$\quad\quad\quad$ = 1 if alcohol intake $\geq$ 30 g/day

$\quad\quad\quad$ = 0 if alcohol intake = 0 g/day

$\quad\quad e_i \sim N(0, \sigma^2)$

We wish to test the hypothesis H_0: $\beta_2 = 0$, $\beta_1 \neq 0$ versus H_1: $\beta_2 \neq 0$, $\beta_1 \neq 0$. β_2 is the partial-regression coefficient for alcohol intake and represents the difference in mean SBP between the high-intake and no-intake groups for subjects of the same age.

11.51 We can use a multiple-regression model with two indicator variables for alcohol intake of the form

$$y_i = \alpha + \beta_1 x_{1i} + \beta_2 x_{2i} + \beta_3 x_{3i} + e_i$$

where $\quad y_i$ = SBP for the ith subject

$\quad\quad x_{1i}$ = age for the ith subject

$\quad\quad x_{2i}$ = 1 if alcohol intake $\geq$ 30 g/day
$\quad\quad\quad\quad$ for the ith subject

$\quad\quad\quad$ = 0 otherwise

$\quad\quad x_{3i}$ = 1 if alcohol intake > 0 and < 30 g/day
$\quad\quad\quad\quad$ for the ith subject

$\quad\quad\quad$ = 0 otherwise

$\quad\quad e_i \sim N(0, \sigma^2)$

We wish to test the hypothesis H_0: $\beta_2 = \beta_3 = 0$, $\beta_1 \neq 0$ versus H_1: at least one of $(\beta_2, \beta_3) \neq 0$, $\beta_1 \neq 0$. β_2 = the mean difference in SBP between the high-intake and the no-intake group for subjects of the same age. β_3 = the mean difference in SBP between the low-intake and the no-intake group for subjects of the same age. The models in this problem and in Problem 11.50 are also referred to as *analysis-of-covariance models*.

11.52 We first run a one-way ANOVA based on the manual-method replicates. The results, using MINITAB, are given as follows:

```
---------------------------------------------------
ANALYSIS OF VARIANCE ON MANUAL
SOURCE    DF        SS       MS       F       P
PERSON     9  11.71510  1.30168  452.76   0.000
ERROR     10   0.02875  0.00287
TOTAL     19  11.74385
---------------------------------------------------
```

From Equation **(9.22)** (text, p. 325), we have $\hat{\sigma}^2 = 0.00287$, $\hat{\sigma}_A^2 = (1.30168 - 0.00287)/2 = 0.64941$. The estimated intraclass correlation = $\hat{\rho}_{I,\text{manual}} = 0.64941/(0.64941 + 0.00287) = 0.64941/0.65228 = 0.996$. From Equation **(11.47)** (text, p. 519), a 95% CI for $\hat{\rho}_{I,\text{manual}}$ is given by (c_1, c_2), where

$$c_1 = \max[(F/F_{9,10,.975} - 1)/(2 + F/F_{9,10,.975} - 1), 0]$$
$$c_2 = \max[(F/F_{9,10,.025} - 1)/(2 + F/F_{9,10,.025} - 1), 0]$$

Using MINITAB, we estimate that $F_{9,10,.975} = 3.78$ and $F_{9,10,.025} = 0.253$. Therefore,

$$c_1 = \max[(452.76/3.78 - 1)/(452.76/3.78 + 1), 0]$$
$$= 118.78/120.78 = .983$$
$$c_2 = \max[(452.76/0.253 - 1)/(452.76/0.253 + 1), 0]$$
$$= 1788.57/1790.57 = .999$$

Therefore, the 95% CI for $\hat{\rho}_{I,\text{manual}} = (.983, .999)$.

11.53 We run a one-way ANOVA based on the digitizer-method replicates as follows:

```
------------------------------------------------
ANALYSIS OF VARIANCE ON DIGITIZER
SOURCE  DF      SS        MS       F       P
PERSON   9   12.67584  1.40843  3658.26  0.000
ERROR   10    0.00385  0.00038
TOTAL   19   12.67969
------------------------------------------------
```

We have $\hat{\sigma}^2 = 0.00038$, $\hat{\sigma}_A^2 = (1.40843 - 0.00038)/2 = 0.704025$. Thus, $\hat{\rho}_{I,digitizer} = 0.704025/(0.704025 + 0.00038) = 0.704025/0.704405 = 0.9995$. From Equation

(11.47) (text, p. 519) a 95% CI for $\hat{\rho}_{I,digitizer} = (c_1, c_2)$, where

$c_1 = \max[(F/F_{9,10,.975} - 1)/(F/F_{9,10,.975} + 1), 0]$

$c_2 = \max[(F/F_{9,10,.025} - 1)/(F/F_{9,10,.025} + 1), 0]$

Thus,

$c_1 = \max[(3658.26/3.78 - 1)/(3658.26/3.78 + 1), 0]$
$\quad = 966.79/968.79 = .9979$

$c_2 = \max[(3658.26/0.253 - 1)/(3658.26/0.253 + 1), 0]$
$\quad = 14,458.53/14,460.53 = .9999$

Therefore, the 95% CI for $\hat{\rho}_{I,digitizer} = (.9979, .9999)$.

REFERENCES

[1] Dec, G. W., Palacios, I. F., Fallon, J. T., Aretz, H. T., Mills, J., Lee, D. C. S., & Johnson, R. A. (1985). Active myocarditis in the spectrum of acute dilated cardiomyopathies: Clinical features, histologic correlates and clinical outcome. *New England Journal of Medicine, 312*(14), 885–890.

[2] Linn, S., Schoenbaum, S. C., Monson, R. R., Rosner, B., Stubblefield, P. C., & Ryan, K. J. (1983). The association of marijuana use with outcome of pregnancy. *American Journal of Public Health, 73*(10), 1161–1164.

[3] Stein, A. D., Shea, S., Basch, C. E., Contento, I. R., & Zybert, P. (1991, December). Variability and tracking of nutrient intakes of preschool children based on multiple administrations of the 24-hour dietary recall. *American Journal of Epidemiology, 134*(12), 1427–1437.

CHAPTER TWELVE

NONPARAMETRIC METHODS

SECTION 12.1   Types of Data

(1) Cardinal data Data on a scale where addition is meaningful (e.g., change in 3 inches for height). There are two types of cardinal data:

    (a) Ratio-scale data Cardinal data on a scale where ratios between values are meaningful (e.g., serum-cholesterol levels).

    (b) Interval-scale data Cardinal data where the zero point is arbitrary. For such data, ratios are not meaningful (e.g., Julian dates; we can calculate the number of days between two dates, but we can't say that one date is twice as large as another date).

(2) Ordinal data Categorical data that can be ordered, but the increment between specific values is arbitrary (e.g., degree of redness of an eye as determined on a photograph in comparison with reference photographs).

(3) Nominal data Categorical data where the categories are not ordered (e.g., ethnic group).

In this chapter, we are concerned with nonparametric methods (methods that do not assume a specific underlying distribution) and will apply these methods to either cardinal or ordinal data.

SECTION 12.2   The Sign Test

12.2.1   Large-Sample Test

Suppose we measure the change in periodontal status among 28 adults with periodontal disease after a dental-education program. The distribution of change scores is as follows:

	Category	Frequency
+3	greatest improvement	4
+2	moderate improvement	5
+1	slight improvement	6
0	no change	5
−1	slight decline	4
−2	moderate decline	2
−3	greatest decline	2

If we eliminate the "no change" people, then 15 people improved and 8 declined. Suppose we wish to determine whether more people improved than declined with the program. We wish to test the hypothesis

$$H_0: p = \frac{1}{2}$$

$$H_1: p \neq \frac{1}{2}$$

where p = probability of improvement given that a person has changed. We use the test statistic

$$z = \frac{\left| \hat{p} - \frac{1}{2} \right| - \frac{1}{2n}}{\sqrt{\frac{1}{4n}}} \sim N(0, 1) \text{ under } H_0$$

The p-value $= 2 \times \Phi(z)$ if $z < 0$, or $= 2 \times [1 - \Phi(z)]$ if $z \geq 0$. This is a special case of the one-sample binomial test, where we test the hypothesis $H_0: p = p_0$ versus H_1: $p \neq p_0$ and $p_0 = .5$. The criterion for using this test is that n = number of people who have changed is ≥ 20.

In this example, $\hat{p} = 15/23$ and the test statistic is

$$z = \frac{\left| \frac{15}{23} - \frac{1}{2} \right| - \frac{1}{46}}{\sqrt{\frac{1}{4(23)}}} = \frac{|.652 - .500| - .022}{0.104}$$

$$= \frac{0.130}{0.104} = 1.251 \sim N(0, 1)$$

$$p = 2 \times [1 - \Phi(1.251)]$$

$$= .21$$

Thus, there is no significant change in periodontal status after the program.

12.2.2 Small-Sample Test

If $n < 20$, then the following small-sample test must be used. If x = number of people who improve among n people who have changed, then

$$p\text{-value} = 2 \times \sum_{k=0}^{x} \binom{n}{k} \left(\frac{1}{2}\right)^n \qquad \text{if } x < \frac{n}{2}$$

$$= 2 \times \left[\sum_{k=x}^{n} \binom{n}{k} \left(\frac{1}{2}\right)^n \right] \qquad \text{if } x > \frac{n}{2}$$

$$= 1 \qquad\qquad\qquad\qquad \text{if } x = \frac{n}{2}$$

The Wilcoxon Signed-Rank Test

Suppose we want to take account of the magnitude of change as well as its direction. In particular, the positive-change scores seem greater than the negative-change scores.

Procedure

Order the change scores by absolute value and rank the absolute values:

Absolute value of change score	Positive	Negative	Total	Rank	Average rank
1	6	4	10	1–10	5.5
2	5	2	7	11–17	14.0
3	4	2	6	18–23	20.5

Assign the average rank to each observation in a group. Compute the rank sum in the 1st group $\equiv R_1$.

$$R_1 = 6 \times 5.5 + 5 \times 14.0 + 4 \times 20.5 = 33 + 70 + 82 = 185$$

Compare the observed rank sum for people with positive scores with the expected rank sum under the null hypothesis that the rank sum is the same for people with positive and negative scores. If most of the changes are in the positive direction, particularly for change scores with large absolute value, then the rank sum for people with positive scores will be large; if most of the changes are in the negative direction, particularly for change scores with large absolute value, then it will be small; if the changes are equally dispersed between positive and negative change scores, particularly for change scores with large absolute value, then the rank sum will be close to $(1 + 2 + \ldots + n)/2 = n(n + 1)/4$. Under H_0,

$$E(R_1) = \frac{n(n +1)}{4}$$

$$Var(R_1) = \frac{n(n + 1)(2n + 1)}{24} - \sum_{i=1}^{g}\left(\frac{t_i^3 - t_i}{2}\right)$$

where t_i = number of observations in the ith tied group. We use the test statistic

$$z = \frac{|R_1 - E(R_1)| - .5}{sd(R_1)} \sim N(0, 1) \quad \text{under } H_0$$

The p-value $= 2 \times \Phi(z)$ if $z < 0$, or $= 2 \times [1 - \Phi(z)]$ if $z \geq 0$. We will only use this test if n = number of nonzero change scores is ≥ 16. In this case,

$$n = 23$$

$$E(R_1) = \frac{23(24)}{4} = 138$$

$$Var(R_1) = \frac{23(24)(47)}{24} - \left[\frac{(10^3 - 10) + (7^3 - 7) + (6^3 - 6)}{2}\right]$$

$$= 1081 - \frac{990 + 336 + 210}{2} = 1081 - \frac{1536}{2}$$

$$= 1081 - 768 = 313$$

$$z = \frac{\left| 185 - 138 \right| - \frac{1}{2}}{\sqrt{313}} = \frac{46.5}{17.69}$$

$$= 2.628 \sim N(0, 1) \quad \text{under } H_0, \text{ two-tailed } p = .009$$

Thus, periodontal status has significantly improved.

Notice that we reached a different conclusion using the signed-rank test, where both the magnitude and the direction of the difference scores are taken into account (viz., a significant benefit of the treatment program) than with the sign test, where only the direction, but not the magnitude of the difference scores are taken into account and no significant benefit of the treatment program was found. The signed-rank test is more appropriate in this case.

If the number of nonzero change scores is ≤ 15, then we refer to Table 12 (Appendix, text). Upper and lower two-tailed critical values are provided for $\alpha = .10$, .05, .02, and .01, for each $n \le 15$. Results are significant at level α if $R_1 \ge$ upper critical value or $R_1 \le$ lower critical value. For example, if we have 10 untied pairs and $R_1 = 49$, then since the upper critical value at $\alpha = .05$ is 47 and the upper critical value at $\alpha = .02$ is 50, it follows that $.02 \le p < .05$.

SECTION 12.4   The Wilcoxon Rank-Sum Test

Since plasma aldosterone has been related to blood-pressure level in adults, and black adults generally have higher mean BP levels than white adults, it is of interest to compare the aldosterone distributions of black and white children.

In Figure 5.21 (p. 138, text), we present the distribution of plasma-aldosterone concentrations for 53 white and 46 black children. The distributions are very skewed, and we are reluctant to use a t test to compare the distributions. Instead, we use the Wilcoxon rank-sum test. For simplicity, we will group the distributions in 200 pmol/liter groups as follows:

Aldosterone group[a]	White	Black	Total	Rank range	Average rank
0–199	12	28	40	1–40	20.5
200–399	17	10	27	41–67	54.0
400–599	15	5	20	68–87	77.5
600+	9	3	12	88–99	93.5
Total	53	46	99		

[a] pmol/liter.

We compute the rank range and average rank in each group over the combined sample (white + black children). We compute the rank sum in the 1st sample (white children). If the rank sum is large, then white children have higher levels than black children; if it is small, then white children have lower levels. We have

$$R_1 = 12(20.5) + \ldots + 9(93.5) = 3168$$

Under the null hypothesis that the median aldosterone level is the same in each group,

$$E(R_1) = n_1\left(\frac{1 + n_1 + n_2}{2}\right)$$

$$Var(R_1) = \frac{n_1 n_2}{12}\left[n_1 + n_2 + 1 - \frac{\Sigma(t_i^3 - t_i)}{(n_1 + n_2)(n_1 + n_2 - 1)}\right]$$

where t_i = number of tied observations in the ith tied group. We have the test statistic

$$T = \frac{|R_1 - E(R_1)| - .5}{sd(R_1)} \sim N(0, 1) \quad \text{under } H_0$$

The p-value $= 2 \times \Phi(T)$ if $T < 0$, or $= 2 \times [1 - \Phi(T)]$ if $T \geq 0$. This test should only be used if n_1 and n_2 are each ≥ 10.

In this case,

$$E(R_1) = 53(1 + 53 + 46)/2 = 2,650$$

$$Var(R_1) = [53(46)/12]\left[100 - \frac{(40^3 - 40) + \ldots + (12^3 - 12)}{99(98)}\right]$$

$$= 203.167(100 - 93,312/9702)$$

$$= 203.167(90.382) = 18,362.65$$

$$sd(R_1) = \sqrt{18,362.65} = 135.51$$

$$T = \frac{|3,168 - 2,650| - .5}{135.51}$$

$$= 517.5/135.51 = 3.82 \sim N(0, 1) \quad \text{under } H_0$$

The p-value $= 2 \times [1 - \Phi(3.82)] < .001$. Thus, since $R_1 > E(R_1)$, white children have significantly higher aldosterone levels than black children.

If either n_1 or n_2 are < 10, then we refer to Table 13 (Appendix, text). For each combination of N_1 = minimum of the two sample sizes $= 4, \ldots, 9$ and N_2 = maximum of the two sample sizes $= 4, \ldots, 50$, two-tailed critical values are provided for $\alpha = .10, .05, .02$, and $.01$. The results are significant at level α if $R_1 \leq$ lower critical value or $R_1 \geq$ upper critical value. Suppose $n_1 = 5$, $n_2 = 10$ and $R_1 = 54$. The critical values for $\alpha = .05$ are 23 and 57. Since $R_1 > 23$ and $R_1 < 57$, it follows that $p > .05$ and there is no significant difference between the two groups.

SECTION 12.5   The Kruskal-Wallis Test

In the Data Set BETACAR.DAT (p. 341, text) are data on the bioavailability of four different preparations of beta carotene (Solatene, Roche, BASF-30, and BASF-60). Serum beta-carotene levels were measured at baseline (two replicates) and at 6, 8, 10, and 12 weeks of follow-up. There were 5–6 subjects for each of the four preparations. To measure bioavailability, we compute

$$x = \text{mean (6-, 8-, 10-, and 12-week levels)/mean (2 baseline levels)}$$

The results are displayed here:

```
------------------------------------------------
ROW          SOL       ROCHE     BASF_30   BASF_60

  1       0.91787     1.98214    2.45614    1.27528
  2       1.99630     1.12273    1.39323    2.33745
  3       1.60106     1.38500    2.09024    2.14773
  4       1.58108     1.44565    1.68466    1.15605
  5       1.91837     1.79009    2.22024    3.43939
  6       2.62264     1.82796               2.68493
------------------------------------------------
```

The distributions are somewhat skewed. If we wish to compare the four preparations, then we could transform x to a different scale (e.g., the ln scale) and perform the fixed-effects one-way ANOVA on the transformed scale. Another approach is to leave the data in the original scale, but perform a nonparametric test to compare the four groups. The nonparametric analogue to the one-way ANOVA is called the *Kruskal-Wallis test*. We wish to test the hypothesis H_0: median $(x$, group 1$)$ = ... = median $(x$, group 4$)$ versus H_1: at least two of the medians are different. The test statistic (in the absence of ties) is

$$H = \frac{12}{N(N+1)} \times \sum_{i=1}^{k} \frac{R_i^2}{n_i} - 3(N+1) \sim \chi^2_{k-1} \quad \text{under } H_0$$

where

R_i is the rank sum in the ith group, where the ranks are based on the pooled sample over all groups

n_i = sample size in the ith group

$N = \Sigma n_i$

If there are ties, then the test statistic is divided by

$$1 - \frac{\sum_{j=1}^{g} (t_j^3 - t_j)}{N^3 - N}$$

where t_j = the number of observations in the jth tied group. The p-value = $Pr(\chi^2_{k-1} > H)$. This test should only be used if the minimum group size is ≥ 5.

We have applied this procedure to the beta-carotene data, with results given as follows:

```
------------------------------------------------
MTB > Kruskal-Wallis C9 C1.

LEVEL       NOBS      MEDIAN    AVE. RANK
  1          6        1.760       11.2
  2          6        1.618        8.5
  3          5        2.090       14.0
  4          6        2.243       14.7
OVERALL     23                    12.0

H = 3.05   d.f. = 3   p = 0.384
------------------------------------------------
```

We see that the beta-carotene level has increased in all groups (median increase = 76%, solatene group (group 1); 62%, Roche group (group 2); 109%, BASF-30 group

(group 3); 124%, BASF-60 group (group 4)). The test statistic $= H = 3.05 \sim \chi_3^2$. The p-value $= Pr(\chi_3^2 > 3.05) = .384$. Thus, there was no significant difference in bio-availability between the four preparations.

If the Kruskal-Wallis test had been significant, then we could compare specific groups using the Dunn procedure [see Equation **(12.11)**, p. 573, text]. Finally, if at least one of the group sizes is < 5, then special small-sample tables must be used to evaluate significance. In Table 14 (Appendix, text) we provide such a table for selected group sizes in the special case of 3 groups.

SECTION 12.6   Rank Correlation

In the Data Set VALID.DAT (p. 41, text), we provide data from a validation study based on 173 women in the Nurses' Health Study of two methods for assessing dietary intake: a food-frequency questionnaire (FFQ) and a diet record (DR). The DR is considered the gold standard, but is very expensive to administer; the FFQ is inexpensive, but has substantial measurement error. Data are presented on four nutrients. Here we focus on the correlation between alcohol intake as measured by the FFQ versus the DR. Since alcohol intake is very skewed and nonnormal, with a substantial fraction of women reporting no intake, we are reluctant to use the ordinary Pearson correlation coefficient. Instead, we will use a nonparametric analogue to the Pearson correlation known as the *Spearman rank-correlation coefficient.* To compute the rank correlation, we first rank each of FFQ and DR and then compute an ordinary Pearson correlation based on the ranked data. Once the correlation is obtained we can test the null hypothesis of no association by using the test statistic

$$t_s = \frac{r_s\sqrt{n-2}}{\sqrt{1 - r_s^2}} \sim t_{n-2} \quad \text{under } H_0$$

The p-value $= 2 \times Pr(t_{n-2} < t_s)$ if $t_s < 0$, or $= 2 \times Pr(t_{n-2} > t_s)$ if $t_s \geq 0$. This test should only be used if $n \geq 10$.

The computations were performed using MINITAB. A sample of the ranked data for DR (labeled ALCD_R) and FFQ (labeled ALCF_R) are shown on p. 207, together with the corresponding raw data (labeled ALCD and ALCF).

The calculation of the rank correlation is given as follows:

```
CORRELATION OF ALCD_R AND ALCF_R = 0.899
```

The rank correlation $= .899$, which is quite high. To perform the significance test, we compute the test statistic

$$t_s = \frac{.899\sqrt{171}}{\sqrt{1 - (.899)^2}} = 26.8 \sim t_{171} \quad \text{under } H_0$$

The p-value $= 2 \times Pr(t_{171} > 26.8) < .001$. Clearly, there is a strong correlation between alcohol intake as assessed by DR and FFQ.

For smaller n ($n \leq 9$), a small-sample table should be used (see Table 15, Appendix, text) to assess significance. To use this table, we reject H_0 if $|r_s| \geq c$, where $c =$ critical value for a specific two-sided α level (e.g., $\alpha = .05$). Suppose $n = 6$ and $r_s = .850$. Since the critical value $= .829$ for $\alpha = .10$ and $.886$ for $\alpha = .05$, it follows that $.05 < p < .10$.

```
              ALCD     ALCF    ALCD_R   ALCF_R
    155       8.18     6.63     107.0    102.0
    156       0.51     0.00      25.0     17.5
    157       7.82    12.08     102.0    134.5
    158       6.60    11.14      90.0    124.5
    159      28.79    52.07     165.0    171.0
    160       3.82     1.81      69.0     61.0
    161       0.00     0.00       9.0     17.5
    162       8.31     2.91     110.0     79.0
    163       8.22    13.76     108.0    142.5
    164       0.96     0.76      38.0     42.5
    165       0.63     0.00      29.0     17.5
    166      15.72    16.81     140.0    151.0
    167      23.97    12.43     158.0    136.0
    168       7.28     6.63      98.0    102.0
    169       4.60     7.25      75.5    108.5
    170       2.56     1.81      56.0     61.0
    171      13.91     8.64     134.0    112.5
    172       5.39     7.19      86.0    107.0
    173       0.60     3.49      27.5     80.0
```

PROBLEMS

Ophthalmology

Suppose an ophthalmologist reviews fundus photographs of 30 patients with macular degeneration both before and 3 months after receiving a laser treatment. To assess the efficacy of treatment, each patient is rated as improved, remained the same, or declined.

12.1 If 20 patients improved, 7 declined, and 3 remained the same, then assess whether or not patients undergoing this treatment are showing significant change from baseline to 3 months afterward. Report a p-value.

Suppose that the patients are divided into two groups according to initial visual acuity (VA). Of the 14 patients with VA 20–40 or better, 8 improved, 5 declined, and 1 stayed the same. Of the 16 patients with VA worse than 20–40, 12 improved, 2 declined, and 2 stayed the same.

12.2 Assess the results in the subgroup of patients with VA of 20–40 or better.

12.3 Assess the results in the subgroup of patients with VA worse than 20–40.

Diabetes

An experiment was conducted to study responses to different methods of taking insulin in patients with type I diabetes. The percentages of glycosolated hemoglobin initially and 3 months after taking insulin by nasal spray are given in Table 12.1 [1].

12.4 Perform a t test to compare the percentages of glycosolated hemoglobin before and 3 months after treatment.

TABLE 12.1 Percentages of glycosolated hemoglobin before and 3 months after taking insulin by nasal spray

Patient number	Before	3 months after
1	11.0	10.2
2	7.7	7.9
3	5.9	6.5
4	9.5	10.4
5	8.7	8.8
6	8.6	9.0
7	11.0	9.5
8	6.9	7.6

Source: Reprinted with permission of the *New England Journal of Medicine, 312*(17), 1078–1084, 1985.

12.5 Suppose normality is not assumed. Perform a nonparametric test corresponding to the t test in Problem 12.4.

12.6 Compare your results in Problems 12.4 and 12.5.

Pathology

Refer to Table 6.1 (p. 69, Study Guide).

12.7 Suppose normality is not assumed. What nonparametric test can be used to compare total heart weight of males with left-heart disease with that of normal males?

12.8 Implement the test in Problem 12.7 and report a *p*-value.

Cardiology

Propranolol is a standard drug given to ease the pain of patients with episodes of unstable angina. A new drug for the treatment of this disease is tested on 30 pairs of patients who are matched on a one-to-one basis according to age, sex, and clinical condition and are assessed as to the severity of their pain. Suppose that in 15 pairs of patients, the patient with the new drug has less pain; in 10 pairs of patients, the patient with propranolol has less pain; and in 5 pairs of patients, the pain is about the same with the two drugs.

12.9 What is the appropriate test to use here?

12.10 Perform the test in Problem 12.9 and report a *p*-value.

Ophthalmology

Table 8.15 (p. 290, text) presents data giving the median gray levels in the lens of the human eye for 6 cataractous and 6 normal people.

12.11 What nonparametric test could be used to compare the median gray levels of normal and cataractous eyes?

12.12 Carry out the test in Problem 12.11 and report a *p*-value.

Psychiatry

Suppose we are conducting a study of the effectiveness of lithium therapy for manic-depressive patients. The study is carried out at two different centers, and we want to determine if the patient populations are comparable at baseline. A self-rating questionnaire about their general psychological well-being is administered to the prospective patients at the two different centers. The outcome measure on the questionnaire is a four-category scale: (1) = feel good; (2) = usually feel good, once in a while feel nervous; (3) = feel nervous half the time; (4) = usually feel nervous. Suppose the data at the two different centers are as follows:

Center 1	3, 4, 1, 1, 3, 2, 3, 4, 4, 3, 2, 4, 4, 4
Center 2	1, 2, 1, 3, 2, 4, 1, 2, 1, 3, 1, 2, 2, 2, 1, 3

12.13 What type of data does this type of scale represent?

12.14 Why might a parametric test not be useful with this type of data?

12.15 Assess if there is any significant difference in the responses of the two patient populations using a nonparametric test.

Psychiatry

Much attention has been given in recent years to the role of transcendental meditation in improving health, particularly in lowering blood pressure. One hypothesis that emerges from this work is that transcendental meditation might also be useful in treating psychiatric patients with symptoms of anxiety. Suppose that a protocol of meditational therapy is administered once a day to 20 patients with anxiety. The patients are given a psychiatric exam at baseline and at a follow-up exam 2 months later. The degree of improvement is rated on a 10-point scale, with 1 indicating the most improvement and 10 the least improvement. Similarly, 26 comparably affected patients with anxiety are given standard psychotherapy and are asked to come back 2 months later for a follow-up exam. The results are given in Table 12.2.

TABLE 12.2 Degree of improvement in patients with anxiety who are treated with transcendental meditation or psychotherapy

Meditation		Psychotherapy	
d^a	f^b	d^a	f^b
1	3	1	0
2	4	2	2
3	7	3	5
4	3	4	3
5	2	5	8
6	1	6	4
7	0	7	2
8	0	8	1
9	0	9	1
10	0	10	0
	20		26

[a]*d* = degree of improvement.
[b]*f* = frequency.

12.16 Why might a parametric test not be useful here?

12.17 What nonparametric test should be used to analyze these data?

12.18 Compare the degree of improvement in the two groups using the test in Problem 12.17.

Pulmonary Disease

Twenty-two young asthmatic volunteers were studied to assess the short-term effects of sulfur dioxide (SO_2) exposure under various conditions [2]. The baseline data in Table 12.3 were presented regarding bronchial reactivity to SO_2 stratified by lung function (as defined by FEV_1/FVC) at screening.

TABLE 12.3 Relationship of bronchial reactivity to SO_2 (cm H_2O/s) grouped by lung function at screening among 22 asthmatic volunteers

Lung function group		
Group A FEV₁/FVC ≤74%	Group B FEV₁/FVC 75–84%	Group C FEV₁/FVC ≥85%
20.8	7.5	9.2
4.1	7.5	2.0
30.0	11.9	2.5
24.7	4.5	6.1
13.8	3.1	7.5
	8.0	
	4.7	
	28.1	
	10.3	
	10.0	
	5.1	
	2.2	

Source: Reprinted with permission of the *American Review of Respiratory Disease, 131*(2), 221–225, 1985.

12.19 Suppose we do not wish to assume normality. What nonparametric test can be used to compare the three groups?

12.20 Implement the test in Problem 12.19 and report a *p*-value.

Pharmacology

Refer to Table 9.18 (p. 340, text) where data are presented on the reduction in fever for patients getting different doses of aspirin.

12.21 Use nonparametric methods to test for significant differences among groups. Identify which groups are significantly different.

12.22 Compare your results for this problem using parametric (see Problems 9.21–9.23, text) and nonparametric methods. Which do you think is the more appropriate method here?

Hypertension

The variability of blood pressure is important in planning screening programs for detecting people with high blood pressure. There are two schools of thought concerning this variability: Some researchers feel that a subgroup of people have extremely variable blood pressures, whereas most people have relatively stable blood pressures; other researchers feel that this variability is common to all people. A study was set up to answer this question. A group of 15 people had their blood pressures measured on three separate visits in each of 2 years, and the between-visit variance of blood pressure was measured each year. The results for systolic blood pressure (SBP) are given in Table 12.4. We wish to assess if there is any relationship between the year 1 and year 2 variances.

12.23 Why might a rank correlation be a useful method for expressing such a relationship?

12.24 Test for the significance of the rank correlation based on the previous data.

12.25 What are your conclusions based on these data?

TABLE 12.4 Blood-pressure variability measured at two points in time

Person	Between-visit variance, year 1	Between-visit variance, year 2
1	8.8	19.0
2	10.0	25.7
3	6.7	17.9
4	13.3	95.7
5	10.0	16.7
6	6.2	15.2
7	35.0	3.9
8	51.2	11.9
9	30.9	14.6
10	61.0	21.6
11	5.4	31.3
12	46.6	21.4
13	37.0	10.1
14	2.6	18.7
15	2.0	22.7

SOLUTIONS

12.1 Use the sign test. If the people who have remained the same are ignored, then 27 people have either improved or declined. Thus, the normal-theory version of the test can be used. We have that C = number of patients improved = 20, n = 27. Reject H_0 if either

$$C > \frac{n}{2} + \frac{1}{2} + z_{1-\alpha/2}\sqrt{n/4} = c_2$$

or

$$C < \frac{n}{2} - \frac{1}{2} - z_{1-\alpha/2}\sqrt{n/4} = c_1$$

and accept H_0 otherwise. We have that α = .05. Therefore,

$$c_2 = \frac{27}{2} + \frac{1}{2} + z_{.975}\sqrt{27/4}$$

$$= 14 + 1.96(2.598) = 14 + 5.09$$

$$= 19.09$$

$$c_1 = 13 - 5.09 = 7.91$$

Since $C = 20 > 19.09$, reject H_0 at the 5% level. The exact p-value is given by

$$p = 2 \times \left\{1 - \Phi\left[\frac{C - n/2 - 0.5}{\sqrt{n/4}}\right]\right\}$$

$$= 2 \times \left[1 - \Phi\left(\frac{20 - 27/2 - 0.5}{\sqrt{27/4}}\right)\right]$$

$$= 2 \times \left[1 - \Phi\left(\frac{6.5 - 0.5}{2.598}\right)\right]$$

$$= 2 \times \left[1 - \Phi\left(\frac{6.0}{2.598}\right)\right] = 2 \times [1 - \Phi(2.309)]$$

$$= 2 \times (1 - .9895) = .021$$

12.2 In the subgroup of patients with better visual acuity, 13 patients have either improved or declined. Thus, the exact binomial test must be used. We have $C = 8, n = 13$. Refer to the exact binomial tables (Table 1, Appendix, text) under $n = 13, p = .5$ to compute

$$p = 2 \times \sum_{k=8}^{13} \binom{13}{k}\left(\frac{1}{2}\right)^{13}$$

$$= 2 \times (.1571 + .0873 + .0349 + .0095$$

$$+ .0016 + .0001)$$

$$= 2 \times .2905 = .581$$

Thus, there is no significant change in the fundus photographs of the subgroup of patients with VA 20-40 or better.

12.3 Fourteen patients have either improved or declined. Thus, the exact binomial test must be used. We have $C = 12, n = 14$. Refer to the exact binomial tables under $n = 14, p = .5$ to compute

$$p = 2 \times \sum_{k=12}^{14} \binom{14}{k}\left(\frac{1}{2}\right)^{14}$$

$$= 2 \times (.0056 + .0009 + .0001)$$

$$= 2 \times .0066 = .013$$

Thus, the fundus photographs of persons with VA of worse than 20-40 at baseline have significantly improved.

12.4 We use a paired t test, where $\bar{d} = -0.075, s_d = 0.821, n = 8$. We have the test statistic

$$t = \frac{\bar{d}}{s_d/\sqrt{n}} = \frac{-0.075}{0.821/\sqrt{8}}$$

$$= \frac{-0.075}{0.290} = -0.26 \sim t_7 \quad \text{under } H_0$$

Since $t_{7,.975} = 2.365 > |t|$, we accept H_0 at the 5% level.

12.5 Since each patient is being used as his or her own control, we use the Wilcoxon signed-rank test. First, rank the difference scores by absolute value as follows:

$|d_i|$	Neg d_i	f_i	Pos d_i	f_i	Number of patients with absolute value	Range of ranks	Average rank
1.5	−1.5	0	1.5	1	1	8	8.0
0.9	−0.9	1	0.9	0	1	7	7.0
0.8	−0.8	0	0.8	1	1	6	6.0
0.7	−0.7	1	0.7	0	1	5	5.0
0.6	−0.6	1	0.6	0	1	4	4.0
0.4	−0.4	1	0.4	0	1	3	3.0
0.2	−0.2	1	0.2	0	1	2	2.0
0.1	−0.1	1	0.1	0	1	1	1.0
		6		2			

We then compute the rank sum of the positive differences $= R_1 = 6 + 8 = 14$. Since there are fewer than 16 non-zero differences, we refer to the special table for the signed-rank test (Table 12, Appendix, text). We note that for $n = 8, \alpha = .10$, the two-tailed upper and lower critical values are 31 and 5, respectively. Since $5 < 14 < 31$, it follows that $p > .10$ and there is no significant difference in the % of glycosolated hemoglobin before and 3 months after taking insulin by nasal spray.

12.6 The same conclusions were reached in Problems 12.4 and 12.5.

12.7 Since the groups are not matched, we can use the Wilcoxon rank-sum test.

12.8 We combine the data from the two groups and assign ranks based on the combined sample.

Value	Left heart disease males	Normal males	Combined sample	Range of ranks	Average rank
245	0	1	1	1	1.0
270	0	1	1	2	2.0
285	1	0	1	3	3.0
290	0	1	1	4	4.0
300	0	2	2	5–6	5.5
310	1	1	2	7–8	7.5
325	1	0	1	9	9.0
340	0	1	1	10	10.0
350	0	1	1	11	11.0
360	0	1	1	12	12.0
375	1	0	1	13	13.0
405	0	1	1	14	14.0
425	1	0	1	15	15.0
450	2	0	2	16–17	16.5
460	1	0	1	18	18.0
495	1	0	1	19	19.0
615	1	0	1	20	20.0
760	1	0	1	21	21.0

We compute the rank sum in the left heart disease group. This is given by $R_1 = 3.0 + 7.5 + \ldots + 21.0 = 158.5$. Since both n_1 and n_2 are ≥ 10, we can use the normal approximation method. We have

$$T = \frac{\left|158.5 - 11(22)/2\right| - .5}{\sqrt{\frac{11(10)}{12}\left[22 - \frac{2(2^2 - 1) + 2(2^2 - 1) + 2(2^2 - 1)}{21(20)}\right]}}$$

$$= \frac{\left|158.5 - 121\right| - .5}{\sqrt{9.167(22 - 18/420)}} = \frac{37.0}{\sqrt{201.274}}$$

$$= \frac{37.0}{14.187} = 2.608 \sim N(0, 1) \quad \text{under } H_0$$

Thus, the p-value $= 2 \times [1 - \Phi(2.608)] = 2 \times (1 - .9954) = .009$. Therefore, the mean total heart weight for left heart disease males is significantly higher than for normal males. This is the same conclusion we reached in Problem 8.7 using the independent-samples t test on these data.

12.9 The appropriate test to use here is the sign test because we only know which treated patient in a pair has less severe pain, but not how much less severe it is.

12.10 We test the hypothesis H_0: $p = 1/2$ versus H_1: $p \neq 1/2$, where p = probability that the patient in a matched pair taking the new drug has less severe pain than the patient taking propranolol. Since the number of nonzero differences $= 25 \geq 20$, we can use the normal-theory method given in **(12.1)** and **(12.2)** (text, p. 554). We have the test statistic

$$z = \frac{\left|C - n/2\right| - .5}{\sqrt{n/4}} = \frac{(\left|15 - 25/2\right| - .5)}{\sqrt{25/4}}$$

$$= \frac{2.0}{2.50} = 0.80 \sim N(0, 1) \quad \text{under } H_0$$

The two-sided p-value is given by $p = 2 \times [1 - \Phi(0.80)]$ $= 2 \times (1 - .7881) = .424$, which is not statistically significant. Thus, there is no significant difference between the two treatments.

12.11 The Wilcoxon rank-sum test.

12.12 We combine the two samples and rank-order the observations in the combined sample. We have

Value	Cataractous people	Normal people	Total	Rank
106	1	0	1	1.0
136	1	0	1	2.0
140	1	0	1	3.0
145	0	1	1	4.0
149	1	0	1	5.0
158	0	1	1	6.0
161	1	0	1	7.0
167	0	1	1	8.0
171	1	0	1	9.0
177	0	1	1	10.0
182	0	1	1	11.0
185	0	1	1	12.0
	6	6	12	

The rank sum in the cataractous group is given by $R_1 = 1.0 + 2.0 + \ldots + 9.0 = 27.0$. We use the exact test due to the small sample sizes. We refer to Table 13 (Appendix, text) under $n_1 = 6$, $n_2 = 6$, $\alpha = .05$ and note that $T_l = 26$, $T_r = 52$. Since $26 < 27 < 52$, it follows that $p \geq .05$ and there is no significant difference between the two groups.

12.13 This type of scale is an example of ordinal data because we can only say that patients with a rating of 1 feel better than those with a rating of 2, but not how much better they feel.

12.14 A parametric test such as the t test would not be useful here because we would have to assume specific cardinal values for the four ratings that would then be added and multiplied.

12.15 We have the following grouped data:

Center	Score 1	2	3	4	Total
1	2	2	4	6	14
2	6	6	3	1	16

We will use the Wilcoxon rank-sum test. First, we pool the data from both samples together and assign a rank to each individual based on his or her rank in the combined sample. We have

Score	Frequency	Rank range	Average rank
1	8	1–8	4.5
2	8	9–16	12.5
3	7	17–23	20.0
4	7	24–30	27.0

We compute the rank sum for center 1 as follows: $R_1 = 2 \times 4.5 + 2 \times 12.5 + 4 \times 20.0 + 6 \times 27.0 = 276$. We use the normal approximation method. We have

$$E(R_1) = \frac{14(31)}{2} = 217$$

$$Var(R_1) = \frac{14(16)}{12}$$
$$\times \left[31 - \frac{8(63) + 8(63) + 7(48) + 7(48)}{30(29)} \right]$$

$$= 18.667(29.069) = 542.621$$

$$sd(R_1) = 23.294$$

Thus, $\quad T = \dfrac{|R_1 - E(R_1)| - .5}{sd(R_1)}$

$$= \frac{|276 - 217| - .5}{23.294}$$

$$= \frac{58.5}{23.294} = 2.511 \sim N(0, 1)$$

Thus, $p = 2 \times [1 - \Phi(2.511)] = .012$. Therefore, there is a significant difference between the patient populations at the two centers, with the patients at center 1 having significantly more anxiety at baseline.

12.16 A parametric test would not be useful here because this is an ordinal scale, with 1 indicating more improvement than 2 but not 1 unit more, so that addition and subtraction are not meaningful with this scale.

12.17 The Wilcoxon rank-sum test should be used to analyze these data because we are comparing two independent groups.

12.18 We will perform the Wilcoxon rank-sum test based on these data. We compute ranks as follows:

Value	Meditation	Psycho-therapy	Total	Rank range	Average rank
1	3	0	3	1–3	2.0
2	4	2	6	4–9	6.5
3	7	5	12	10–21	15.5
4	3	3	6	22–27	24.5
5	2	8	10	28–37	32.5
6	1	4	5	38–42	40.0
7	0	2	2	43–44	43.5
8	0	1	1	45	45.0
9	0	1	1	46	46.0
	20	26	46		

We compute the rank sum in the meditation group: $R_1 = 3 \times 2 + 4 \times 6.5 + 7 \times 15.5 + 3 \times 24.5 + 2 \times 32.5 + 1 \times 40 = 319$. The corresponding expected value and variance of the rank sum are given as follows:

$$E(R_1) = \frac{20(47)}{2} = 470$$

$$Var(R_1) = \frac{20(26)}{12}$$

$$\times \left[47 - \frac{3(3^2 - 1) + \ldots + 2(2^2 - 1)}{46(45)} \right]$$

$$= \frac{520}{12} \times \left(47 - \frac{3,276}{2,070} \right) = 1968.087$$

$$sd(R_1) = 44.36$$

The test statistic is obtained as follows:

$$T = \frac{|319 - 470| - .5}{44.36}$$

$$= \frac{150.5}{44.36} = 3.392 \sim N(0, 1) \quad \text{under } H_0$$

The p-value is given by $2 \times [1 - \Phi(3.392)] = 2 \times (1 - .9997) < .001$. This result is very highly significant and indicates that there is a significant difference in improvement between the two treatment groups, with the meditation group showing more improvement.

12.19 The Kruskal-Wallis test.

12.20 We have the test statistic

$$H = H^* / [1 - \Sigma(t_j^3 - t_j)/(N^3 - N)] \sim \chi^2_{k-1} \quad \text{under } H_0$$

where $\quad H^* = \frac{12}{N(N + 1)} (\Sigma R_i^2/n_i) - 3(N + 1)$

and $\quad R_i = $ rank sum in the ith group.

We pool the three samples and compute ranks in the combined sample as follows:

Value	Frequency	Rank range	Average rank
2.0	1	1	1.0
2.2	1	2	2.0
2.5	1	3	3.0
3.1	1	4	4.0
4.1	1	5	5.0
4.5	1	6	6.0
4.7	1	7	7.0
5.1	1	8	8.0
6.1	1	9	9.0
7.5	3	10–12	11.0
8.0	1	13	13.0
9.2	1	14	14.0
10.0	1	15	15.0
10.3	1	16	16.0
11.9	1	17	17.0
13.8	1	18	18.0
20.8	1	19	19.0
24.7	1	20	20.0
28.1	1	21	21.0
30.0	1	22	22.0

We now compute the rank sum in the three samples:

$$R_1 = 19.0 + \ldots + 18.0 = 84.0$$
$$R_2 = 11.0 + \ldots + 2.0 = 131.0$$
$$R_3 = 14.0 + \ldots + 11.0 = 38.0$$

Thus, the test statistic is given by

$$H^* = \frac{12}{22(23)} \times (84^2/5 + 131^2/12 + 38^2/5) - 3(23)$$

$$= 0.0237(1411.2 + 1430.1 + 288.8) - 69$$

$$= 0.0237(3130.1) - 69 = 5.231$$

Finally, we apply the correction for ties and obtain

$$H = 5.231/[1 - (3^3 - 3)/(22^3 - 22)]$$

$$= 5.231/(1 - 24/10,626)$$

$$= 5.231/.9977 = 5.24 \sim \chi_2^2 \quad \text{under } H_0$$

Since $\chi_{2,.95}^2 = 5.99 > 5.24$, it follows that $p > .05$ and there is no significant difference among the three means.

12.21 We use the Kruskal-Wallis test to assess overall differences between groups. We combine the three samples and compute ranks in the pooled sample as follows:

Value	Rank	Value	Rank	Value	Rank
−1.0	1.0	0.3	6.5	1.2	11.0
−0.4	2.0	0.3	6.5	1.3	12.0
−0.2	3.0	0.5	8.0	1.6	13.0
0.2	4.5	0.6	9.0	2.0	14.0
0.2	4.5	1.1	10.0	2.1	15.0

We now compute the rank sum in each of the three groups:

Drug A		Drug B		Drug C	
Value	Rank	Value	Rank	Value	Rank
2.0	14.0	0.5	8.0	1.1	10.0
1.6	13.0	1.2	11.0	−1.0	1.0
2.1	15.0	0.3	6.5	−0.2	3.0
0.6	9.0	0.2	4.5	0.2	4.5
1.3	12.0	−0.4	2.0	0.3	6.5
Rank sum	63.0		32.0		25.0

The test statistic is given by

$$H = \frac{1}{C}\left[\frac{12}{15(16)}\left(\frac{63.0^2}{5} + \frac{32.0^2}{5} + \frac{25.0^2}{5}\right) - 3(16)\right]$$

$$= \frac{1}{C}\left[0.05(793.8 + 204.8 + 125.0) - 48\right]$$

$$= \frac{1}{C}\left[0.05(1,123.6) - 48\right] = \frac{1}{C}\left(56.18 - 48\right) = \frac{8.18}{C}$$

The correction for ties (C) is

$$C = 1 - \frac{(2^3 - 2) + (2^3 - 2)}{15^3 - 15}$$

$$= 1 - \frac{12}{3360} = .9964$$

Thus, $H = 8.18/.9964 = 8.21 \sim \chi_2^2$ under H_0. Since $\chi_{2,.975}^2 = 7.38$, $\chi_{2,.99}^2 = 9.21$ and $7.38 < 8.21 < 9.21$, it follows that $.01 < p < .025$ and some of the groups are significantly different.

To identify which groups are different, we use the Dunn procedure (text, p. 573). The test statistics and p-values for the comparison of each pair of groups are given as follows:

Groups	Test statistic	Critical value	p-value
A, B	$z = \dfrac{(63.0 - 32.0)/5}{\sqrt{\dfrac{15(16)}{12} \times \left(\dfrac{1}{5} + \dfrac{1}{5}\right)}}$		
	$= \dfrac{6.2}{2.83} = 2.19$	2.394[a]	NS
A, C	$z = \dfrac{(63.0 - 25.0)/5}{2.83}$		
	$= \dfrac{7.6}{2.83} = 2.69$	2.394	$p < .05$
B, C	$z = \dfrac{(32.0 - 25.0)/5}{2.83}$		
	$= \dfrac{1.4}{2.83} = 0.49$	2.394	NS

[a]$z_{1-.05/6} = z_{.9917} = 2.394$

Thus, under this procedure only groups A and C are significantly different.

12.22 Using parametric methods, group A was significantly different from both groups B and C (text, answers, p. 666). Using nonparametric methods, group A was only significantly different from group C. A stem-and-leaf plot is given as follows:

−1	0
−0	42
0	653223
1	6321
2	01

From the plot, we see that the normality assumption seems

to be reasonably satisfied with these data and thus, the parametric ANOVA procedure is probably more appropriate here.

12.23 The rank correlation is a useful measure here, because the distribution of the variances is quite skewed and very far from being normal.

12.24 We first rank the two sets of scores as follows:

Between-visit variance (year 1)	Frequency	Rank range	Average rank
2.0	1	1	1.0
2.6	1	2	2.0
5.4	1	3	3.0
6.2	1	4	4.0
6.7	1	5	5.0
8.8	1	6	6.0
10.0	2	7–8	7.5
13.3	1	9	9.0
30.9	1	10	10.0
35.0	1	11	11.0
37.0	1	12	12.0
46.6	1	13	13.0
51.2	1	14	14.0
61.0	1	15	15.0

Between-visit variance (year 2)	Frequency	Rank range	Average rank
3.9	1	1	1.0
10.1	1	2	2.0
11.9	1	3	3.0
14.6	1	4	4.0
15.2	1	5	5.0
16.7	1	6	6.0
17.9	1	7	7.0
18.7	1	8	8.0
19.0	1	9	9.0
21.4	1	10	10.0
21.6	1	11	11.0
22.7	1	12	12.0
25.7	1	13	13.0
31.3	1	14	14.0
95.7	1	15	15.0

The ranks are then displayed for each person as follows:

Person number	Between-visit variance (year 1 rank)	Between-visit variance (year 2 rank)
1	6.0	9.0
2	7.5	13.0
3	5.0	7.0
4	9.0	15.0
5	7.5	6.0
6	4.0	5.0
7	11.0	1.0
8	14.0	3.0
9	10.0	4.0
10	15.0	11.0
11	3.0	14.0
12	13.0	10.0
13	12.0	2.0
14	2.0	8.0
15	1.0	12.0

We now compute the Spearman rank-correlation coefficient as follows:

$$r_s = \frac{L_{xy}}{(L_{xx}L_{yy})^{1/2}} = \frac{-91.5}{[279.5(280)]^{1/2}}$$

$$= \frac{-91.5}{279.75} = -0.327$$

We then have the following test statistic:

$$t_s = \frac{r_s\sqrt{n-2}}{\sqrt{1-r_s^2}}$$

$$= \frac{-.327\sqrt{13}}{\sqrt{1-(-.327)^2}}$$

$$= \frac{-1.179}{0.945} = -1.248 \sim t_{13} \quad \text{under } H_0$$

Since $t_{13,.975} = 2.160 > |t_s|$, it follows that the results are *not* statistically significant.

12.25 The results are consistent with the theory that blood-pressure variability is not a characteristic of specific people, but rather is common among all people. Otherwise, there should have been a strong positive rank correlation. Indeed, the estimated correlation in this small group of people is negative.

REFERENCES

[1] Salzman, R., Manson, J. E., Griffing, G. T., Kimmerle, R., Ruderman, N., McCall, A., Stoltz, E. I., Mullin, C., Small, D., Armstrong, J., & Melby, J. C. (1985). Intranasal aerosolized insulin: Mixed-meal studies and long term use in type I diabetes. *New England Journal of Medicine, 312*(17), 1078–1084.

[2] Linn, W. S., Shamoo, D. A., Anderson, K. R., Whynot, J. D., Avol, E. L., & Hackney, J. D. (1985). Effects of heat and humidity on the responses of exercising asthmatics to sulfur dioxide exposure. *American Review of Respiratory Disease, 131*(2), 221–225.

CHAPTER THIRTEEN

HYPOTHESIS TESTING: PERSON-TIME DATA

SECTION 13.1 **Measures of Effect for Person-Time Data**

(1) **Incidence Density (λ)**   Number of events/amount of person-time. It represents the number of disease events per unit time. In subsequent discussion, we will use the term *incidence rate* synonymously with *incidence density.*

(2) **Cumulative Incidence [CI(t)]**   The probability of developing a disease over a time period (t). The relationship between cumulative incidence and incidence density is

$$CI(t) = 1 - \exp(-\lambda t) \cong \lambda t, \text{ for small } \lambda t$$

Approximately 100,000 women in the Nurses' Health Study, ages 30–64, were followed for 1,140,172 person-years from 1976 to 1990, during which time 2214 new breast cancers occurred. The incidence density for breast cancer was 2214/1,140,172 = .00194 events per person-year or 194 events per 100,000 (10^5) person-years. The cumulative incidence of breast cancer over 5 years $= 1 - \exp[-5(.00194)] = .00966 = 966$ events per 10^5 person-years.

SECTION 13.2 **Comparison of Incidence Rates**

Women were asked whether their mothers or any sisters had ever had breast cancer. If they responded yes, then they were considered to have a family history of breast cancer. From 1976 to 1990, there were 295 cases of breast cancer over 80,539 person-years among women with a family history of breast cancer and 1919 cases over 1,059,633 person-years among women without a family history. How can we compare the two incidence rates? We wish to test the hypothesis H_0: $ID_1 = ID_2$ versus H_1: $ID_1 \neq ID_2$ (where ID = incidence density). We use the test statistic

$$z = \frac{|a_1 - E_1| - .5}{\sqrt{V_1}} \sim N(0, 1) \text{ under } H_0$$

where

a_i = observed number of cases in group i, $i = 1, 2$

t_i = number of person-years in group i, $i = 1, 2$

E_1 = expected number of cases in group 1
(family history = yes) under H_0

$= (a_1 + a_2)t_1/(t_1 + t_2)$

V_1 = variance of number of cases in group 1 under H_0

$= (a_1 + a_2)t_1 t_2/(t_1 + t_2)^2$

The p-value $= 2 \times \Phi(z)$ if $z < 0$, or $= 2 \times [1 - \Phi(z)]$ if $z \geq 0$. This test should only be used if $V_1 \geq 5$.

In this case, we have $a_1 = 295$, $a_2 = 1919$, $t_1 = 80,539$, $t_2 = 1,059,633$. Therefore,

$$E_1 = 2214(80,539/1,140,172) = 156.4$$

$$V_1 = 2214(80,539)(1,059,633)/(1,140,172)^2$$

$$= 145.3$$

The test statistic is

$$z = \frac{|295 - 156.4| - .5}{\sqrt{145.3}}$$

$$= 138.1/12.1 = 11.46 \sim N(0, 1) \text{ under } H_0$$

The p-value $= 2 \times [1 - \Phi(11.46)] < .001$

Thus, women with a family history have a significantly higher incidence rate of breast cancer than women without a family history.

If $V_1 < 5$, then exact methods based on the binomial distribution must be used [see Equation (**13.6**), p. 590, text].

SECTION 13.3   Estimation of the Rate Ratio

The *rate ratio* (*RR*) is defined by

RR = incidence rate in an exposed group/incidence rate in an unexposed group

It is estimated by $(a_1/t_1)/(a_2/t_2) = \widehat{RR}$.

To obtain confidence limits for *RR*, we first obtain confidence limits for $\ln(RR) = (c_1, c_2)$ given by

$$c_1 = \ln(\widehat{RR}) - z_{1-\alpha/2}\sqrt{1/a_1 + 1/a_2}$$

$$c_2 = \ln(\widehat{RR}) + z_{1-\alpha/2}\sqrt{1/a_1 + 1/a_2}$$

The corresponding $100\% \times (1 - \alpha)$ CI for $RR = [\exp(c_1), \exp(c_2)]$.

The estimated incidence rate for women with a family history = 295/80,539 =

$366/10^5$, and for women without a family history $= 1919/1,059,633 = 181/10^5$. The estimated RR for women with a family history of breast cancer versus women without a family history $= (366/10^5)/(181/10^5) = 2.02$. A 95% CI for RR is given by $[\exp(c_1), \exp(c_2)]$, where

$$c_1 = \ln(2.02) - 1.96\sqrt{1/295 + 1/1919}$$

$$= 0.704 - 0.123 = 0.582$$

$$c_2 = 0.704 + 0.123 = 0.827$$

Thus, the 95% CI for $RR = [\exp(0.582), \exp(0.827)] = (1.79, 2.29)$.

SECTION 13.4   Inference for Stratified Person-Time Data

In the breast-cancer data, it is possible that women who report a family history (Hx) of breast cancer are older than those women who do not report such a history, since their mothers and sisters are likely to be older. Since in other data sets breast-cancer incidence rises with age, it is important to control for age in the analysis. To accomplish this, we obtain age-specific incidence rates of breast cancer for women with and without a family history as shown here:

Age	Family Hx = yes			Family Hx = no		
	Cases	Person-years	Incidence rate $(\times 10^{-5})$	Cases	Person-years	Incidence rate $(\times 10^{-5})$
30–39	12	8,564	140	124	173,104	72
40–44	31	12,157	255	268	200,916	133
45–49	54	15,422	350	387	215,329	180
50–54	68	18,110	375	461	214,921	214
55–59	86	16,384	525	416	165,380	252
60–64	44	9,902	444	263	89,983	292
Total	295	80,539	366	1,919	1,059,633	181

We wish to test the hypothesis $H_0: RR = 1$ versus $H_1: RR \neq 1$, where RR = rate ratio for women with a family history versus women without a family history within a specific age group. We use the test statistic

$$X^2 = \frac{(|A - E(A)| - .5)^2}{Var(A)} \sim \chi_1^2 \text{ under } H_0$$

where

$A =$ total observed number of cases over all strata $= \sum_{i=1}^{k} a_{1i}$

$E(A) =$ total expected number of cases over all strata

$$= \sum_{i=1}^{k} (a_{1i} + a_{2i})t_{1i}/(t_{1i} + t_{2i})$$

$Var(A)$ = variance of total number of cases over all strata

$$= \sum_{i=1}^{k} (a_{1i} + a_{2i})t_{1i}t_{2i}/(t_{1i} + t_{2i})^2$$

a_{1i}, t_{1i} = number of cases and person-years for women with a family history in the ith age group, $i = 1, \ldots, k$

a_{2i}, t_{2i} = number of cases and person-years for women without a family history in the ith age group, $i = 1, \ldots, k$

The p-value = $Pr(\chi_1^2 > X^2)$. We should only use this test if $Var(A) \geq 5$. In this case,

$$A = 295$$

$$E(A) = (12 + 124)(8564)/(8564 + 173{,}104)$$
$$+ \ldots + (44 + 263)(9902)/(9902 + 89{,}983)$$
$$= 6.41 + \ldots + 30.43 = 169.74$$

$$Var(A) = (12 + 124)(8564)(173{,}104)/(8564 + 173{,}104)^2$$
$$+ \ldots + (44 + 263)(9902)(89{,}983)/(9902 + 89{,}983)^2$$
$$= 6.11 + \ldots + 27.42 = 156.20$$

$$X^2 = \frac{(|295 - 169.74| - .5)^2}{156.20}$$

$$= 124.76^2/156.20 = 99.65$$

$$p\text{-value} = Pr(\chi_1^2 > 99.65) < .001$$

Therefore, women with a family history of breast cancer have significantly higher breast-cancer incidence rates than women without a family history, even after controlling for age.

13.4.1 Estimation of the Rate Ratio for Stratified Data

To estimate the rate ratio for stratified data, we compute a weighted average of the log rate ratios within specific strata and then take the antilog of the weighted average. This estimate is referred to as an *age-adjusted rate ratio*. Specifically, $\widehat{RR}$ = estimated age-adjusted rate ratio = $\exp(c)$, where

$$c = \sum_{i=1}^{k} w_i \ln(\widehat{RR}_i) / \sum_{i=1}^{k} w_i$$

and

$$\widehat{RR}_i = (a_{1i}/t_{1i})/(a_{2i}/t_{2i})$$

$$= \text{estimated rate ratio in the } i\text{th stratum}$$

$$w_i = (1/a_{1i} + 1/a_{2i})^{-1}$$

A $100\% \times (1 - \alpha)$ CI for RR is given by $[\exp(c_1), \exp(c_2)]$, where

$$c_1 = c - z_{1-\alpha/2}\sqrt{1/\sum_{i=1}^{k} w_i}$$

$$c_2 = c + z_{1-\alpha/2}\sqrt{1/\sum_{i=1}^{k} w_i}$$

For the breast-cancer data, we have

$$RR_1 = (140/10^5)/(72/10^5) = 1.96$$
$$w_1 = (1/12 + 1/124)^{-1} = 10.94$$

etc.

Age group	$\widehat{RR}_i$	w_i
30–39	1.96	10.94
40–44	1.91	27.79
45–49	1.95	47.39
50–54	1.75	59.26
55–59	2.09	71.27
60–64	1.52	37.69

Therefore,

$$c = \frac{10.94 \ln(1.96) + \ldots + 37.69 \ln(1.52)}{10.94 + \ldots + 37.69}$$

$$= 158.35/254.33 = 0.623$$
$$\widehat{RR} = \exp(0.623) = 1.86$$
$$c_1 = 0.623 - 1.96\sqrt{1/254.33}$$
$$= 0.623 - 0.123 = 0.500$$
$$c_2 = 0.623 + 0.123 = 0.745$$

The 95% CI for the age-adjusted RR is given by $[\exp(0.500), \exp(0.745)] = (1.65, 2.11)$.

We have assumed in the breast-cancer analyses that the underlying rate ratio is the same in all age groups. To check this assumption, a test for the homogeneity of the rate ratios over different strata can be performed [see Equation **(13.18)**, p. 600, text].

Sample size estimation for stratified (e.g., by age) incidence rate data depends on (1) the age-specific incidence rates of disease in the unexposed group, (2) the distribution of person-years within age-exposure-specific groups, (3) the true rate ratio under H_1, and (4) type I and type II errors. The appropriate formula is given in Equation **(13.19)**, p. 601, text. Similarly, estimation of power for incidence rate data with a specified sample size is given in Equation **(13.20)**, p. 603, text.

SECTION 13.5   Testing for Trend–Incidence Rate Data

It appears from Section 13.4 that the incidence rate of breast cancer increases with age. However, the rate of increase is greater prior to age 45 than after age 45. Suppose we wish to estimate the rate of increase after age 45 and wish to control for the confounding effect of family history. We assume a model of the form

$$\ln(p_{ij}) = \alpha_i + \beta S_j$$

where $\quad p_{ij}$ = incidence rate for the ith family history (Hx) group and jth age group, $i = 1, \dots, s, j = 1, \dots, k$

S_j = score variable for the jth age group

$$\hat{\beta} = L_{xy}^{\cdot}/L_{xx}$$

where $\quad L_{xy} = \sum_{i=1}^{s}\sum_{j=1}^{k} w_{ij} S_j \ln(\hat{p}_{ij}) - \left(\sum_{i=1}^{s}\sum_{j=1}^{k} w_{ij} S_j\right)\left[\sum_{i=1}^{s}\sum_{j=1}^{k} w_{ij} \ln(\hat{p}_{ij})\right]\bigg/\sum_{i=1}^{s}\sum_{j=1}^{k} w_{ij}$

$$L_{xx} = \sum_{i=1}^{s}\sum_{j=1}^{k} w_{ij} S_j^2 - \left(\sum_{i=1}^{s}\sum_{j=1}^{k} w_{ij} S_j\right)^2 \bigg/\sum_{i=1}^{s}\sum_{j=1}^{k} w_{ij}$$

$w_{ij} = a_{ij}$ = number of cases for the ith family Hx group and jth age group.

We will assume that $S_j = 1$ for age 45–49, 2 for age 50–54, 3 for age 55–59, and 4 for 60–64. In this case, $i = 1$ corresponds to family Hx = yes and $i = 2$ to family Hx = no. Therefore,

$$\sum_{i=1}^{2}\sum_{j=1}^{4} w_{ij} = 54 + \dots + 263 = 1779$$

$$\sum_{i=1}^{2}\sum_{j=1}^{4} w_{ij} S_j = 54(1) + \dots + 263(4) = 4233$$

$$\sum_{i=1}^{2}\sum_{j=1}^{4} w_{ij} S_j^2 = 54(1) + \dots + 263(16) = 11{,}987$$

$$L_{xx} = 11{,}987 - 4233^2/1779 = 1914.9$$

$$\sum_{i=1}^{2}\sum_{j=1}^{4} w_{ij} \ln(\hat{p}_{ij}) = 54 \ln(350 \times 10^{-5})$$
$$+ \dots + 263 \ln(292 \times 10^{-5})$$
$$= -10{,}678.5$$

$$\sum_{i=1}^{2}\sum_{j=1}^{4} w_{ij} S_j \ln(\hat{p}_{ij}) = 54(1) \ln(350 \times 10^{-5})$$
$$+ \dots + 263(4) \ln(292 \times 10^{-5})$$
$$= -25{,}092.7$$

$$L_{xy} = -25{,}092.7 - 4233(-10{,}678.5)/1779 = 316.2$$

Therefore, $\qquad \hat{\beta} = 316.2/1914.9 = 0.165$

Thus, breast-cancer incidence increases by $e^{0.165} = 18\%$ per 5-year increase in age, or roughly $e^{0.165/5} = 3.4\%$ per year after age 45. To obtain a standard error for the slope, we use

$$se(\hat{\beta}) = \sqrt{1/L_{xx}}$$
$$= \sqrt{1/1914.9} = .023$$

Therefore, a 95% CI for β is $0.165 \pm 1.96(0.023) = 0.165 \pm 0.045 = (0.120, 0.210)$. The corresponding 95% CI for the rate of increase is $[\exp(0.120), \exp(0.210)] = (13\%, 23\%)$ per 5-year increase in age or $(2.4\%, 4.3\%)$ per year after menopause. If we wish to test the hypothesis H_0: $\beta = 0$ versus H_1: $\beta \neq 0$, we can use the test statistic

$$z = \hat{\beta}/se(\hat{\beta}) = 0.165/0.023 = 7.23 \sim N(0, 1) \text{ under } H_0$$

The p-value $= 2 \times \Phi(z)$ if $z < 0$ or $2 \times [1 - \Phi(z)]$ if $z \geq 0$. In this case, $p = 2 \times [1 - \Phi(7.23)] < .001$. Thus, breast-cancer incidence significantly increases with age after age 45.

SECTION 13.6   Introduction to Survival Analysis

For incidence rates that remain constant over time, the methods in Sections 13.1–13.5 are applicable. If incidence rates vary over time, then methods of survival analysis are more appropriate. In the latter case, the hazard function plays a similar role to that of incidence density except that it is allowed to vary over time.

The **hazard function h(t)** is the instantaneous probability of having an event at time t (per unit time) given that one has not had an event up to time t.

A group of 1676 children were followed from birth to the 1st year of life for the development of otitis media (OTM), a common ear condition characterized by fluid in the middle ear (either the right or the left ear) and some clinical symptoms (e.g., fever, irritability). The presence of OTM was determined at every clinic visit. For simplicity, the data are reported here in 3-month intervals (i.e., it is assumed that the episodes of OTM occur exactly at 3 months, 6 months, 9 months, and 12 months).

Time (t)	Fail	Censored	Survive	Total	$h(t)$	$S(t)$
3	121	32	1523	1676	.0722	.9278
6	250	53	1220	1523	.1641	.7755
9	283	41	896	1220	.2320	.5956
12	182	714	0	896	.2031	.4746

The hazard at 3 months = the number of children who failed at 3 months/number of children available for examination at 3 months = $121/1676 = .0722$. Similarly, the hazards at 6, 9, and 12 months are .1641, .2320, and .2031, where the denominators at 6, 9, and 12 months are the number of children available for examination at these times and who had not had an episode prior to these times. Clearly, the hazard function appears to vary over time.

13.6.1   Censored Data

In a clinical study, some patients have not failed up to a specific point in time. If these patients are not followed any longer, they are referred to as *censored observations*. For example, in the OTM study, 32 children were censored at 3 months; these children were followed for 3 months, did not develop OTM at that time, but were not followed any further. An additional 1523 children also did not fail at 3 months but were followed further and are listed as *surviving*. The two groups are distinguished because they are treated differently in the estimation of *survival curves*.

SECTION 13.7 Estimation of Survival Curves

Suppose there are S_{i-1} subjects who survive up to time t_{i-1} and are not censored at that time. Of these subjects, S_i survive, d_i fail, and l_i are censored at time t_i. The Kaplan-Meier estimator of survival at time t_i is

$$S(t_i) = \left(1 - \frac{d_1}{S_0}\right) \times \left(1 - \frac{d_2}{S_1}\right) \times \ldots \times \left(1 - \frac{d_i}{S_{i-1}}\right)$$

For the OTM data, the estimated survival probability at 3 months = $(1 - 121/1676)$ = .9278. The estimated survival probability at 6 months = $(1 - 121/1676) \times (1 - 250/1523)$ = .7755, ... , etc. It is estimated that 47% of children "survive" (i.e., do not have OTM for 1 year). Thus, 53% of children have a 1st episode of OTM during the 1st year of life.

SECTION 13.8 Comparison of Survival Curves

It is interesting to compare survival curves between different subgroups. For example, one hypothesis is that infants of smoking mothers have more OTM than infants of nonsmoking mothers. How can we test this hypothesis? We have computed separate survival curves for infants of smoking and nonsmoking mothers as follows:

Time (months)	Fail	Censored	Survive	Total	h(t)	S(t)
Mother smokes = yes						
3	22	8	159	189	.1164	.8836
6	36	5	118	159	.2264	.6835
9	30	5	83	118	.2542	.5098
12	21	62	0	83	.2530	.3808
Mother smokes = no						
3	99	24	1364	1487	.0666	.9334
6	214	48	1102	1364	.1569	.7870
9	253	36	813	1102	.2296	.6063
12	161	652	0	813	.1980	.4862

Children of smoking mothers have a consistently higher hazard function and a lower survival probability than children of nonsmoking mothers. To compare these groups, we use the *log-rank test*. Using this test, we can take into account not only whether an event occurs, but when it occurs. We will test the hypothesis $H_0: h_1(t)/h_2(t) = 1$ versus $H_1: h_1(t)/h_2(t) = \beta \neq 1$, where $h_1(t)$ = hazard at time t for children of smoking mothers and $h_2(t)$ = hazard at time t for children of nonsmoking mothers. Thus, the hazards are allowed to vary over time, but the hazard ratio is assumed to be constant.

To implement this test, we construct separate 2×2 tables for each 3-month time interval of the following form:

OTM

		+	−	Total
Smoke	+	a_i	b_i	$a_i + b_i$
	−	c_i	d_i	$c_i + d_i$
		$a_i + c_i$	$b_i + d_i$	N_i

where

a_i = number of children of smoking mothers who had their 1st episode of OTM during the ith time interval (actually at time i, since we assume that events only occur exactly at 3, 6, 9, and 12 months of life, respectively)

b_i = number of children of smoking mothers who did not have a 1st episode of OTM at time i

c_i, d_i are defined similarly for children of nonsmoking mothers

We then perform the Mantel-Haenszel test over the collection of all such 2×2 tables using the test statistic

$$X_{LR}^2 = \frac{(|O - E| - .5)^2}{V} \sim \chi_1^2 \text{ under } H_0$$

where

$$O = \sum_{i=1}^{k} a_i$$

$$E = \sum_{i=1}^{k} E_i = \sum_{i=1}^{k} (a_i + b_i)(a_i + c_i)/N_i$$

$$V = \sum_{i=1}^{k} V_i$$

$$= \sum_{i=1}^{k} (a_i + b_i)(c_i + d_i)(a_i + c_i)(b_i + d_i)/[N_i^2(N_i - 1)]$$

The p-value $= Pr(\chi_1^2 > X_{LR}^2)$. We should only use this test if $V \geq 5$.

We construct the 2×2 tables for the OTM data as follows:

3 months
- -

OTM

		+	−	Total
Smoke	+	22	167	189
	−	99	1388	1487
Total		121	1555	1676

6 months

```
                      OTM
                  +       −      Total

Smoke    +      36      123      159

         −     214     1150     1364

       Total   250     1273     1523
```

9 months

```
                      OTM
                  +       −      Total

Smoke    +      30       88      118

         −     253      849     1102

       Total   283      937     1220
```

12 months

```
                      OTM
                  +       −      Total

Smoke    +      21       62       83

         −     161      652      813

       Total   182      714      896
```

Thus,

$$O = 22 + \ldots + 21 = 109$$

$$E = 189(121)/1676 + \ldots + 83(182)/896$$
$$= 13.64 + \ldots + 16.86 = 83.98$$

$$V = 121(1555)(189)(1487)/[1676^2(1675)]$$
$$+ \ldots + 182(714)(83)(813)/[896^2(895)]$$
$$= 11.24 + \ldots + 12.20 = 62.00$$

The test statistic is

$$X_{LR}^2 = (|109 - 83.98| - .5)^2/62.00$$
$$= 601.41/62.00 = 9.70 \sim \chi_1^2 \quad \text{under } H_0$$

The p-value $= Pr(\chi_1^2 > 9.70) = .002$. Thus, children of smoking mothers have significantly more OTM during the 1st year of life than children of nonsmoking mothers.

The Proportional-Hazards Model

The log-rank test is useful for comparing survival curves between two groups. It can be extended to allow one to control for a limited number of other covariates as well. However, if there are many covariates to be controlled for, then use of a *proportional-hazards model* is preferable. The *Cox proportional-hazards model* is a regression-type approach for modeling survival data. If we have k variables of interest, $x_1, \ldots, x_k$, then the hazard at time t is modeled as

$$h(t) = h_0(t) \exp(\beta_1 x_1 + \ldots + \beta_k x_k)$$

where $\quad h_0(t) =$ hazard function at time t for a subject with all covariate values $= 0$

13.9.1 Interpretation of Parameters

(1) Dichotomous predictors If x_i is a dichotomous predictor of OTM, (e.g., sex, where 1 = female and 0 = male), then $\exp(\beta_i)$ = ratio of hazards for OTM for females versus males, all other factors held constant.

(2) Continuous predictors If x_i is a continuous predictor variable [e.g., weight (lbs)], then $\exp(\beta_i)$ = hazard ratio for OTM for the heavier versus the lighter of two infants that are 1 lb apart in weight, all other variables held constant.

13.9.2 Hypothesis Testing

To test the hypothesis H_0: $\beta_i = 0$, all other $\beta_j \neq 0$, versus H_1: all $\beta_j \neq 0$, we use the test statistic $z = \hat{\beta}_i / se(\hat{\beta}_i) \sim N(0, 1)$ under H_0. The p-value $= 2 \times \Phi(z)$ if $z < 0$, or $= 2 \times [1 - \Phi(z)]$ if $z \geq 0$.

We have run the Cox proportional-hazards model with the OTM data, where the risk factors are

(1) Sex (coded as 1 if female and 0 if male)

(2) At least one sibling with Hx (history) of ear disease (coded as 1 if yes and 0 if no)

(3) Some sibs, but none with a Hx of ear disease (coded as 1 if yes and 0 if no)

(4) Ever breastfed during the 1st year of life (coded as 1 if yes and 0 if no)

(5) Mother smoking (ever during the 1st year of life) (coded as 1 if yes and 0 if no)

The results are as follows:

Variable	Regression coefficient	Standard error	z	p-value	Relative risk
Sex	−0.080	0.069	−1.16	.25	0.92
Sibs with Hx ear disease	0.331	0.073	4.53	< .001	1.39
Sibs, no Hx ear disease	0.207	0.120	1.73	.085	1.23
Ever breastfed	0.047	0.075	0.63	.53	1.05
Mother smoking	0.223	0.106	2.10	.035	1.25

There are significant effects of maternal smoking, even after controlling for the other covariates ($p = .035$). The relative risk = hazard ratio = 1.25; thus, children whose mothers smoke are 25% more likely to develop a 1st episode of OTM at any point in time in the 1st year of life compared with children whose mothers do not smoke, all other risk factors being equal. In addition, there was a significant effect of having a sibling with a Hx of ear disease ($p < .001$) versus not having any sibs ($RR = 1.39$, $p < .001$). The other covariates were not statistically significant.

In addition, it is possible to obtain confidence limits for both the true regression coefficients and RR's. Specifically, a 95% CI for the regression coefficient for maternal smoking = $0.223 \pm 1.96(0.106) = (0.015, 0.431)$. The corresponding 95% CI for RR is $[\exp(0.015), \exp(0.431)] = (1.02, 1.54)$.

PROBLEMS

Cancer

Refer to the data in Table 13.4 (text, p. 594).

13.1 Obtain an estimate of the rate ratio of breast cancer for past postmenopausal hormone users vs. never users, and provide a 95% CI about this estimate (control for age in your analysis).

13.2 Test the hypothesis that past use of postmenopausal hormones is associated with an increased (or decreased) risk of breast cancer compared with never users.

13.3 Suppose we plan a new study and assume that the true rate ratio comparing breast-cancer incidence of past to never users is 1.2. How many subjects do we need to study if (a) each subject is followed for 10 years, (b) the distribution of person-years among past and never users is the same as in Table 13.4 (p. 594, text), (c) the age-specific incidence rate of breast cancer among never users is the same as in Table 13.4, and (d) we wish to conduct a two-sided test with $\alpha = .05$ and power = 80%?

13.4 Suppose that the planners of the study expect to enroll 10,000 postmenopausal women who are past or never PMH users and follow each of them for 10 years. How much power would such a study have under the assumptions given in Problem 13.3?

Cardiovascular Disease

A proportional-hazards model was obtained using SAS PROC PHGLM based on the Framingham Heart Study data presented in Example 11.65 (p. 531, text). The exam when the 1st event occurred for an individual was used as the time of failure (exams 5, 6, 7, 8, or 9, where the exams are approximately 2 years apart). The results are presented in Table 13.1.

13.5 Assess the statistical significance of each of the variables in Table 13.1 and report a p-value.

13.6 Estimate the hazard ratio corresponding to each of the risk factors, and obtain a 95% CI corresponding to each point estimate. For cholesterol, glucose, and systolic blood pressure, compare people who differ by twofold. For number of cigarettes per day, compare people who differ by 1 pack (20 cigarettes) per day. Interpret each of the hazard ratios in words.

13.7 Compare your results with the corresponding results using multiple logistic-regression methods in Table 11.25 (p. 533, text). Which do you think is more informative for this data set?

TABLE 13.1 Proportional-hazards regression model, based on Framingham Heart Study data described in Example 11.65 (p. 531, text)

```
-----------------------------------------------
  PROPORTIONAL HAZARDS GENERAL LINEAR MODEL
                 PROCEDURE


          DEPENDENT VARIABLE: YRCOMB

            EVENT INDICATOR: EVT

1731 OBSERVATIONS
 163 UNCENSORED OBSERVATIONS
   0 OBSERVATIONS DELETED DUE TO MISSING
     VALUES
```

VARIABLE	BETA	STD. ERROR
AGE4554	0.68142175	0.23561312
AGE5564	1.09884661	0.24006788
AGE6569	1.39473508	0.33602235
LCHLD235	1.71310056	0.46486115
LSUGRD82	0.51543162	0.27821210
SMOKEM13	0.01689457	0.00580539
LBMID26	1.36816659	0.64236298
LMSYD132	2.61028377	0.51597055

```
-----------------------------------------------
```

Accident Epidemiology

An article was published in the *New York Post* [1] listing the five most dangerous New York City subway stations based on the reported number of felonies in 1992.

TABLE 13.2

- -

Five most dangerous stations in the New York City subway system
(based on number of felonies)

Station	Train line	1992 felonies	1992 paying patrons per avg. wkday
Times Square	1	289	105,826
42nd St./8th Ave.	A	254	105,826
34th St.	1	174	70,279
Grand Central	4	169	105,113
72nd St.	1	127	24,286

Source: Transit Police

- -

13.8 Estimate the incidence rate of the number of felonies per 1,000,000 paying patrons at the Times Square station. (Assume that there are 260 weekdays and 105 weekend days in a year and that the number of paying patrons per weekend day $= .5 \times$ number of paying patrons per weekday.)

13.9 Answer the question posed in Problem 13.8 for the Grand Central subway station.

13.10 Compare the incidence rates at the Times Square and Grand Central stations using hypothesis-testing methods and report a *p*-value.

13.11 Estimate the rate ratio of the true felony rate per paying patron at the Times Square station vs. the Grand Central station and provide a 95% CI about this estimate.

13.12 Suppose you use the Times Square station twice per day for 5 working days per week and 50 working weeks for 1 year. What is the probability that you will be a felony victim?

13.13 Answer the same question posed in Problem 13.12 for the Grand Central station.

TABLE 13.3

- -

Five most dangerous stations in the New York City subway system
(based on crime-to-1,000 passenger ratio)

Station	Train line	1992 felonies	1992 paying patrons per avg. wkday	Ratio per 1,000
Bowery	J	38	233	163
Dean St.	S	8	89	90
Livonia Ave.	L	52	627	83
Broad Channel	A	12	199	60
Atlantic Ave.	L	21	375	56

Source: Transit Police

- -

In the same article, the five stations with the highest felony rates per 1000 paying patrons were also given (see Table 13.3).

13.14 Compare the incidence rates at the Times Square and Bowery subway stations using hypothesis-testing methods and provide a *p*-value.

13.15 Estimate the rate ratio of the true felony rate at the Bowery station vs. the Times Square station. Suppose you use the Bowery station twice per weekday for 1 year (50 weeks). What is the probability that you will be a felony victim? Which station do you think is the more dangerous?

Cancer

The data in Table 13.4 provides the relationship between breast-cancer incidence rate and age at 1st birth by age, based on the Nurses' Health Study data from 1976–1990.

13.16 For parous women, assess if there is a trend relating age at 1st birth to breast-cancer incidence rate while controlling for age. Please report a *p*-value.

13.17 For parous women, estimate the percentage of increase (or decrease) in breast-cancer incidence rate for every 5-year increase in age at 1st birth for women of a given age. Provide a 95% CI for the percentage of increase (or decrease).

TABLE 13.4 Relationship between breast-cancer incidence rate and age at 1st birth after controlling for age, Nurses' Health Study, 1976–1990

				Age at 1st birth			
	Nulliparous	**< 20**	**20–24**	**25–29**	**30–34**	**35–39**	**40+**
Age	Cases/ person-years (Incidence rate)[a]	Cases/ person-years (Incidence rate)[a]	Cases/ person-years (Incidence rate)[a]	Cases/ person-years (Incidence rate)[a]	Cases/ person-years (Incidence rate)[a]	Cases/ person-years (Incidence rate)[a]	Cases/ person-years (Incidence rate)[a]
30–39	13/15,265 (85)	3/1,067 (281)	62/97,140 (64)	73/82,959 (88)	11/11,879 (93)	1/937 (107)	——
40–49	44/30,922 (142)	3/2,381 (126)	349/218,239 (160)	290/173,714 (167)	73/32,895 (222)	16/6264 (255)	3/813 (369)
50–59	102/35,206 (290)	3/1,693 (177)	327/156,907 (208)	454/181,244 (250)	133/43,721 (304)	43/11,291 (381)	13/2,341 (555)
60–69	32/11,594 (276)	3/261 (1149)	72/30,214 (238)	148/53,486 (277)	67/15,319 (437)	21/4,057 (518)	1/877 (114)

[a]Per 100,000 person-years.

SOLUTIONS

13.1 We estimate the rate ratio and obtain 95% CI's using the method in Equation (**13.17**) (text, p. 598). The point estimate is given $\exp(c)$, where

$$c = \frac{\sum_{i=1}^{5} w_i \ln(\widehat{RR}_i)}{\sum_{i=1}^{5} w_i}$$

where $\widehat{RR}_i$ = estimated rate ratio in the ith age group

$$= (a_{1i}/t_{1i})/(a_{2i}/t_{2i})$$

and $$w_i = (1/a_{1i} + 1/a_{2i})^{-1}$$

The 95% CI for $\ln(RR)$ is given by

$$\ln(\widehat{RR}) \pm 1.96\sqrt{1/\sum_{i=1}^{5} w_i} = (c_1, c_2)$$

The corresponding 95% CI for RR is given by $[\exp(c_1), \exp(c_2)] = (RR_1, RR_2)$. We compute $\widehat{RR}_i$, w_i, c, c_1, c_2, RR_1, and RR_2 using MINITAB, as shown on page 231. The point estimate = $\widehat{RR}$ = 0.99 with 95% CI = (0.82, 1.18).

13.2 We use the test procedure in Equation (**13.12**) (text, p. 596). For each age group, we compute E_i = the expected number of cases among past users = $(a_{1i} + a_{2i})t_{1i}/(t_{1i} + t_{2i})$ and the variance of the number of cases

among past users conditional on the total observed number of cases among past or never users in the ith age stratum given by

$$V_i = (a_{1i} + a_{2i})t_{1i}t_{2i}/(t_{1i} + t_{2i})^2$$

We then compute the test statistic

$$X^2 = \frac{[|A - E(A)| - .5]^2}{Var(A)} \sim \chi_1^2 \text{ under } H_0$$

where $A = \sum_{i=1}^{5} a_{1i}, E(A) = \sum_{i=1}^{5} E_i, Var(A) = \sum_{i=1}^{5} V_i$

We compute E_i, V_i, A, $E(A)$, $Var(A)$, and X^2 using MINITAB as shown on p. 232. The chi-square statistic = 0.011 with 1 df. The p-value = $Pr(\chi_1^2 > 0.011)$ = .92. Thus, past users do not have a significantly different incidence of breast cancer than never users.

13.3 We use the sample-size formula given in Equation (**13.19**) (text, p. 601). The total number of person-years required is

$$T = \frac{c_1 RR}{c_2(RR - 1)^2}\left(\frac{z_{1-\alpha/2}}{\sqrt{c_1}} + \frac{z_{1-\beta}}{\sqrt{c_2}}\right)^2$$

where

$$c_1 = RR \sum_{i=1}^{5} \frac{\lambda_{1i}\lambda_{2i}p_i}{RR\lambda_{1i} + \lambda_{2i}} = RR \sum_{i=1}^{5} c_{1i}$$

Solution 13.1

```
MTB > LET C6=(C3/C4)/(C1/C2)
MTB > LOGE C6 C7.
MTB > LET C8=1/(1/C1+1/C3)
MTB > LET C9=C7*C8
MTB > SUM C8 K1.
    SUM     =     114.30
MTB > SUM C9 K2.
    SUM     =     -1.5880

MTB > LET K3=K2/K1
MTB > PRINT K3
K3        -0.0138933
MTB > LET K4=EXP(K3)
MTB > PRINT K4
K4        0.986203
MTB > LET K5=K3-1.96*SQRT(1/K1)
MTB > LET K6=K3+1.96*SQRT(1/K1)
MTB > LET K7=EXP(K5)
MTB > LET K8=EXP(K6)

MTB > PRINT C1-C9 K1-K8.
```

K1	114.300	$=\Sigma W_i$
K2	-1.58801	$=\Sigma W_i LN(\hat{RR}_i)$
K3	-0.0138933	$=K2/K1$
K4	0.986203	$=\hat{RR}$
K5	-0.197223	$=C_1$
K6	0.169436	$=C_2$
K7	0.821008	$=RR_1$
K8	1.18464	$=RR_2$

ROW	NEVR_CAS	NEVR_PY	PAST_CAS	PAST_PY	AGE	RR	LN(RR)
1	5	4722	4	3835	39-44	0.98503	-0.0150805
2	26	20812	12	8921	45-49	1.07673	0.0739318
3	129	71746	46	26256	50-54	0.97440	-0.0259334
4	159	73413	82	39785	55-59	0.95163	-0.0495738
5	35	15773	29	11965	60-64	1.09227	0.0882616

ROW	W	W*LN(RR)
1	2.2222	-0.03351
2	8.2105	0.60702
3	33.9086	-0.87936
4	54.0996	-2.68192
5	15.8594	1.39977

$$c_2 = \sum_{i=1}^{5} \frac{\lambda_{1i}\lambda_{2i}p_i}{\lambda_{1i} + \lambda_{2i}} = \sum_{i=1}^{5} c_{2i}$$

$$\lambda_{1i} = t_{1i}/\sum_{j=1}^{2}\sum_{i=1}^{5} t_{ji}$$

$$\lambda_{2i} = t_{2i}/\sum_{j=1}^{2}\sum_{i=1}^{5} t_{ji}$$

$$p_i = a_{2i}/t_{2i}$$

The computations are performed using MINITAB as shown on p. 233. We see that 584,803 person-years are required. Since each woman is followed for 10 years, it follows that we need 58,480 subjects who are past or never users in the study.

13.4 We use the power formula given in Equation **(13.20)** (text, p. 603). The power is given by $\Phi(z)$, where

$$z = \frac{c_2\sqrt{T}\,|RR - 1|}{\sqrt{c_1}RR} - \sqrt{\frac{c_2}{c_1}}z_{1-\alpha/2}$$

We have that $T = 10,000 \times 10 = 100,000$ person-years, $z_{1-\alpha/2} = 1.96$, $RR = 1.2$. From the MINITAB output in Problem 13.3, $c_1 = 0.000468914$, $c_2 = 0.000416923$. Therefore,

$$z = \frac{0.000416923\sqrt{100,000}(0.2)}{\sqrt{0.000468914(1.2)}}$$

$$- \sqrt{\frac{0.000416923}{0.000468914}}(1.96)$$

$$= 1.112 - 0.943(1.96) = -0.737$$

The power $= \Phi(-0.737) = .231$. Thus, the study would have only 23% power. This makes sense since from Problem 13.3, 58,480 subjects were needed to achieve 80% power.

13.5 We compute the test statistic $z_i = b_i/se(b_i)$ and the p-value $= 2 \times [1 - \Phi(z_i)]$ for each variable in Table 13.1 as follows:

Test statistics and p-values for Framingham Heart Study data in Table 13.1

Variable	z_i	p-value
Age 45–54	2.89	.004
Age 55–64	4.58	< .001
Age 65–69	4.15	< .001
ln(cholesterol/235)	3.69	< .001
ln(glucose/82)	1.85	.064
Number of cigarettes per day − 13	2.91	.004
ln(BMI/26)	2.13	.033
ln(SBP/132)	5.06	< .001

Solution 13.2

```
MTB > LET C10=(C1+C3)*C4/(C2+C4)
MTB > NAME C11='V'
MTB > LET 'V'=(C1+C3)*C2*C4/(C2+C4)**2
MTB > NAME C10='E'


MTB > LET K9=SUM(C3)
MTB > LET K10=SUM(C10)
MTB > LET K11=SUM(C11)
MTB > LET K12=(ABS(K9-K10)-.5)**2/K11


MTB > PRINT C1-C11 K9-K12.
K9        173.000=A
K10       174.629=E(A)
K11       115.161=VAR(A)
K12       0.0110754=X2


ROW  NEVR_CAS  NEVR_PY  PAST_CAS  PAST_PY   AGE      RR       LN(RR)

 1         5     4722        4     3835   39-44  0.98503  -0.0150805
 2        26    20812       12     8921   45-49  1.07673   0.0739318
 3       129    71746       46    26256   50-54  0.97440  -0.0259334
 4       159    73413       82    39785   55-59  0.95163  -0.0495738
 5        35    15773       29    11965   60-64  1.09227   0.0882616

ROW        W   W*LN(RR)       E        V

 1    2.2222   -0.03351   4.0335   2.2258
 2    8.2105    0.60702  11.4014   7.9806
 3   33.9086   -0.87936  46.8848  34.3237
 4   54.0996   -2.68192  84.7028  54.9328
 5   15.8594    1.39977  27.6069  15.6984
```

Solution 13.3

```
MTB > LET C12=C1/C2
MTB > NAME C12='PI'
MTB > LET C13=C4/(SUM(C2)+SUM(C4))
MTB > NAME C13='LAM1I'
MTB > LET C14=C2/(SUM(C2)+SUM(C4))
MTB > NAME C14='LAM2I'
MTB > LET C15=C12*C13*C14/(1.2*C13+C14)
MTB > NAME C15='C1I'
MTB > LET C16=C12*C13*C14/(C13+C14)
MTB > NAME C16='C2I'
MTB > LET K13=1.2*SUM(C15)
MTB > LET K14=SUM(C16)
MTB > LET K15=K13*1.2/(K14*.04)*(1.96/SQRT(K13)+0.84/SQRT(K14))**2
MTB > SAVE 'PR1313.MTW'.

WORKSHEET SAVED INTO FILE: PR1313.MTW
MTB > PRINT C1-C16 K13-K15
K13      0.000468914=C1
K14      0.000416923=C2
K15      584803     =T
```

ROW	NEVR_CAS	NEVR_PY	PAST_CAS	PAST_PY	AGE	RR	LN(RR)
1	5	4722	4	3835	39-44	0.98503	-0.0150805
2	26	20812	12	8921	45-49	1.07673	0.0739318
3	129	71746	46	26256	50-54	0.97440	-0.0259334
4	159	73413	82	39785	55-59	0.95163	-0.0495738
5	35	15773	29	11965	60-64	1.09227	0.0882616

ROW	W	W*LN(RR)	E	V	PI	LAM1I	LAM2I
1	2.2222	-0.03351	4.0335	2.2258	0.0010589	0.013833	0.017033
2	8.2105	0.60702	11.4014	7.9806	0.0012493	0.032179	0.075072
3	33.9086	-0.87936	46.8848	34.3237	0.0017980	0.094709	0.258798
4	54.0996	-2.68192	84.7028	54.9328	0.0021658	0.143510	0.264811
5	15.8594	1.39977	27.6069	15.6984	0.0022190	0.043159	0.056895

ROW	C1I	C2I
1	0.0000074	0.0000081
2	0.0000265	0.0000281
3	0.0001183	0.0001247
4	0.0001883	0.0002016
5	0.0000501	0.0000545

13.6 The estimated hazard ratio $= RR_i = \exp(b_i\Delta_i)$ with 95% CI $= [\exp(c_1), \exp(c_2)]$, where $c_1 = b_i\Delta_i - 1.96\Delta_i se(b_i)$, $c_2 = b_i\Delta_i + 1.96\Delta_i se(b_i)$. The results are given as follows:

Variable	Δ_i	RR_i	95% CI
Age 45–54	1	1.98	(1.25, 3.14)
Age 55–64	1	3.00	(1.87, 4.80)
Age 65–69	1	4.03	(2.09, 7.79)
ln(cholesterol/235)	ln(2)	3.28	(1.74, 6.17)
ln(glucose/82)	ln(2)	1.43	(0.98, 2.09)
Number of cigarettes per day − 13	20	1.40	(1.12, 1.76)
ln(SBP/132)	ln(2)	6.11	(3.03, 12.31)

The relative risk for age 45–54 represents the hazard for coronary heart disease (CHD) for a 45–54-year-old man compared with a 35–44-year-old man. Thus, the 45–54-year-old man is twice as likely to have a first CHD event over the short term as a 35–44-year-old man, all other factors being equal. Similarly, a 55–64-year-old man and a 65–69-year-old man are 3 and 4 times as likely, respectively, to have a first CHD event over the short term, as a 35–44-year-old man (all other factors being equal). The relative risk of 3.28 for cholesterol represents the hazard ratio for subject A versus subject B, where subject A has double the cholesterol of subject B, all other factors being equal. Since cholesterol is represented on the ln scale, the difference between subject A versus subject B's ln(cholesterol) level $= \ln(2)$. Similarly, the relative risks for glucose and SBP are 1.43 and 6.11, respectively, and represent the comparative hazards for two subjects whose glucose or SBP values are in the ratio of 2:1, all other factors being equal. Finally, the relative risk for number of cigarettes per day $= 1.40$ and represents the comparative hazard for two subjects who differ by 1 pack (i.e., 20 cigarettes) per day (e.g., a smoker of 1 pack per day versus a nonsmoker), all other factors being equal.

13.7 The results are very similar to those given in Table 11.25 (text, p. 533). A proportional-hazards model is probably more appropriate because it takes into account when an event occurs versus a multiple logistic regression model, which only takes into account whether an event occurs, but not when it occurs. For a relatively rare event such as CHD (163 events out of 1731 subjects), there is not much difference between the two methods. Another advantage of a proportional-hazards model is the ability to use time-dependent covariates; i.e., covariates that change over time. This is particularly important for risk factors that change a lot over time, such as cigarette-smoking habits. In this example, for simplicity, only fixed covariates defined at baseline were used. In the Framingham Heart Study data set, risk factors are updated every 2 years, and therefore it is possible to include time-dependent covariates.

13.8 The number of patrons at the Times Square station in 1992 $= 260(105{,}826) + 105(105{,}826/2) = 33{,}070{,}625$. The incidence rate per paying patron $= 289/33{,}070{,}625 = 8.74/10^6$.

13.9 The number of paying patrons at the Grand Central station in 1992 $= 260(105{,}113) + 105(105{,}113/2) = 32{,}847{,}813$. The incidence rate per paying patron $= 169/32{,}847{,}813 = 5.14/10^6$.

13.10 We will use the large-sample procedure given in Equation **(13.5)** (text, p. 588), since the variance $(V_1) \geq 5$. We use the test statistic

$$z = \frac{|a_1 - E_1| - .5}{\sqrt{V_1}} \sim N(0, 1) \quad \text{under } H_0$$

where $a_1 = 289$

$$\begin{aligned} E_1 &= (289 + 169)(33{,}070{,}625)/ \\ &\quad (33{,}070{,}625 + 32{,}847{,}813) \\ &= 458(.502) = 229.8 \end{aligned}$$

$$V_1 = 458(.501)(.499) = 114.5$$

Thus, $z = (|289 - 229.8| - .5)/\sqrt{114.5} = 58.73/10.70 = 5.49 \sim N(0, 1)$ under H_0. The p-value $= 2 \times [1 - \Phi(5.49)] < .001$. Thus, there is a significant difference between the felony rates at the two stations.

13.11 We have that $\widehat{RR} = (8.74/10^6)/(5.14/10^6) = 1.70$. Furthermore, $Var[\ln(\widehat{RR})] = 1/289 + 1/169 \doteq 0.00938$. Thus, a 95% CI for $\ln(RR) = \ln(1.70) \pm 1.96\sqrt{0.00938} = 0.530 \pm 0.190 = (0.340, 0.720)$. The corresponding 95% CI for $RR = [\exp(0.340), \exp(0.720)] = (1.40, 2.05)$.

13.12 You use the station $50 \times 5 \times 2 = 500$ times per year. We will assume that the number of felonies (X) is Poisson-distributed with parameter $\mu = 500(8.74/10^6) = 0.0044$. We wish to compute $Pr(X \geq 1) = 1 - Pr(X = 0) = 1 - \exp(-0.0044) = .0044$.

13.13 We will assume that the number of felonies (X) is Poisson-distributed with parameter $\mu = 500(5.14/10^6) = 0.0026$. We wish to compute $Pr(X \geq 1) = 1 - Pr(X = 0) = 1 - \exp(-0.0026) = .0026$.

13.14 The number of paying patrons in 1992 at the Bowery station $= 233(250) + (233/2)(105) = 70{,}483$. The incidence rate per paying patron $= 38/70{,}483 = 539.1/10^6$. We cannot use the large-sample test procedure because $V_1 = (289 + 38)[70{,}483(33{,}070{,}625)/(70{,}483 + 33{,}070{,}625)^2] = 327(.0021)(.9979) = 0.69 < 5$. Instead, we must use the exact test given in Equation **(13.6)** (text, p. 590). We have

$$p\text{-value} = 2 \sum_{k=38}^{327} \binom{327}{k}(.0021)^k(.9979)^{327-k}$$

$$= 2\left[1 - \sum_{k=0}^{37} \binom{327}{k}(.0021)^k(.9979)^{327-k}\right]$$

This is tedious to compute, but fortunately is not necessary. From the recursion rule for binomial probabilities,

$$Pr(0) = (.9979)^{327} = .4985$$

$$Pr(1) = (21/9979)(327)(.4985) = .3474$$

$$Pr(2) = (21/9979)(326/2)(.3474) = .1207$$

$$Pr(3) = (21/9979)(325/3)(.1207) = .0279$$

Thus, $Pr(X \geq 4) = 1 - Pr(X \leq 3) = 1 - (.4985 + .3474 + .1207 + .0279 = .0056$. Since $Pr(X \geq 38) \leq Pr(X \geq 4) = .0056$, it follows that we can reject H_0 and conclude that the incidence rate of felonies is significantly higher at the Bowery station.

13.15 The estimated rate ratio $(RR) = (539.1/10^6)/(8.74/10^6) = 61.7$. Thus, a patron is about 60 times as likely to be a felony victim at the Bowery station than at the Times Square station. Indeed, if a patron uses the Bowery station every workday, twice a day, for a year (i.e., $50 \times 5 \times 2 = 500$ times, allowing for a 2-week vacation), then the probability of at least one felony $= Pr(X \geq 1) = 1 - Pr(X = 0) = 1 - \exp(-\mu)$. In this case, $\mu = 500(539.1/10^6) = .2696$. Thus, $Pr(X \geq 1) = 1 - \exp(-0.2696) = .236$! Thus, there is a 24% probability of being a felony victim over a 1-year period at the Bowery station versus a 0.4% probability at the Times Square station. Clearly, the Bowery station is more dangerous.

13.16 We use the test for trend for incidence-rate data given in Equation **(13.23)** (text, p. 605). We assume a model of the form

$$\ln(p_{ij}) = \alpha_i + \beta S_j$$

where

p_{ij} = incidence rate in the ith age group and the jth age at 1st birth group

S_j = score for the jth age at 1st birth group, $j = 1, \ldots, 6$

We will use scores of $1, \ldots, 6$ for the age at 1st birth groups $= < 20$, 20–24, 25–29, 30–34, 35–39, and 40+, respectively. We wish to test the hypothesis $H_0: \beta = 0$ versus $H_1: \beta \neq 0$. We use the test statistic $z = \hat{\beta}/se(\hat{\beta}) \sim N(0, 1)$ under H_0, where $\hat{\beta} = L_{xy}/L_{xx}$.

$$L_{xy} = \sum_{i=1}^{4} \sum_{j=1}^{6} a_{ij} S_j \ln(\hat{p}_{ij})$$

$$- \left(\sum_{i=1}^{4} \sum_{j=1}^{6} a_{ij} S_j \right) \times \left[\sum_{i=1}^{4} \sum_{j=1}^{6} a_{ij} \ln(\hat{p}_{ij}) \right] / \sum_{i=1}^{4} \sum_{j=1}^{6} a_{ij}$$

$$L_{xx} = \sum_{i=1}^{4} \sum_{j=1}^{6} a_{ij} S_j^2 - \left(\sum_{i=1}^{4} \sum_{j=1}^{6} a_{ij} S_j \right)^2 / \sum_{i=1}^{4} \sum_{j=1}^{6} a_{ij}$$

$$se(\hat{\beta}) = 1/\sqrt{L_{xx}}$$

We perform the computations using MINITAB as shown on p. 236. We see that $\hat{\beta} = 0.243$, $se(\hat{\beta}) = 0.025$, $z = 9.63$. Clearly, the p-value $= 2[1 - \Phi(z)] < .001$. Thus, among parous women, there is a significant relationship between breast-cancer incidence and age at 1st birth.

13.17 The estimated percentage increase in breast-cancer incidence for every 5-year increase in age at 1st birth $= \hat{\Delta} = \exp(0.243) = 27.5\%$. A 95% CI for $\Delta = [\exp(c_1), \exp(c_2)]$ where

$$c_1 = \hat{\beta} - 1.96 \, se(\hat{\beta})$$

$$= 0.243 - 1.96(0.025) = 0.194$$

$$c_2 = 0.243 + 1.96(0.025) = 0.293$$

Thus, the 95% CI for $\Delta = [\exp(0.194), \exp(0.293)] = (21.4\%, 34.0\%)$.

REFERENCE

[1] *New York Post*, August 30, 1993, p. 4.

Solution 13.16

```
MTB > LET C14 = C1+2*C3+3*C5+4*C7+5*C9+6*C11
MTB > LET C15 = C1+4*C3+9*C5+16*C7+25*C9+36*C11
MTB > LET C16 = C1+C3+C5+C7+C9+C11

MTB > LET K1=SUM(C15)-SUM(C14)**2/SUM(C16)
MTB > PRINT K1
K1        1566.64=Lxx
MTB > LET C17 = LOGE(C1/C2)
MTB > LET C18 = LOGE(C3/C4)
MTB > LET C19 = LOGE(C5/C6)
MTB > LET C20 = LOGE(C7/C8)
MTB > LET C21 = LOGE(C9/C10)
MTB > LET C22 = LOGE(C11/C12)
MTB > LET C22 = LOGE(C11/C12)
                            J
*** VALUES OUT OF BOUNDS DURING OPERATION AT J
    MISSING RETURNED 1 TIMES

 NOTE * THE DATA SCREEN WAS USED TO CHANGE THE WORKSHEET
MTB > LET C23=C1*C17+C3*C18+C5*C19+C7*C20+C9*C21+C11*C22
MTB > LET C24=C1*C17+2*C3*C18+3*C5*C19+4*C7*C20+5*C9*C21+6*C11*C22
MTB > LET LET K2=SUM(C24)-SUM(C14)*SUM(C23)/SUM(C16)
MTB > PRINT K2
K2         381.004=Lxy
MTB > LET C25=K2/K1
MTB > LET C26=1/SQRT(K1)
MTB > LET C27=C25/C26
MTB > CDF C27;
SUBC>    NORMAL 0.0 1.0.
    9.6260    1.0000
```

ROW	<20_CA	<20_PY	20-24_CA	20-24_PY	25-29_CA	25-29_PY	30-34_CA
1	3	1067	62	97140	73	82959	11
2	3	2381	349	218239	290	173714	73
3	3	1693	327	156907	454	181244	133
4	3	261	72	30214	148	53486	67

ROW	30-34_PY	35-39_CA	35-39_PY	40-44_CA	40-44_PY	AGE GRP	C14	C15
1	11879	1	937	0	0	30-39	395	1109
2	32895	16	6264	3	813	40-49	1961	5685
3	43721	43	11291	13	2341	50-59	2844	9068
4	15319	21	4057	1	877	60-69	970	3256

ROW	C16	C17	C18	C19	C20	C21	C22
1	150	-5.87399	-7.35677	-7.03564	-6.98463	-6.84268	0.00000
2	734	-6.67666	-6.43827	-6.39528	-6.11062	-5.96999	-5.60212
3	973	-6.33565	-6.17345	-5.98950	-5.79523	-5.57056	-5.19338
4	312	-4.46591	-6.03939	-5.88996	-5.43216	-5.26368	-6.77651

ROW	C23	C24	BETA	SE(BETA)	Z
1	-1071.02	-2812.2	0.243198	0.0252648	9.62597
2	-4680.02	-12440.6			
3	-5834.77	-16900.0			
3	-1801.22	-5547.4			

```
K1        1566.64=Lxx
K2         381.004=Lxy
```

CHAPTER FOURTEEN
MISCELLANEOUS PROBLEMS

SECTION 14.1 Introduction

The purpose of this chapter is to give the student practice in choosing the appropriate statistical analysis for a particular dataset. Problems are selected from the material in all the chapters and are arranged in random order. Although this is not a substitute for experience in "real" data analysis, it should increase confidence in applying the knowledge obtained from this text in new situations. If you are unsure as to which method to use for a particular problem, consult the Flow Chart of Methods of Statistical Inference (p. 671–675, text).

PROBLEMS

Hypertension

Suppose we wish to design an experiment to assess the effectiveness of a new drug for treating hypertensive patients. We decide to declare the drug "effective" if the mean diastolic blood pressure (DBP) of the 50 patients participating in the study drops by at least 2 mm Hg after using the drug daily for 1 month.

Let us set up this study as a hypothesis-testing problem and assume that the population standard deviation of the before–after blood-pressure differences is *known* to be 10 mm Hg.

14.1 If we identify declaring the drug effective with rejecting the null hypothesis, then what is the significance level of the test?

14.2 Suppose that the underlying population before–after DBP decrease from using this drug is 3 mm Hg. What is the power of the test in Problem 14.1?

Gynecology

In a 1985 study, 89 of 283 women with primary tubal infertility (cases) and 640 of 3833 control women reported ever having used an IUD [1].

14.3 What is the best point estimate of the rate of IUD use among case and control women, respectively?

14.4 Provide a 95% CI for the estimates in Problem 14.3.

Emergency Medicine

Mannitol and Decadron are drugs that are often administered to patients with severe head injury when they are admitted to the emergency room of a hospital. One hypothesis is that this type of treatment would be more beneficial if administered by paramedics to patients in the field before they are transported to the hospital. To plan such a study, a pilot study is performed, whereby 4 of 10 patients with field treatment and 6 of 10 patients with no field treatment die before discharge from the hospital.

14.5 If a clinical trial is planned based on a two-tailed test with $\alpha = .05$, assuming that the pilot study results are valid, how much power would such a study have if 20 patients are randomized to each of the two groups?

14.6 How many patients are needed in each group to achieve an 80% power with the preceding study design if the differences found in the pilot study are assumed to be valid?

14.7 If the true mortality rate in the field-treated group were actually .50 rather than .40, how would these new data affect the power estimate in Problem 14.5 and the sample-size estimate in Problem 14.6? (Provide only a qualitative answer.)

Cardiovascular Disease

Much controversy has arisen recently on the possible association of myocardial infarction (MI) and coffee drinking. Suppose the information in Table 14.1 on coffee drinking and prior MI status is obtained from 200 60–64-year-old males in the general population.

TABLE 14.1 Coffee drinking and prior MI status

Coffee drinking (cups/day)	MI in last 5 years	Number of people
0	Yes	3
0	No	57
1	Yes	7
1	No	43
2	Yes	8
2	No	42
3 or more	Yes	12
3 or more	No	28
	Total yes	30
	Total no	170

14.8 Test for the association between history of MI and coffee-drinking status, which is categorized as follows: 0 cups, 1 or more cups.

14.9 Suppose coffee drinking is categorized as follows: 0 cups, 1 cup, 2 cups, 3 or more cups. Perform a test to investigate whether or not there is a "dose–response" relationship between these two variables (i.e., does the prevalence of prior MI increase or decrease as the number of cups of coffee per day increases?).

Cardiology

A study was conducted among a group of people who underwent coronary angiography at Baptist Memorial Hospital, Memphis, Tennessee, between January 1, 1972, and December 31, 1986 [2]. A group of 1493 people with coronary-artery disease were identified and were compared with a group of 707 people without coronary-artery disease (the controls). Both groups were age 35–49 years. Risk-factor information was collected on each group. Among cases, the mean serum cholesterol was 234.8 mg/dL with standard deviation = 47.3 mg/dL. Among controls, the mean serum cholesterol was 215.5 mg/dL with standard deviation = 47.3 mg/dL.

14.10 What test is appropriate to determine if the true mean serum cholesterol is different between the two groups?

14.11 Implement the test in Problem 14.10 and report a two-sided p-value.

14.12 What power did the study have to detect a significant difference using a two-sided test with $\alpha = 0.05$ if the true mean difference is 10 mg/dL between the two groups and the true standard deviations are the same as the sample standard deviations in the study?

Obstetrics

A new drug therapy is proposed for the prevention of low-birthweight deliveries. A pilot study undertaken, using the drug on 20 pregnant women, found that the mean birthweight in this group is 3500 g with a standard deviation of 500 g.

14.13 What is the standard error of the mean in this case?

14.14 What is the difference in interpretation between the standard deviation and standard error in this case (in words)?

14.15 Suppose $\dfrac{x - \mu_0}{s/\sqrt{n}} = 2.73$ and a one-sample t test is performed based on 20 subjects. What is the two-tailed p-value?

Health-Services Administration

A comparison is made between demographic characteristics of patients using fee-for-service practices and prepaid group health plans. Suppose the data presented in Table 14.2 are found.

TABLE 14.2 Characteristics of patients using fee-for-service practices and prepaid group health plans

Characteristic	Fee-for-service			Prepaid group health plans		
	Mean	sd	n	Mean	sd	n
Age (years)	58.1	6.2	57	52.6	4.3	48
Education (years)	11.8	0.7	57	12.7	0.8	48

14.16 Test for a significant difference in the variance of age between the two groups.

14.17 What is the appropriate test to compare the mean ages of the two groups?

14.18 Perform the test in Problem 14.17 and report a p-value.

14.19 Compute a 95% CI for the mean age difference between the two groups.

14.20 Answer Problem 14.16 for number of years of education.

14.21 Answer Problem 14.17 for number of years of education.

14.22 Answer Problem 14.18 for number of years of education.

14.23 Answer Problem 14.19 for number of years of education.

Hypertension

An investigator wishes to determine if sitting upright in a chair versus lying down on a bed will affect a person's blood pressure. The investigator decides to use each of 10 patients as his or her own control and collects systolic blood-pressure (SBP) data in both the sitting and lying positions, as given in Table 14.3.

TABLE 14.3 Effect of position on SBP level (mm Hg)

Patient	Sitting upright	Lying down
1	142	154
2	100	106
3	112	110
4	92	100
5	104	112
6	100	100
7	108	120
8	94	90
9	104	104
10	98	114

14.24 What is the distinction between a one-sided and a two-sided hypothesis test in this problem?

14.25 Which hypothesis test is appropriate here? Why?

14.26 Using an α level of .05, test the hypothesis that position affects the level of blood pressure.

14.27 What is the p-value of the preceding test? Compute either the specific p-value or a range within which the p-value lies.

Hypertension

A frequently encountered phenomenon in Western society is the positive correlation of blood pressure and age. One finding of recent interest is the apparent absence of this correlation in many underdeveloped countries. Suppose that in a sample of 903 U.S. males, the observed correlation is .402, whereas in a sample of 444 Polynesian males, the observed correlation is .053.

14.28 Test for whether or not a significant difference exists between the two underlying correlations.

Cardiovascular Disease

Many studies have demonstrated that the level of HDL (high-density lipoprotein) cholesterol is positively related to alcohol consumption. This relationship has intrigued researchers because the level of HDL cholesterol is inversely correlated with the incidence of heart disease.

One possible mechanism was explored by Kuller et al. in an analysis of participants in the MRFIT study relating the level of HDL cholesterol to the level of SGOT, a parameter commonly used to assess liver function [3]. The data in Table 14.4 were presented.

TABLE 14.4 The relationship of HDL cholesterol to SGOT in the MRFIT population

SGOT	Mean HDL cholesterol (mg/dL)
≤ 9	40.0
10–12	41.2
13–14	42.3
15–16	42.8
17–18	43.8
19–20	43.6
≥ 21	46.5

Source: Reprinted with permission of the *American Journal of Epidemiology,* *117*(4), 406–418, 1983.

14.29 Fit a regression line predicting mean HDL cholesterol as a function of SGOT. For this purpose assume that all levels of SGOT ≤ 9 and ≥ 21 are 9.5 and 20.5, respectively, and that levels of SGOT within all other groups occur at the midpoint of the group. Assume also that group assignments are made after rounding; for example, 10–12 represents an actual range from 9.5 to 12.5. Using this convention, if x = SGOT and y = mean HDL cholesterol, then

$$\sum_{i=1}^{7} x_i = 107 \qquad \sum_{i=1}^{7} x_i^2 = 1740.5 \qquad \sum_{i=1}^{7} y_i = 300.2$$

$$\sum_{i=1}^{7} y_i^2 = 12{,}900.2 \qquad \sum_{i=1}^{7} x_i y_i = 4637.6$$

14.30 Test the line fitted in Problem 14.29 for statistical significance and report a *p*-value.

14.31 What is the best estimate of the HDL cholesterol for an average person with SGOT level of 11, and what is the standard error of this estimate?

14.32 Suppose that a mistake was made in Problem 14.29 and the SGOT groups were assigned by truncation rather than rounding (e.g., 10–12 would represent 10.0–12.99 rather than 9.5–12.5). It would follow that average levels of SGOT within each group would increase by 0.5 (assume this is also the case for the groups ≤ 9 and ≥ 21, respectively). What effect does the truncation-assignment rule have on the estimates of the regression parameters in Problem 14.29? (No further calculation should be necessary.)

As part of the same study, a multiple-regression analysis was performed relating the change in HDL cholesterol to baseline levels and changes in other risk factors. The results are presented in Table 14.5.

14.33 Interpret the coefficients for change in SGOT and drinks/week in the multiple-regression model.

TABLE 14.5 Multiple regression of change in HDL cholesterol from the second screening visit to the 48-month follow-up visit (follow-up minus baseline) among participants receiving standard medical care (*n* = 5,112)

Variable	Regression coefficient	Regression coefficient/ standard error
Age (years)	−0.0054	−0.3
Screen 2 HDL cholesterol (mg/dL)	−0.3109	−29.3
Screen 1 diastolic blood pressure (mm Hg)	−0.0322	−1.7
Screen 1 cigarettes/day	−0.0324	−4.6
Change in diastolic blood pressure (mm Hg)	0.1233	9.1
Change in thiocyanate[a] (mg/L)	−0.0106	−4.4
Change in body mass index (kg/m^2)	−1.2702	−18.0
Change in drinks/week	0.1053	9.5
Change in SGOT (IU/L)	0.1160	8.4

[a] A biochemical marker used to measure the actual number of cigarettes smoked recently.

Nutrition

In a dietary reproducibility study, sucrose intake was assessed on the 1st day of each of 4 weeks separated in time by approximately 3 months. A one-way ANOVA was run with results given in Table 14.6.

TABLE 14.6 One-way ANOVA results for sucrose data, dietary reproducibility study (*n* = 173)

	SS	df	MS
Between	473,103.2	172	2750.6
Within	366,725.4	519	706.6

14.34 Is there a significant difference between mean sucrose intake for different individuals?

14.35 Estimate the between-person and within-person components of variation for sucrose intake.

14.36 What is the intraclass correlation for sucrose intake? What does it mean, in words?

Cardiology

Suppose a drug to relieve anginal pain is effective within 8 hours in 30% of 100 patients studied.

14.37 Derive an upper one-sided 95% CI for the percentage of patients who could get pain relief from the drug within 8 hours.

14.38 Suppose that if the patients are untreated, 15% will be free from pain within 8 hours. Assuming that the drug either benefits the patients or has no effect at all, what is your opinion on the effectiveness of the drug?

Suppose that in the same study 20% of the patients become pain-free within 4 hours of administration of the drug.

14.39 Derive an upper one-sided CI for the percentage of patients who could get pain relief from the drug within 4 hours.

14.40 Suppose 10% of untreated patients will be pain-free within 4 hours. Assuming that the drug either benefits the patients or has no effect at all, what is your opinion on the effectiveness of the drug?

Ophthalmology

A new drug is developed to relieve the ocular symptoms of hay fever. The drug is composed of two components, A and B: A is supposed to relieve itching of the eye and B is supposed to prevent redness; the combination is supposed to relieve both itching and redness. Federal regulations require that each component be proven effective both separately and in combination. Three experiments are performed on a group of 25 patients with hay fever. In the

TABLE 14.7 Comparison of drug A versus placebo for the relief of redness and itching in hay-fever patients

Subject	Redness	Itching	Subject	Redness	Itching
1	0	+	13	+	+
2	0	+	14	0	0
3	−	−	15	0	0
4	+	−	16	−	0
5	0	+	17	−	+
6	+	+	18	−	0
7	+	+	19	0	0
8	−	0	20	−	+
9	−	+	21	−	0
10	0	−	22	−	+
11	+	+	23	0	0
12	0	+	24	−	0
			25	−	0

first experiment, drug A is administered to a randomly selected eye and a placebo is administered to the other eye, and the change from baseline is noted for each eye. The data are given in Table 14.7. A plus sign represents more improvement in the drug-treated eye; a minus sign represents more improvement in the placebo-treated eye; 0 represents equal improvement in both eyes.

14.41 Why might a nonparametric statistical test be useful in comparing drug A with placebo for this experiment?

14.42 Why is it important to administer the placebo to the second eye of the same person rather than to a different group of people with hay fever?

14.43 What nonparametric statistical test would you use to compare drug A with placebo for redness or itching? Why?

14.44 Compare redness in the drug-A and placebo eyes and report a p-value.

14.45 Compare itching in the drug-A and placebo eyes and report a p-value.

In the second experiment, drug B is administered to a randomly selected eye and a placebo is administered to the other eye. The data are given in Table 14.8.

TABLE 14.8 Comparison of drug B versus placebo for the relief of redness and itching in hay-fever patients

Subject	Redness	Itching	Subject	Redness	Itching
1	+	−	13	+	0
2	+	−	14	+	0
3	+	+	15	+	0
4	+	+	16	0	0
5	+	+	17	+	−
6	0	−	18	+	−
7	+	0	19	+	+
8	+	+	20	+	+
9	+	0	21	+	0
10	0	0	22	+	0
11	+	−	23	+	0
12	+	+	24	+	0
			25	+	0

TABLE 14.9 Comparison of combination of drugs A and B versus placebo for the relief of redness and itching in hay-fever patients

Subject	Redness	Itching	Subject	Redness	Itching
1	+	+	13	+	+
2	+	0	14	+	0
3	+	+	15	+	0
4	+	+	16	+	+
5	+	+	17	+	+
6	+	−	18	+	+
7	+	+	19	0	+
8	+	0	20	0	+
9	+	+	21	+	0
10	+	+	22	+	0
11	+	−	23	+	0
12	+	−	24	+	0
			25	+	+

14.46 Compare redness in the drug-B and placebo eyes and report a p-value.

14.47 Compare itching in the drug-B and placebo eyes and report a p-value.

In the third experiment, the combination of drugs A and B is administered to a randomly selected eye, and a placebo is administered to the other eye. The data are given in Table 14.9.

14.48 Compare redness in the combination and placebo eyes and report a p-value.

14.49 Compare itching in the combination and placebo eyes, and report a p-value.

14.50 Summarize the results of the three experiments as concisely as possible.

Cancer

Suppose a clinical trial is performed to assess the effect of a new treatment for cancer of the esophagus. No attempt is made to match the patients in any way because the number of cases is too small. It is found that of 100 patients who were given the standard treatment, 6 lived for 3 years and 5 lived for 5 years. Correspondingly, of 47 patients who were given the new treatment, 10 survived for 3 years, whereas 2 survived for 5 years.

14.51 Is there any evidence that the new treatment is helpful for the 3-year prognosis of the patient?

14.52 Is there any evidence that the new treatment is helpful for the 5-year prognosis?

14.53 Suppose a person has survived for 3 years. Is

there any evidence for a treatment effect on the prognosis for the next 2 years?

14.54 Suppose we want to use the Wilcoxon signed-rank test and have a rank sum of 27 based on 7 untied pairs. Evaluate the significance of the results.

14.55 Answer Problem 14.54 for a rank sum of 65 based on 13 untied pairs.

14.56 Answer Problem 14.54 for a rank sum of 90 with 15 untied pairs.

Cancer

A study was conducted to look at the relationship between self-reports of mole count and prevalence of malignant melanoma [4]. 98 cases of malignant melanoma occurring between 1976 and 1982 and 190 age-matched controls among members of the Nurses' Health Study (age 30–54) were selected for analysis. The total number of moles in the left and right arms combined was compared in cases and controls. The data are given in Table 14.10.

14.57 Test the hypothesis that cases, on average, have more moles (or less moles) than controls.

14.58 Estimate the odds ratio relating specific mole-count levels (1–5, 6–15, 16–30, 31+, respectively) versus having no moles to malignant melanoma. Provide a 95% CI for each odds-ratio estimate.

The prevalence in 1973 of a rare disease in two communities, A and B, was compiled and it was found that of 100,000 people in community A, 5 have the disease,

TABLE 14.10 Number of moles in left and right arm (combined)

	0	1-5	6-15	16-30	31+	Total
Cases	16	20	20	12	30	98
Controls	57	40	45	20	28	190

Source: Reprinted with permission of the *American Journal of Epidemiology, 127*(4), 703–712, 1988.

whereas of 200,000 people in community B, only 1 has the disease.

14.59 Test whether the underlying prevalence rates are significantly different.

Hypertension
Refer to Table 9.20 (p. 340, text).

14.60 Compute the intraclass correlation between replicate SBP measurements, and provide a 95% CI about this estimate.

Hypertension
Suppose we are interested in investigating the effect of race on level of blood pressure. A study was conducted in Evans County, Georgia, comparing the mean level of blood pressure among whites and blacks for different age–sex groups [5]. The mean and standard deviation of systolic blood pressure (SBP) among 25–34-year-old white males were reported as 128.6 mm Hg and 11.1 mm Hg, respectively, based on a large sample.

14.61 Suppose the mean SBP of 38 25–34-year-old black males is reported as 135.7 mm Hg. If the standard deviation of white males is assumed to hold for black males as well, then test if the underlying mean SBPs are the same in the two groups.

14.62 The actual reported standard deviation among 25–34-year-old black males is 12.5 mm Hg. Test the hypothesis that the underlying variance of blood pressure for white and black males is the same.

14.63 Suppose the actual underlying mean for black males is 135 mm Hg. What is the power of the test in Problem 14.61 in this case if we assume that the standard deviations for white males and black males are the same?

Suppose we do *not* assume that the standard deviation of 11.1 mm Hg is correct for 25–34-year-old black males.

14.64 Test the hypothesis that the underlying mean SBPs are the same in the two groups.

14.65 Derive a 95% CI for the mean SBP for 25–34-

year-old black males under the assumptions in Problem 14.64.

14.66 Relate your answers to Problems 14.64 and 14.65.

14.67 Suppose we have two normally distributed samples of sizes 12 and 16 with sample standard deviations of 12.6 and 6.2, respectively. Test if the variances are significantly different in the two samples.

Endocrinology, Bone and Joint Disease
A study was conducted to relate serum estrogens to bone metabolism.

14.68 Suppose that a correlation of .52 is found between serum estradiol and the bone density of the lumbar region of the spine, as computed by CT scan among 23 postmenopausal women. Test for the statistical significance of these results and report a *p*-value.

14.69 Similarly, a correlation of .34 is found between serum estradiol and the bone density of the distal radius of the nondominant arm among the same 23 postmenopausal women. Test for the statistical significance of these results and report a *p*-value.

14.70 Finally, a correlation coefficient of .395 was computed between the two preceding measures of bone metabolism in Problems 14.68 and 14.69 based on the same group of women. Test for the statistical significance of these results and report a *p*-value.

14.71 Suppose in a previous large study, other investigators found a correlation of .5 between the two measures of bone metabolism in Problems 14.68 and 14.69. Test for whether or not the sample results in Problem 14.70 are compatible with these previous results. Report a *p*-value.

Cancer
A case–control study was performed to look at the effect of cigarette smoking on the risk of cervical cancer [6]. There were 230 cervical-cancer cases and 230 controls consisting of women ages 22–74 matched to the cases for age within 5-year intervals and admitted to the hospital for acute conditions other than malignancy, hormonal, or gynecologic disorders. The smoking habits of these women are given in Table 14.11. Suppose we wish to compare each smoking group versus nonsmokers.

14.72 Compute the odds ratio and associated 95% CI's comparing ex-smokers, current smokers of < 15 cigarettes per day, and current smokers of ≥ 15 cigarettes per day, respectively, versus nonsmokers.

14.73 What is your overall interpretation of the data?

TABLE 14.11 Association between cigarette smoking and the risk of cervical cancer

	Nonsmokers	Ex-smokers	Current smokers < 15 cigarettes/day	Current smokers ≥15 cigarettes/day
Cervical-cancer cases	155	13	33	29
Controls	169	21	22	18

Source: Reprinted with permission by the *American Journal of Epidemiology, 123*(1), 22–29, 1986.

TABLE 14.12 Relationship of type of treatment for Hodgkin's disease and incidence of leukemia

Treatment procedure	Number of leukemias	Number of person-years of follow-up after treatment
(A) No intensive therapy	2	3322.8
(B) Either intensive radiotherapy or chemotherapy	4	1858.2

Source: Reprinted with permission of the *Journal of the National Cancer Institute, 67*(4), 751–760, 1981.

Cancer

An investigation was conducted looking at the incidence of leukemia after treatment among people with Hodgkin's disease according to the type of treatment received [7]. The data in Table 14.12 were reported.

Suppose we assume under the null hypothesis that the type of treatment has no effect on the incidence of leukemia.

14.74 What is the expected proportion of leukemias in the no-intensive-therapy group under the null hypothesis?

14.75 What test procedure can be used to test if there is a different incidence density of leukemia in group A than expected under the null hypothesis, if the total number of leukemias (= 6) are considered fixed?

14.76 Implement the test procedure in Problem 14.75 and report a *p*-value.

Infectious Disease, Pulmonary Disease, Hospital Epidemiology

A study was performed to relate reactivity to tuberculin to job activity within a hospital. In particular, suppose that according to statewide, age-specific rates, 31% of nurses in a hospital are expected to have positive tuberculin skin tests. In the hospital under study, 93 out of 221 nurses are tuberculin positive.

14.77 Give a 95% CI for the proportion of positive tuberculin skin tests among nurses.

14.78 How does the nurses' observed rate compare with the expected rate of 31%?

Occupational Health

To assess quantitatively the association between benzene exposure and cancer mortality, a study was conducted among 1165 men who worked in a rubber-hydrochloride plant with occupational exposure to benzene. A comparison was made between observed cancer mortality during the period 1950–1981 and expected cancer mortality as derived from U.S. white male mortality rates over the same time period. The data in Table 14.13 were obtained [8].

TABLE 14.13 Mortality data for workers exposed to benzene

	Number of deaths	
Cause of death	Observed	Expected
Total cancer	69	66.8
Multiple myeloma	4	1.0

14.79 What are the standardized mortality ratios (SMRs) for total cancer and multiple myeloma in this group?

14.80 Perform a test to compare observed and expected total cancer mortality in this cohort. Report a *p*-value.

14.81 Perform a test comparing observed and expected multiple myeloma rates in this cohort. Report a *p*-value.

Infectious Disease

Refer to the lymph-node data in Table 8.18 (p. 292, text).

14.82 Use nonparametric methods to assess whether there are systematic differences between the assessments of Doctor A versus Doctor B regarding the number of palpable lymph nodes among sexual contacts of AIDS or ARC patients.

14.83 Are there any differences between your conclusions in Problem 14.82 and your conclusions using parametric methods in Problem 8.69 (p. 292, text)?

14.84 Use nonparametric methods to estimate the between-observer reproducibility in the assessment of the number of palpable lymph nodes.

Cardiovascular Disease, Nutrition

A study was performed of the effects of walnut consumption on serum lipids and blood pressure [9]. The same subjects were administered a regular cholesterol-lowering diet (called the *reference diet*) for 4 weeks and an experimental diet substituting 3 servings of walnuts per day for portions of some foods in the reference diet for 4 weeks. The total calories, protein, carbohydrates, and fat were kept the same during each period. The order in which the diets were administered was randomized for individual subjects. 18 subjects completed the protocol. Table 14.14 shows the values obtained for mean HDL cholesterol at the end of each 4-week dietary period.

14.85 What does it mean that the "order in which the diets were administered was randomized"?

14.86 From the information provided, estimate the standard error of the difference in mean HDL cholesterol between the two diets.

14.87 Perform a statistical test to test if the HDL cholesterol was the same for the reference versus walnut diets. Report a p-value as precisely as possible using the tables in the text.

14.88 Which of the following statements, if any, are inconsistent with the data in Table 14.14 and why? (no additional calculation is necessary): **(a)** p-value in Problem 14.87 $< .05$; **(b)** 90% CI for mean difference =

$(-4.1, -0.5)$, **(c)** 98% CI for mean difference = $(-4.2, -0.4)$.

Pulmonary Disease

A study was performed comparing radiologic findings between bronchitic and normal individuals. Eighteen randomly selected 18–29-year-old bronchitic women and 17 randomly selected 18–29-year-old normal women were chosen for study. It was found that 7 out of 18 bronchitic women had abnormal chest x-ray findings, while 0 out of 17 normal women had abnormal findings.

14.89 What test procedure can be used to compare the percentage of abnormal findings between bronchitic and normal women?

14.90 Implement the test procedure and report a p-value (versus the one-sided alternative that bronchitic women have more abnormal radiologic findings than normal women).

14.91 Is the preceding design a paired- or independent-samples design?

Cardiovascular Disease

Some reports in the literature have noted that there is an inverse relationship between moderate alcohol consumption and cholesterol levels. To test this hypothesis, a questionnaire is administered to a group of workers at a particular company as to their alcohol consumption, and blood samples are taken to measure their cholesterol levels. The workers are subdivided into those who report no alcohol consumption, those who drink ≤ 2 oz of alcohol, and those who drink more than 2 oz of alcohol on an average day. The 23 nondrinkers had average cholesterol levels of 205.6 mg/dL with a standard deviation of 25.3 mg/dL. The 15 light drinkers had average cholesterol levels of 182.7 mg/dL with a standard deviation of 21.9 mg/dL. The 12 heavy drinkers had average cholesterol levels of 199.8 mg/dL with a standard deviation of 30.3 mg/dL.

14.92 Test the hypothesis that there is an overall difference in underlying mean cholesterol levels among these three groups.

TABLE 14.14 Comparison of HDL cholesterol while on walnut diet versus reference diet

	Reference diet	Walnut diet	Mean difference (walnut diet- reference diet)	95% CI for mean difference
HDL cholesterol (mg/dL)	47 ± 11[a]	45 ± 10[a]	-2.3	$(-3.9, -0.7)$

[a]Mean ± sd.

14.93 Use the method of multiple comparisons to test for significant differences in mean cholesterol levels between each pair of groups.

Obstetrics

Assume that the distribution of birthweights in the general population is normal with mean 120 oz and standard deviation 20 oz. We wish to test a drug that, when administered to mothers in the prenatal period, will reduce the number of low-birthweight infants. We anticipate that the mean birthweight of the infants whose mothers are on the drug will be $1/2$ lb heavier than that for the general newborn population.

14.94 How large a sample is needed to have an 80% chance of finding a significant difference if a one-sided test with significance level .05 is used?

14.95 Answer Problem 14.94 for a 90% chance.

SOLUTIONS

14.1 Under H_0, we have $\bar{d} \sim N(0, \sigma^2/n)$ where $\bar{d}$ = mean difference of before–after, $\sigma = 10$, $n = 50$. Therefore,

$$\alpha = Pr[\bar{d} > 2.0 \,|\, \bar{d} \sim N(0, \sigma^2/n)]$$

$$= 1 - \Phi\left(\frac{2}{\sigma/\sqrt{n}}\right) = 1 - \Phi\left(\frac{2\sqrt{n}}{\sigma}\right)$$

$$= 1 - \Phi\left(\frac{2\sqrt{50}}{10}\right)$$

$$= 1 - \Phi(1.414) = .079$$

14.2 We have

$$Power = Pr[\bar{d} > 2.0 \,|\, \bar{d} \sim N(3, \sigma^2/n)]$$

$$= 1 - \Phi\left[\frac{(2 - 3)\sqrt{n}}{\sigma}\right]$$

$$= 1 - \Phi\left(\frac{-\sqrt{n}}{\sigma}\right) = \Phi\left(\frac{\sqrt{n}}{\sigma}\right)$$

$$= \Phi\left(\frac{\sqrt{50}}{10}\right) = \Phi(0.71) = .760$$

Thus, if the underlying reduction in blood pressure is 3 mm Hg, then there is a 76% chance of showing that the drug is effective, based on a study with 50 subjects.

14.3 Case women, $89/283 = .314$; control women, $640/3833 = .167$.

14.4 A 95% CI for case women is given by

$$.314 \pm 1.96\sqrt{.314(.686)/283}$$

$$= .314 \pm 1.96(.0276) = .314 \pm .054$$

$$= (.260, .369)$$

A 95% CI for control women is given by

$$.167 \pm 1.96\sqrt{.167(.833)/3833}$$

$$= .167 \pm 1.96(.0060)$$

$$= .167 \pm .012 = (.155, .179)$$

14.5 The power is given in **(10.27)** (text, p. 385) by

$$Power = \Phi\left[\frac{\Delta}{\sqrt{(p_1q_1 + p_2q_2)/n}}\right.$$

$$\left. - z_{1-\alpha/2}\frac{\sqrt{2\bar{p}\bar{q}}}{\sqrt{p_1q_1 + p_2q_2}}\right]$$

In this case, $p_1 = .6$, $p_2 = .4$, $\Delta = |.6 - .4| = .2$, $\alpha = .05$, $n = 20$, $\bar{p} = (.6 + .4)/2 = .5$. We have

$$Power = \Phi\left\{\frac{.2}{\sqrt{[.6(.4) + .4(.6)]/20}}\right.$$

$$\left. - z_{.975}\frac{\sqrt{2(.5)(.5)}}{\sqrt{.6(.4) + .4(.6)}}\right\}$$

$$= \Phi\left(\frac{.2}{.155} - 1.96 \times \frac{.707}{.693}\right)$$

$$= \Phi(1.291 - 2.000) = \Phi(-0.709) = .239$$

Thus, such a study would only have a 24% chance of detecting a significant difference.

14.6 We use the sample-size formula in **(10.26)** (text, p. 384) as follows:

$$n = \frac{(\sqrt{2\bar{p}\bar{q}}\,z_{1-\alpha/2} + \sqrt{p_1q_1 + p_2q_2}\,z_{1-\beta})^2}{\Delta^2}$$

$$= \frac{[\sqrt{2(.5)(.5)}\,z_{.975} + \sqrt{.6(.4) + .4(.6)}\,z_{.8}]^2}{.2^2}$$

$$= \frac{[.707(1.96) + 0.693(0.84)]^2}{.04}$$

$$= \frac{1.968^2}{.04} = 96.8$$

Thus, we need to enroll 97 patients in each group in order to achieve an 80% power.

14.7 The estimated power in Problem 14.5 would decrease and the required sample size in Problem 14.6 would increase.

14.8 We form the following 2×2 table:

Coffee drinking

		0	1+	
MI status	Yes	3	27	30
	No	57	113	170
		60	140	200

We use the chi-square test for 2×2 tables since all expected values are ≥ 5 (the smallest expected value $= (60 \times 30)/200 = 9.0$). We have the following test statistic:

$$X^2 = \frac{n(|ad - bc| - n/2)^2}{(a + b)(c + d)(a + c)(b + d)}$$

$$= \frac{200(|3(113) - 57(27)| - 100)^2}{30 \times 170 \times 60 \times 140}$$

$$= 5.65 \sim \chi^2_1 \text{ under } H_0$$

From the chi-square table (Table 6, Appendix, text), we see that $\chi^2_{1,.975} = 5.02$, $\chi^2_{1,.99} = 6.63$, and thus $.01 < p < .025$. Therefore, there is a significant association between prior MI status and coffee drinking, with coffee drinkers having a higher incidence of prior MI.

14.9 The results in Problem 14.8 would be more convincing if we were able to establish a "dose–response" relationship between coffee drinking and MI status with the risk of MI increasing as the number of cups per day of coffee consumption increases. For this purpose, we form the following 2×4 table:

Number of cups of current coffee consumption

		0	1	2	3+	
MI status	Yes	3	7	8	12	30
	No	57	43	42	28	170
		60	50	50	40	200

We perform the chi-square test for trend in binomial proportions using the score statistic 1, 2, 3, 4 for the coffee-consumption groups 0, 1, 2, 3+ respectively. From (10.36) (text, p. 397), we have the test statistic $X^2_1 = A^2/B$, where

$$A = \sum_{i=1}^{k} x_i S_i - \bar{x}S = 3(1) + 7(2) + 8(3) + 12(4)$$

$$- 30 \times \left[\frac{60(1) + \dots + 40(4)}{200} \right]$$

$$= 89 - \frac{30(470)}{200} = 89 - 70.5 = 18.5$$

$$B = \bar{p}\,\bar{q}\left[\sum_{i=1}^{k} n_i S_i^2 - \left(\sum_{i=1}^{k} n_i S_i \right)^2 / N \right]$$

$$= \frac{30}{200} \times \frac{170}{200} \times \left[60(1^2) + \dots + 40(4^2) - \frac{470^2}{200} \right]$$

$$= .1275(1350 - 1104.50) = .1275(245.5)$$

$$= 31.30$$

Therefore, $X^2_1 = 18.5^2/31.30 = 10.93 \sim \chi^2_1$ under H_0. Since $\chi^2_{1,.999} = 10.83 < X^2_1$, it follows that $p < .001$. Thus, there is a significant linear trend, with the rate of prior MI increasing as the number of cups/day of coffee consumed increases.

14.10 Let x_i be the serum cholesterol for the ith case and y_j be the serum cholesterol for the jth control. We assume $x_i \sim N(\mu_1, \sigma_1^2)$, $y_j \sim N(\mu_2, \sigma_2^2)$. We wish to test the hypothesis H_0: $\mu_1 = \mu_2$ versus H_1: $\mu_1 \neq \mu_2$. Since $s_1 = s_2$, we will assume equal variances, i.e., $\sigma_1^2 = \sigma_2^2$. Thus, we use the two-sample t test for independent samples with equal variances.

14.11 We have the test statistic

$$t = \frac{\bar{x} - \bar{y}}{\sqrt{s^2(1/n_1 + 1/n_2)}}$$

$$= \frac{234.8 - 215.5}{\sqrt{47.3^2(1/1493 + 1/707)}}$$

$$= \frac{19.3}{2.159} = 8.94 \sim t_{1493+707-2} = t_{2198}$$

Since $t > t_{120,.9995} = 3.373 > t_{2198,.9995}$, it follows that $p < 2 \times (1 - .9995)$ or $p < .001$.

14.12 We use the power formula

$$\text{Power} = \Phi\left[-z_{1-\alpha/2} + \frac{\sqrt{n_1}\,|\Delta|}{\sqrt{\sigma_1^2 + \sigma_2^2/k}} \right]$$

In this case $\alpha = .05$, $z_{1-\alpha/2} = z_{.975} = 1.96$, $\Delta = 10$, $n_1 = 1493$, $k = n_2/n_1 = 707/1493 = 0.474$, $\sigma_1^2 = \sigma_2^2 = 47.3^2$. Thus, we have

$$\text{Power} = \Phi\left[-1.96 + \frac{\sqrt{1493}(10)}{\sqrt{47.3^2 + 47.3^2/0.474}} \right]$$

$$= \Phi\left(-1.96 + \frac{386.4}{83.44} \right) = \Phi(-1.96 + 4.63)$$

$$= \Phi(2.671) = .996$$

Thus, there is a 99.6% chance of finding a significant difference.

14.13 sem $= 500/\sqrt{20} = 111.8$

14.14 The standard deviation is a measure of variability for the birthweight of *one* infant. The standard error of the mean is a measure of variability for the *mean* birth-weight of a group of n infants (in this case $n = 20$). The standard error will always be smaller than the standard deviation because a mean of more than one birthweight will be less variable in repeated samples than an individual birthweight.

14.15 $p = 2 \times Pr(t_{19} > 2.73)$. We refer to Table 5 (Appendix, text) and note that $t_{19,.99} = 2.539$, $t_{19,.995} = 2.861$. Since $2.539 < 2.73 < 2.861$, it follows that $2 \times (1 - .995) < p < 2 \times (1 - .99)$ or $.01 < p < .02$. The exact p-value obtained by computer is $p = .013$.

14.16 We assume that the distribution of age is normally distributed in each group. We have the test statistic $F = s_1^2/s_2^2 = 6.2^2/4.3^2 = 2.08 \sim F_{56,47}$ under H_0. Since $F = 2.08 > F_{24,40,.975} = 2.01 > F_{56,47,.975}$, it follows that $p < .05$ and there are significant differences between the variances.

14.17 A two-sample t test with unequal variances.

14.18 We have the test statistic

$$t = \frac{\bar{x}_1 - \bar{x}_2}{\sqrt{s_1^2/n_1 + s_2^2/n_2}} = \frac{58.1 - 52.6}{\sqrt{6.2^2/57 + 4.3^2/48}}$$

$$= \frac{5.5}{1.029} = 5.34$$

We determine the effective df from **(8.22)** (text, p. 272) as follows:

$$d' = \frac{(6.2^2/57 + 4.3^2/48)^2}{(6.2^2/57)^2/56 + (4.3^2/48)^2/47} = 99.5$$

Therefore, $t = 5.34 \sim t_{99}$ under H_0. Since $t > t_{60,.9995} = 3.460 > t_{99,.9995}$, it follows that $p < .001$ and there are significant differences in age between the two groups.

14.19 The 95% CI is given by

$$\bar{x}_1 - \bar{x}_2 \pm t_{99,.975}\sqrt{s_1^2/n_1 + s_2^2/n_2}$$

$$= 5.5 \pm t_{99,.975}(1.029)$$

We estimate $t_{99,.975}$ using the computer and obtain 1.984. Therefore, the 95% CI is given by

$$5.5 \pm 1.984(1.029)$$

$$= 5.5 \pm 2.0 = (3.5, 7.5)$$

14.20 We assume that number of years of education is normally distributed in each group. We have the test statistic $F = s_2^2/s_1^2 = 0.8^2/0.7^2 = 1.31 \sim F_{47,56}$ under H_0. Since $F < F_{\infty,60,.975} = 1.48 < F_{47,56,.975}$, it follows that $p > .05$ and there is no significant difference between the variances.

14.21 The two-sample t test with equal variances.

14.22 The pooled-variance estimate is given by

$$s^2 = \frac{0.7^2(56) + 0.8^2(47)}{103} = 0.558$$

Therefore, the test statistic is obtained from

$$t = \frac{\bar{x}_1 - \bar{x}_2}{\sqrt{s^2(1/n_1 + 1/n_2)}}$$

$$= \frac{11.8 - 12.7}{\sqrt{.558(1/57 + 1/48)}}$$

$$= \frac{-0.9}{0.146} = -6.1 \sim t_{103} \text{ under } H_0$$

Since $|t| > t_{60,.9995} = 3.460 > t_{103,.9995}$, it follows that $p < .001$. Therefore, there are highly significant differences in educational level between the two groups.

14.23 A 95% CI is given by $-0.9 \pm t_{103,.975}(0.146)$. We obtain $t_{103,.975}$ from the computer $= 1.983$. Therefore, the 95% CI $= -0.9 \pm 1.983(0.146) = -0.9 \pm 0.29 = (-1.19, -0.61)$. Thus, in summary, patients in fee-for-service practices are older and have less years of education than patients in prepaid group health plans.

14.24 The distinction between a one-sided and two-sided test in this case is that for a one-sided test we would test the hypothesis H_0: $\mu_1 = \mu_2$ versus H_1: $\mu_1 > \mu_2$ or, alternatively, H_0: $\mu_1 = \mu_2$ versus H_1: $\mu_1 < \mu_2$, where μ_1 represents mean SBP (systolic blood pressure) sitting upright and μ_2 represents mean SBP lying down. For a two-sided test we would test the hypothesis H_0: $\mu_1 = \mu_2$ versus H_1: $\mu_1 \neq \mu_2$.

14.25 A two-sided test is appropriate here, since we have no preconceived notions as to the relative orderings of μ_1 and μ_2 and would be equally interested in the outcomes $\mu_1 < \mu_2$ and $\mu_1 > \mu_2$.

14.26 Since each person is serving as his or her own control, we are dealing with highly dependent samples and must use the paired t test. We test the hypothesis H_0: $\mu_d = 0$ versus H_1: $\mu_d \neq 0$, where $d_i = $ sitting SBP $-$ lying SBP for the ith person and $d_i \sim N(\mu_d, \sigma_d^2)$. We have the following set of within-pair differences: -12, -6, $+2$, -8, -8, 0, -12, $+4$, 0, -16. Compute the test statistic

$$t = \bar{d}/(s_d/\sqrt{n})$$

$$= -5.60/(6.786/\sqrt{10}) = -5.60/2.146$$

$$= -2.61$$

Under H_0, $t \sim t_9$ and we have from Table 5 (Appendix, text) that $t_{9,.975} = 2.262 < |t|$.

Therefore, H_0 would be rejected at the 5% level and the hypothesis that the position affects SBP, with the sitting upright position having the lower blood pressure, would be accepted.

14.27 We have $t_{9,.975} = 2.262$, $t_{9,.99} = 2.821$. Since $t_{9,.99} = 2.821 > |t|$, $.02 < p < .05$. Using the computer, the exact p-value $= p = 2 \times Pr(t_9 > 2.61) = .028$.

14.28 We wish to test the hypothesis $H_0: \rho_1 = \rho_2$ versus $H_1: \rho_1 \neq \rho_2$. We use the test statistic

$$\lambda = \frac{z_1 - z_2}{\sqrt{\dfrac{1}{n_1 - 3} + \dfrac{1}{n_2 - 3}}} \sim N(0, 1) \text{ under } H_0$$

We have

$$z_1 = .5 \times \ln\left(\frac{1 + .402}{1 - .402}\right) = .5 \times \ln(2.344) = 0.426$$

$$z_2 = .5 \times \ln\left(\frac{1 + .053}{1 - .053}\right) = .5 \times \ln(1.112) = 0.053$$

Thus,

$$\lambda = \frac{0.426 - 0.053}{\sqrt{\dfrac{1}{900} + \dfrac{1}{441}}} = \frac{0.373}{0.058}$$

$$= 6.42 \sim N(0, 1) \text{ under } H_0$$

Clearly, $p < .001$ and thus, there is a highly significant difference between the two correlations.

14.29 We fit the least-squares line with $b = L_{xy}/L_{xx}$ and $a = (\Sigma y_i - b\Sigma x_i)/n$. We have that

$$L_{xx} = 1740.5 - \frac{107^2}{7} = 104.93$$

$$L_{xy} = 4637.6 - \frac{107(300.2)}{7} = 48.83$$

$$L_{yy} = 12,900.2 - \frac{(300.2)^2}{7} = 25.93$$

It follows that $b = 48.83/104.93 = 0.465$, $a = [300.2 - 0.465(107)]/7 = 35.77$. Thus, the least-squares line is $y = 35.77 + 0.465x$.

14.30 We use the regression F test. We have that

$$\text{Reg SS} = \frac{L_{xy}^2}{L_{xx}} = \frac{48.83^2}{104.93} = 22.72 = \text{Reg MS}$$

$$\text{Res SS} = \text{Total SS} - \text{Reg SS}$$

$$= 25.93 - 22.72 = 3.21$$

$$\text{Res MS} = \frac{\text{Res SS}}{n - 2} = \frac{3.21}{5} = 0.641$$

Therefore, we have the test statistic $F = \text{Reg MS}/\text{Res MS} = 22.72/0.641 = 35.4 \sim F_{1,5}$ under H_0. Since $F_{1,5,.995} = 22.78$, $F_{1,5,.999} = 47.18$ and $22.78 < F < 47.18$, it follows that $1 - .999 < p < 1 - .995$ or $.001 < p < .005$. Thus, there is a significant association between HDL cholesterol and SGOT.

14.31 We have that $\hat{y} = 35.77 + 0.465(11) = 40.89$ mg/dL. The standard error of this estimate is given by

$$se = \sqrt{\text{Res MS}\left[\frac{1}{n} + \frac{(x - \bar{x})^2}{L_{xx}}\right]}$$

$$= \sqrt{0.641\left[\frac{1}{7} + \frac{(11 - 107/7)^2}{104.93}\right]}$$

$$= \sqrt{0.641(0.143 + 0.175)} = 0.45 \text{ mg/dL}$$

14.32 If we represent the old and new x-values by x_{old} and x_{new}, then we are given that $x_{\text{new}} = x_{\text{old}} + 0.5$. It follows that $\bar{x}_{\text{new}} = \bar{x}_{\text{old}} + 0.5$. This should have no effect on L_{xy} or L_{xx}, since the terms in x in these expressions are of the form $x_i - \bar{x}$, which should be unaffected by a change in origin. Thus $b_{\text{new}} = (L_{xy})_{\text{new}}/(L_{xx})_{\text{new}} = (L_{xy})_{\text{old}}/(L_{xx})_{\text{old}} = b_{\text{old}}$, and the estimate of the slope is unaffected by the change in x. However, the estimate of the intercept will change in the following way:

$$a_{\text{new}} = \bar{y}_{\text{new}} - b_{\text{new}}\bar{x}_{\text{new}}$$

$$= \bar{y}_{\text{old}} - b_{\text{old}}(\bar{x}_{\text{old}} + 0.5)$$

$$= (\bar{y}_{\text{old}} - b_{\text{old}}\bar{x}_{\text{old}}) - 0.5b_{\text{old}}$$

$$= a_{\text{old}} - 0.5b_{\text{old}} < a_{\text{old}}$$

Thus, the estimate of intercept is reduced by $0.5b_{\text{old}}$.

14.33 The results indicate that after simultaneously controlling for the effects of all other risk factors in the multiple-regression model, people with (a) increases in alcohol consumption and (b) increases in SGOT levels, respectively, showed the greatest increases in HDL-cholesterol levels after controlling for the effects of all other risk factors. The results as regards SGOT levels supports the data presented in Table 14.4 relating the HDL-cholesterol levels to SGOT levels at one point in time.

14.34 We perform the F test for one-way ANOVA. We have the F statistic $F = 2750.6/706.6 = 3.89 \sim F_{172,519}$ under H_0. Since $F > F_{24,120,.999} = 2.40 > F_{172,519,.999}$, it follows that $p < .001$. Thus, there is a significant difference between mean sucrose intake for different people.

14.35 The between-person variance $= \sigma_A^2$ is estimated by $(2750.6 - 706.6)/4 = 511.0$. The within-person variance $= \sigma^2$ is estimated by 706.6.

14.36 The intraclass correlation $= 511.0/(511.0 + 706.6) = 511.0/1217.6 = .42$. It represents the proportion of the total variability that is attributable to between-person variation. Sucrose is a much less variable nutrient than dietary cholesterol, where the intraclass correlation was .13 (see Section 9.6, p. 122, Study Guide).

14.37 The upper one-sided 95% CI is obtained from the formula

$$p > \hat{p} - z_{.95}\sqrt{\hat{p}\hat{q}/n}$$

$$= .30 - 1.645\sqrt{.3 \times .7/100}$$

$$= .30 - 1.645 \times .046 = .30 - .075 = .225$$

14.38 Since 15% is not within the one-sided 95% CI computed in Problem 14.37, it follows that the true rate of pain relief is greater with the drug than if the patients were untreated. Therefore, the drug has proven benefit.

14.39 The upper one-sided 95% CI is given by

$$p > \hat{p} - z_{.95}\sqrt{\hat{p}\hat{q}/n}$$

$$= .20 - 1.645\sqrt{.2 \times .8/100}$$

$$= .20 - 1.645 \times .04 = .20 - .066 = .134$$

14.40 Since 10% is not within the one-sided 95% CI computed in Problem 14.39, we conclude that the drug is beneficial as regards pain relief within 4 hours.

14.41 A nonparametric statistical test would be useful because the change in redness or itching cannot be numerically quantified; only which eye has improved more can be assessed.

14.42 The degree of redness or itching is probably similar in two eyes of the same person at baseline. Thus, this procedure would eliminate the substantial between-person variability that can occur if the drug were administered to one group of people and the placebo to another group.

14.43 The sign test should be used here because we are comparing two paired samples consisting of alternate eyes from the same people and we only know which eye did better but not how much better.

14.44 There are 5 +'s and 11 −'s for redness. Since there are < 20 nonzero differences, the exact test must be used. In particular, the p-value is given by

$$p = 2 \times \sum_{k=0}^{5}\binom{16}{k}\left(\frac{1}{2}\right)^{16}$$

From the binomial tables (Table 1, Appendix, text) using $n = 16, p = .50,$

$$Pr(0) = .0000 \quad Pr(1) = .0002$$
$$Pr(2) = .0018 \quad Pr(3) = .0085$$
$$Pr(4) = .0278 \quad Pr(5) = .0667$$

Thus,

$$p = 2 \times (.0000 + .0002 + .0018$$
$$+ .0085 + .0278 + .0667)$$
$$= .210$$

Thus, there is no significant difference in redness between the two sets of treated eyes.

14.45 There are 12 +'s and 3 −'s for itching. Again use the exact test, calculated from the binomial tables, with the p-value given by

$$p = 2 \times \sum_{k=0}^{3}\binom{15}{k}\left(\frac{1}{2}\right)^{15}$$

$$= 2 \times (.0000 + .0005 + .0032 + .0139)$$

$$= .035$$

Thus, there is a significant difference in itching between the drug-treated and placebo-treated eyes. This is what is expected, since drug A is only supposed to be effective against itching.

14.46 For redness there are 22 +'s and 3 0's. We can use the sign test using the normal approximation, since we have 22 untied pairs. The test statistic is given by

$$z = \frac{\left|C - \dfrac{n}{2}\right| - .5}{\sqrt{\dfrac{n}{4}}} = \frac{|22 - 11| - .5}{\sqrt{\dfrac{22}{4}}} = \frac{10.5}{2.345}$$

$$= 4.48 \sim N(0, 1) \text{ under } H_0$$

The p-value is

$$2 \times [1 - \Phi(z)] = 2 \times [1 - \Phi(4.48)] < .001$$

There is clearly overwhelming evidence that drug B is effective against redness.

14.47 For itching there are 7 +'s and 6 −'s. We use the exact binomial version of the sign test. The p-value is obtained from the binomial tables as follows:

$$p = 2 \times \sum_{k=0}^{6}\binom{13}{k}\left(\frac{1}{2}\right)^{13}$$

$$= 2 \times (.0001 + .0016 + .0095 + .0349$$
$$+ .0873 + .1571 + .2095)$$

$$= 1.0$$

There is clearly no significant difference in itching between treated and untreated eyes with drug B. These results are again consistent with our expectations for drug B, which is only supposed to be effective against redness.

14.48 For redness there are 23 +'s and 2 0's. This result is even more extreme than that in Problem 14.46, and if we use the sign test with the normal approximation, we surely will find $p < .001$. Thus, the combination drug is very effective against redness.

14.49 For itching there are 14 +'s and 3 −'s. We use the exact binomial version of the sign test, as follows:

$$p = 2 \times \sum_{k=0}^{3} \binom{17}{k} \left(\frac{1}{2}\right)^{17}$$

$$= 2 \times (.0000 + .0001 + .0010 + .0052)$$

$$= .013$$

Thus, the combination drug is significantly better than the placebo for itching as well, but not as strongly as for redness.

14.50 The results of the three experiments agree with our prior hypotheses. Indeed, drug A is significantly better than the placebo for itching ($p = .035$) but not for redness; drug B is significantly better than the placebo for redness ($p < .001$) but not for itching; the combination drug is significantly better than the placebo for both redness ($p < .001$) and itching ($p = .013$).

14.51 We have the following 2×2 table for the 3-year prognosis:

		Status		
		Alive	**Dead**	
Treatment	**Standard**	6	94	100
	New	10	37	47
		16	131	147

We can use the chi-square test here since the smallest expected value $= (16 \times 47)/147 = 5.12 \geq 5$. The chi-square statistic with Yates' correction is given as follows:

$$X^2 = \frac{n(|ad - bc| - n/2)^2}{(a+b)(c+d)(a+c)(b+d)}$$

$$= \frac{147(|6(37) - 10(94)| - 73.5)^2}{16 \times 131 \times 100 \times 47}$$

$$= \frac{147(718 - 73.5)^2}{16 \times 131 \times 100 \times 47}$$

$$= \frac{147(644.5)^2}{16 \times 131 \times 100 \times 47}$$

$$= 6.20 \sim \chi_1^2 \text{ under } H_0$$

Since $\chi_{1,.975}^2 = 5.02$, $\chi_{1,.99}^2 = 6.63$ and $5.02 < 6.20 < 6.63$, we have $.01 < p < .025$. Therefore, the new treatment is significantly better than the standard treatment as regards the 3-year prognosis.

14.52 We have the following 2×2 table for the 5-year prognosis:

		Status		
		Alive	**Dead**	
Treatment	**Standard**	5	95	100
	New	2	45	47
		7	140	147

We use Fisher's exact test here since the smallest expected value $= (7 \times 47)/147 = 2.24 < 5$. We must enumerate all the possible tables with the same row and column margins as follows:

0	100		1	99		2	98		3	97		4	96
7	40		6	41		5	42		4	43		3	44

5	95		6	94		7	93
2	45		1	46		0	47

Thus, using the recursion rule we have

$$Pr(0) = 1 \, Pr(0)$$

$$Pr(1) = Pr(0) \times \frac{7 \times 100}{1 \times 41} = 17.073 \, Pr(0)$$

$$Pr(2) = Pr(1) \times \frac{6 \times 99}{2 \times 42} = 120.732 \, Pr(0)$$

$$Pr(3) = Pr(2) \times \frac{5 \times 98}{3 \times 43} = 458.593 \, Pr(0)$$

$$Pr(4) = Pr(3) \times \frac{4 \times 97}{4 \times 44} = 1010.990 \, Pr(0)$$

$$Pr(5) = Pr(4) \times \frac{3 \times 96}{5 \times 45} = 1294.067 \, Pr(0)$$

$$Pr(6) = Pr(5) \times \frac{2 \times 95}{6 \times 46} = 890.843 \, Pr(0)$$

$$Pr(7) = Pr(6) \times \frac{1 \times 94}{7 \times 47} = 254.527 \, Pr(0)$$

Thus, $Pr(0) \times (1 + 17.073 + \ldots + 254.527) = 1$ or $Pr(0) = 1/4047.825 = 2.470 \times 10^{-4}$, $Pr(1) = .004$, $Pr(2) = .030$, $Pr(3) = .113$, $Pr(4) = .250$, $Pr(5) = .320$, $Pr(6) = .220$, $Pr(7) = .063$. Since our table is the "5" table, it follows that the two-sided p-value $= 2 \times \min(2.470 \times 10^{-4} + \ldots + .320, .320 + .220 + .063, .5) = 2 \times \min(.717, .603, .5) = 2 \times .50 = 1.0$, which is clearly not statistically significant. Therefore, there is no significant difference between the two treatments as regards the 5-year prognosis.

14.53 We can look at the set of people who are alive after 3 years to answer this question. We have the following 2 × 2 table:

Status after 5 years

		Alive	Dead	
Treatment	**Standard**	5	1	6
	New	2	8	10
		7	9	16

We again use Fisher's exact test, because some of the expected values are < 5. We enumerate all the possible tables with the preceding row and column totals:

0	6		1	5		2	4		3	3		4	2
7	3		6	4		5	5		4	6		3	7

5	1		6	0
2	8		1	9

We use the recursion rule as follows:

$$Pr(0) = 1 \, Pr(0)$$

$$Pr(1) = Pr(0) \times \frac{7 \times 6}{1 \times 4} = 10.50 \, Pr(0)$$

$$Pr(2) = Pr(1) \times \frac{6 \times 5}{2 \times 5} = 31.50 \, Pr(0)$$

$$Pr(3) = Pr(2) \times \frac{5 \times 4}{3 \times 6} = 35.00 \, Pr(0)$$

$$Pr(4) = Pr(3) \times \frac{4 \times 3}{4 \times 7} = 15.00 \, Pr(0)$$

$$Pr(5) = Pr(4) \times \frac{3 \times 2}{5 \times 8} = 2.25 \, Pr(0)$$

$$Pr(6) = Pr(5) \times \frac{2 \times 1}{6 \times 9} = 0.083 \, Pr(0)$$

Therefore, $Pr(0) \times (1 + 10.5 + \ldots + 0.083) = 1$ or $Pr(0) = 1/95.333 = .0105$, $Pr(1) = .110$, $Pr(2) = .330$, $Pr(3) = .367$, $Pr(4) = .157$, $Pr(5) = .024$, $Pr(6) = .001$. Our table is the "5" table and the two-tailed p-value $= 2 \times \min(.0105 + \ldots + .024, .024 + .001) = 2 \times .024 = .049$, which is statistically significant. Thus, if a person has survived for 3 years he or she is significantly more likely to survive for the next 2 years with the standard treatment.

14.54 We refer to Table 12 (Appendix, text). Because the upper two-tailed critical value $= 26$ for $\alpha = .05$ and 28 for $\alpha = .02$, it follows that $.02 \leq p < .05$.

14.55 We refer to Table 12 (Appendix, text) under $n = 13$, $\alpha = .10$. We note that the lower and upper critical values are 21 and 70. Because $21 < 65 < 70$, it follows that $p \geq .10$ and we accept H_0 at the 5% level.

14.56 The exact tables for the signed-rank test must be used, since the number of untied pairs is less than 16. Refer to Table 12 (Appendix, text) under $n = 15$ and note that the upper critical value for $\alpha = .10 = 90$, whereas the upper critical value for $\alpha = .05$ is 95. Since $R_1 = 90 \geq 90$ and $90 < 95$, we have $.05 \leq p < .10$ using a two-sided test.

14.57 We will use the chi-square test for trend. We have the test statistic $X_1^2 = A^2/B$, where we use scores of 1, 2, 3, 4, and 5 for the 5 mole groups. We have

$$A = 1(16) + 2(20) + 3(20) + 4(12) + 5(30)$$
$$- (98/288)[1(73) + 2(60) + 3(65)$$
$$+ 4(32) + 5(58)]$$
$$= 314 - 274.26 = 39.74$$

$$B = [98(190)/288^2]\{1^2(73) + 2^2(60)$$
$$+ 3^2(65) + 4^2(32) + 5^2(58)$$
$$- [1(73) + 2(60) + 3(65)$$
$$+ 4(32) + 5(58)]^2/288\}$$
$$= .2245(2860 - 806^2/288) = 135.66$$

Thus, $X_1^2 = 39.74^2/135.66 = 11.64 \sim \chi_1^2$ under H_0. Since $11.64 > 10.83 = \chi_{1,.999}^2$, it follows that $p < .001$. Therefore, we reject H_0. Since $A > 0$, we conclude that the mole count of cases is significantly higher than for controls.

14.58 The odds ratio comparing the 1–5 mole group to the 0 mole group $= 20(57)/[16(40)] = 1.78$. Using the Woolf approach, $se[\ln(\hat{OR})] = \sqrt{1/20 + 1/57 + 1/16 + 1/40} = 0.3938$. Therefore, a 95% CI for $\ln(OR) = \ln(1.78) \pm 1.96(0.3938) = (-0.1944, 1.3491)$. The corresponding 95% CI for OR $= [\exp(-0.1944), \exp(1.3491)] = (0.82, 3.85)$. Similar methods are used for each of the other 3 mole-count groups versus the 0 mole-count group, with results summarized as follows:

Odds ratio and 95% CI for the 1–5, 6–15, 16–30, and 31+ mole-count groups, respectively, versus the 0 mole-count group

Mole-count group	$\widehat{OR}$	95% CI
1–5 moles	1.78	(0.82, 3.85)
6–15 moles	1.58	(0.74, 3.40)
16–30 moles	2.14	(0.86, 5.29)
31+ moles	3.82	(1.79, 8.14)

Interestingly, there is only a significant difference between the 31+ mole-count group versus the 0 mole-count group. However, the overall trend reveals an increasing odds ratio as the mole count increases. Therefore, the chi-square test for trend is the most appropriate method for analyzing these data.

14.59 The 2 × 2 contingency table is shown as follows:

	Disease status		
Community	+	–	Total
A	5	99,995	100,000
B	1	199,999	200,000
Total	6	299,994	300,000

The expected table is formed as follows:

2.0	99,998
4.0	199,996

Since two of the four cells have expected values < 5, Fisher's exact test rather than the χ^2 test must be used. Thus, the following seven tables are formed:

0	100,000
6	199,994

1	99,999
5	199,995

2	99,998
4	199,996

3	99,997
3	199,997

4	99,996
2	199,998

5	99,995
1	199,999

6	99,994
0	200,000

We have

$$Pr(0) = 1.000\ Pr(0)$$
$$Pr(1) = \frac{6 \times 100,000}{1 \times 199,995}\ Pr(0) = 3.000\ Pr(0)$$
$$Pr(2) = \frac{5 \times 99,999}{2 \times 199,996}\ Pr(1) = 3.750\ Pr(0)$$
$$Pr(3) = \frac{4 \times 99,998}{3 \times 199,997}\ Pr(2) = 2.500\ Pr(0)$$
$$Pr(4) = \frac{3 \times 99,997}{4 \times 199,998}\ Pr(3) = 0.938\ Pr(0)$$
$$Pr(5) = \frac{2 \times 99,996}{5 \times 199,999}\ Pr(4) = 0.187\ Pr(0)$$
$$Pr(6) = \frac{1 \times 99,995}{6 \times 200,000}\ Pr(5) = 0.016\ Pr(0)$$

Thus, $Pr(0)(1 + 3.000 + 3.750 + 2.500 + 0.938 + 0.187 + 0.016) = 1$

$Pr(0) = .0878$	$Pr(1) = .263$	$Pr(2) = .329$
$Pr(3) = .219$	$Pr(4) = .082$	$Pr(5) = .016$
	$Pr(6) = .001$	

The one-sided p-value associated with the "5" table is $.016 + .001 = .018$. The two-sided p-value is .036. Thus, there is a significant difference between the prevalence rates.

14.60 We first perform a one-way ANOVA using MINITAB, as follows:

```
ANALYSIS OF VARIANCE ON SBP
SOURCE   DF      SS      MS     F      P
DAY       9   152.0    16.9   1.16   0.405
ERROR    10   145.0    14.5
TOTAL    19   297.0
```

From Equation (**9.22**) (text, p. 325) we have $\hat{\sigma}^2 = 14.5$, $\hat{\sigma}_A^2 = (16.9 - 14.5)/2 = 1.2$. The estimated intraclass correlation $= \hat{\rho}_I = 1.2/(1.2 + 14.5) = 1.2/15.7 = .076$. From Equation (**11.47**) (text, p. 519) a 95% CI for ρ_I is given by (c_1, c_2), where

$$c_1 = \max[(F/F_{9,10,.975} - 1)/(F/F_{9,10,.975} + 1), 0]$$
$$c_2 = \max[(F/F_{9,10,.025} - 1)/(F/F_{9,10,.025} + 1), 0]$$

We use MINITAB to obtain $F_{9,10,.975} = 3.78$ and $F_{9,10,.025} = 0.253$. Thus,

$$c_1 = \max[(1.16/3.78 - 1)/(1.16/3.78 + 1), 0]$$
$$= \max(-0.692/1.308, 0) = 0$$

$$c_2 = \max[(1.16/0.253 - 1)/(1.16/0.253 + 1), 0]$$
$$= \max(3.604/5.604, 0) = .64$$

Therefore, the 95% CI for $\rho_{\mathrm{I}} = (0, .64)$.

14.61   We use a two-sided one-sample z test here, because the standard deviation is known. We have H_0: $\mu = 128.6$, $\sigma = 11.1$ versus H_1: $\mu \neq 128.6$, $\sigma = 11.1$. It is appropriate to use a two-sided test because it is not known in which direction the mean blood pressure will fall. We base our significance test on

$$z = \frac{\bar{x} - \mu_0}{\sigma/\sqrt{n}}$$

$$= \frac{135.7 - 128.6}{11.1/\sqrt{38}} = 3.94$$

We have that $p = 2 \times Pr[z \geq 3.94 \,|\, z \sim N(0, 1)] < .001$. Thus, this difference is very highly significant.

14.62   We use a one-sample χ^2 test for the variance of a normal distribution here to test the hypothesis H_0: $\sigma^2 = 11.1^2$ versus H_1: $\sigma^2 \neq 11.1^2$. We base our significance test on

$$X^2 = \frac{(n-1)s^2}{\sigma^2}$$

$$= \frac{37(12.5)^2}{(11.1)^2} = 46.92$$

which under H_0 follows a χ^2_{37} distribution. We note from Table 6 (Appendix, text) that $\chi^2_{37, .975} > \chi^2_{30, .975} = 46.98 > X^2$. Thus, it follows that $p > 2(.025) = .05$, and the variances of systolic blood pressure for white and black males are not significantly different.

14.63   The power of a test $= Pr(\text{reject } H_0 \,|\, H_1 \text{ true})$. If we use a two-sided 5% significance test, then the power is given by the formula

$$\Phi[z_{\alpha/2} + |\mu_1 - \mu_0| \sqrt{n}/\sigma]$$

We have $\mu_0 = 128.6$, $\mu_1 = 135$, $\sigma = 11.1$, $n = 38$, and $\alpha = .05$. It follows that

$$\text{Power} = \Phi\left[-1.96 + \frac{(135 - 128.6)\sqrt{38}}{11.1}\right]$$

$$= \Phi(1.594) = .94$$

Thus, we have a 94% chance of finding a significant difference using a two-sided test with an α level $= .05$ and a sample size of 38.

14.64   We have the test statistic

$$t = \frac{\bar{x} - \mu_0}{s/\sqrt{n}} = \frac{135.7 - 128.6}{12.5/\sqrt{38}}$$

$$= \frac{7.1}{2.028} = 3.50 \sim t_{37} \text{ under } H_0$$

We see from Table 5 (Appendix, text) that $t_{30, .995} = 2.750$, $t_{30, .9995} = 3.646$, $t_{40, .995} = 2.704$, $t_{40, .9995} = 3.551$. Thus, since $2.750 < t < 3.646$, and $2.704 < t < 3.551$, it follows that if we had either 30 or 40 df, then we would obtain $.001 < p < .01$. The exact p-value obtained by computer is $p = 2 \times Pr(t_{37} > 3.50) = .0012$. Therefore, we again conclude that there is a significant difference between the mean SBPs of 25–34-year-old black and white males.

14.65   A 95% CI for μ is given by $\bar{x} \pm t_{37, .975} s/\sqrt{n}$. We obtain $t_{37, .975}$ using MINITAB, and obtain 2.026. Thus, the 95% CI for μ is given by $135.7 \pm 2.026(12.5)/\sqrt{38} = 135.7 \pm 4.1 = (131.6, 139.8)$.

14.66   The 95% CI in Problem 14.65 does not contain the mean for white males $= 128.6$. This must be the case, since the p-value for the test in Problem 14.64 was $< .05$.

14.67   We test the hypothesis H_0: $\sigma_1^2 = \sigma_2^2$ versus H_1: $\sigma_1^2 \neq \sigma_2^2$. We compute the F statistic $= s_1^2/s_2^2 = 12.6^2/6.2^2 = 4.13 \sim F_{11, 15}$ under H_0. Since $F > F_{8, 14, .975} = 3.29 > F_{11, 15, .975}$, it follows that $p < 2 \times (1 - .975)$ or $p < .05$, and the variances are significantly different.

14.68   We have the test statistic

$$t = \frac{r\sqrt{n-2}}{\sqrt{1-r^2}} = \frac{.52\sqrt{21}}{\sqrt{1-(.52)^2}}$$

$$= \frac{2.383}{0.854} = 2.79 \sim t_{21} \text{ under } H_0$$

Since $t_{21, .99} = 2.518$, $t_{21, .995} = 2.831$ and $2.518 < 2.79 < 2.831$, it follows that $2 \times (1 - .995) < p < 2 \times (1 - .99)$ or $.01 < p < .02$.

14.69   We have the test statistic

$$t = \frac{.34\sqrt{21}}{\sqrt{1-(.34)^2}}$$

$$= \frac{1.558}{0.940} = 1.66 \sim t_{21} \text{ under } H_0$$

Since $t_{21, .975} = 2.080 > t$, it follows that $p > .05$ and there is no significant association between serum estradiol and the bone density of the distal radius.

14.70   We have the test statistic

$$t = \frac{.395\sqrt{21}}{\sqrt{1-(.395)^2}} = \frac{1.810}{0.919} = 1.970 \sim t_{21} \text{ under } H_0$$

Since $t_{21, .975} = 2.080$, $t_{21, .95} = 1.721$ and $1.721 < 1.970 < 2.080$, it follows that $.05 < p < .10$ and there is a trend toward a significant association between the two measures of bone metabolism.

14.71 We use the one-sample z test to test the hypothesis H_0: $\rho = \rho_0$ versus H_1: $\rho \neq \rho_0$. We have the test statistic

$$\lambda = \sqrt{n-3}(z - z_0) \sim N(0, 1) \text{ under } H_0$$

where $z_0 = .5 \times \ln\left(\dfrac{1 + .5}{1 - .5}\right)$

$= 0.549$ from Table 11 (Appendix, text)

and $z = .5 \times \ln\left(\dfrac{1 + .395}{1 - .395}\right)$

$= .5 \times \ln(2.306) = 0.418$

Thus, $\lambda = \sqrt{20}(0.418 - 0.549) = -0.589 \sim N(0,1)$ under H_0. This yields a p-value of $2 \times \Phi(-0.589) = 2(.2781) = .56$. Thus, our results are compatible with the previous results, because no statistical significance was established.

14.72 The odds ratio for ex-smokers versus nonsmokers $= \hat{OR} = 13(169)/[155(21)] = 0.67$. Using the Woolf approach, $se[\ln(\hat{OR})] = \sqrt{1/13 + 1/21 + 1/155 + 1/169}$ $= 0.3700$. Therefore, a 95% CI for $\ln(OR) = \ln(0.67) \pm 1.96(0.3700) = -0.3931 \pm 0.7252 = (-1.1183, 0.3321)$. The corresponding 95% CI for OR $=$ $[\exp(-1.1183), \exp(0.3321)] = (0.33, 1.39)$.

The odds ratio for current smokers of < 15 cigarettes per day versus nonsmokers $=$ $33(169)/[155(22)] = 1.64$. We have $se[\ln(\hat{OR})] =$ $\sqrt{1/33 + 1/169 + 1/155 + 1/22} = 0.2969$. A 95% CI for $\ln(OR) = \ln(1.64) \pm 1.96(0.2969) = 0.4919 \pm 0.5818 = (-0.0899, 1.0738)$. The corresponding 95% CI for OR $= [\exp(-0.0899), \exp(1.0738)] = (0.91, 2.93)$.

The odds ratio for current smokers of ≥ 15 cigarettes per day versus nonsmokers $=$ $29(169)/[18(155)] = 1.76$. We have $se[\ln(\hat{OR})] =$ $\sqrt{1/29 + 1/169 + 1/18 + 1/155} = 0.3200$. A 95% CI for $\ln(OR) = \ln(1.76) \pm 1.96(0.3200) = 0.5634 \pm 0.6272 = (-0.0638, 1.1906)$. The corresponding 95% CI for OR $= [\exp(-0.0638), \exp(1.1906)] = (0.94, 3.29)$.

14.73 The overall impression is that there is an excess number of current smokers among cervical-cancer cases versus controls. Although the light and heavy current smokers, when considered separately do not differ significantly from the nonsmokers, if one combines them into one group there is a significant excess number of current smokers among cervical-cancer cases versus controls.

14.74 Under the null hypothesis, the type of treatment has no effect on the proportion of leukemias. Therefore, the proportion of leukemias in each group should be in direct proportion to the number of person-years of follow-up after treatment in each group. Thus, $p_0 = 3322.8/(3322.8 + 1858.2) = 3322.8/5181.0 = .641$.

14.75 We wish to compare incidence-density measures in groups A and B. Since $V_1 = 6(.641)(.359) = 1.38 < 5$, we must use the exact method in (13.6) (text, p. 590).

14.76 Since $\hat{p} = 2/6 = .33 < p_0 = .641$, the p-value for the test in Problem 14.75 is given by

$$p = 2 \times \sum_{k=0}^{2} {}_6C_k p_0^k q_0^{6-k}$$

$= 2 \times [(.359)^6 + 6(.641)(.359)^5 + {}_6C_2(.641)^2(.359)^4]$

$= 2 \times (.00213 + .02284 + .10209) = 2 \times .12706$

$= .254$

Thus, there is no significant difference between the incidence density of leukemia in the no-intensive-therapy group versus the intensive-therapy group.

14.77 We have that $\hat{p} = 93/221 = .421$. Thus, a 95% CI is given by

$$\hat{p} \pm 1.96\sqrt{\hat{p}\hat{q}/n}$$

$= .421 \pm 1.96\sqrt{.421(.579)/221}$

$= .421 \pm .065 = (.356, .486)$

14.78 Since 31% does not fall in this interval, we can conclude that the reactivity rate is significantly higher than the statewide rate.

14.79 The SMR for total cancer $= 100\% \times 69/66.8 = 103$. The SMR for multiple myeloma $= 100\% \times 4/1.0 = 400$.

14.80 Since the expected number of deaths $= 66.8 \geq 5$, we can use the large-sample test. We have the test statistic

$$X^2 = \frac{(O - E)^2}{E}$$

$$= \frac{(69 - 66.8)^2}{66.8} = 0.07 \sim \chi_1^2 \text{ under } H_0$$

Since $\chi_{1,.10}^2 = 0.02$, $\chi_{1,.25}^2 = 0.10$, and $0.02 < 0.07 < 0.10$, it follows that $1 - .25 < p < 1 - .10$ or $.75 < p < .90$. Clearly, the results are not statistically significant; i.e., the cancer mortality rate in this cohort is not significantly different from the U.S. white male mortality rates.

14.81 Since the expected number of deaths $= 1.0 < 5.0$, we must use the small-sample test. Since the observed number of deaths is greater than the expected, we compute the p-value using the formula

$$p = 2 \times \sum_{k=4}^{\infty} \frac{e^{-1.0}1.0^k}{k!}$$

$$= 2 \times \left(1 - \sum_{k=0}^{3} \frac{e^{-1.0}}{k!}\right)$$

Referring to the Poisson table (Table 2, Appendix, text) under $\mu = 1.0$, we obtain $p = 2 \times [1 - (.3679 + ... + .0613)] = 2 \times (1 - .9810) = .038$. Thus, there is a significant excess of multiple-myeloma deaths among workers exposed to benzene.

14.82 We will use the Wilcoxon signed-rank test. The difference scores are ranked by absolute value as follows:

	Neg		Pos				
$\|d_i\|$	d_i	f_i	d_i	f_i	Total	Rank Range	Average Rank
10	−10	0	10	1	1	29	29.0
8	−8	0	8	1	1	28	28.0
6	−6	0	6	2	2	26–27	26.5
5	−5	0	5	3	3	23–25	24.0
4	−4	0	4	4	4	19–22	20.5
3	−3	1	3	10	11	8–18	13.0
2	−2	2	2	1	3	5–7	6.0
1	−1	1	1	3	4	1–4	2.5
		4		25	29		
0	0	3					

The rank sum of the positive differences is given by $R_1 = 29(1) + 28(1) + ... + 2.5(3) = 407.5$. Since there are 29 untied pairs, we can use the normal-theory test. The test statistic is

$$T = \frac{|407.5 - 29(30)/4| - .5}{\sqrt{29(30)(59)/24 - [(2^3 - 2) + (3^3 - 3) + ... + (4^3 - 4)]/2}}$$

$$= \frac{189.5}{\sqrt{2138.75 - 1494/2}} = \frac{189.5}{\sqrt{1391.75}}$$

$$= \frac{189.5}{37.31} = 5.08 \sim N(0, 1)$$

The p-value $= 2 \times [1 - \Phi(5.08)] < .001$.

14.83 The results are the same: Doctor A systematically detects more palpable lymph nodes than Doctor B.

14.84 We will compute a Spearman rank correlation between the lymph-node assessments of Doctors A and B. We first rank-order the assessments of Doctors A and B, separately.

Doctor A			
Number of lymph nodes	Frequency	Rank range	Avg rank
19	1	32	32
17	1	31	31
15	1	30	30
13	1	29	29
12	4	25–28	26.5
11	1	24	24
10	1	23	23
9	2	21–22	21.5
8	3	18–20	19
7	1	17	17
6	3	14–16	15
5	7	7–13	10
4	2	5–6	5.5
3	3	2–4	3
1	1	1	1
Total	32		

Doctor B			
Number of lymph nodes	Frequency	Rank range	Avg rank
13	1	32	32
12	1	31	31
11	1	30	30
9	7	23–29	26
7	2	21–22	21.5
6	4	17–20	18.5
4	4	13–16	14.5
3	1	12	12
2	2	10–11	10.5
1	5	5–9	7
0	4	1–4	2.5
Total	32		

We now compile the ranks for Doctors A and B by patient as follows:

Patient	Doctor A	Doctor B
1	5.5	7.0
2	31.0	26.0
3	3.0	10.5
4	24.0	32.0
5	26.5	26.0
6	10.0	10.5
7	10.0	18.5
8	15.0	12.0
9	3.0	2.5
10	10.0	2.5
11	21.5	18.5
12	1.0	7.0
13	10.0	14.5
14	19.0	14.5
15	17.0	21.5
16	19.0	18.5

Patient	Doctor A	Doctor B
17	5.5	7.0
18	26.5	26.0
19	23.0	21.5
20	21.5	30.0
21	10.0	2.5
22	3.0	2.5
23	26.5	31.0
24	10.0	7.0
25	29.0	26.0
26	26.5	18.5
27	15.0	26.0
28	32.0	26.0
29	19.0	14.5
30	30.0	26.0
31	15.0	7.0
32	10.0	14.5

To compute the Spearman rank-correlation coefficient, we must compute the ordinary Pearson correlation between ranks. The Spearman rank-correlation coefficient (r_s) is given by

$$r_s = \frac{2228.5}{\sqrt{2688(2674)}} = \frac{2228.5}{2681.0} = 0.831$$

This correlation is high. In general, the rank correlation and the signed-rank test provide complementary information. It is possible that the signed-rank test is not significant, whereby, on average, Doctors A and B are comparable, yet they may vary substantially for individual subjects, which would result in a low rank correlation. Conversely, the signed-rank test may be highly significant, which would indicate that on average Doctors A and B differ, but if this difference is consistent for each subject, then the rank correlation will be high. This seems to be the situation here.

14.85 This means that a random-number table was used to select the order in which the diets were administered to an individual subject. Each subject had a probability of 1/2 of having the reference diet or the walnut diet first for 4 weeks, followed by the other diet for an additional 4 weeks. This does not guarantee that in a small study there will be exactly an equal number of subjects assigned to each diet first. Indeed, in the actual study, 10 subjects received the walnut diet first, while 8 subjects received the reference diet first.

14.86 The 95% CI for the true mean difference $= \bar{d} \pm t_{n-1,.975}(se) = -2.3 \pm t_{17,.975}(se)$. The width of the 95% CI $= 2t_{17,.975}(se)$. Thus

$$2t_{17,.975}(se) = -0.7 - (-3.9) = 3.2$$

or

$$se = \frac{3.2}{2t_{17,.975}} = \frac{1.6}{2.110} = 0.758$$

14.87 Since we are using each person as his or her own control, we will use the paired t test. We wish to test the hypothesis $H_0: \mu_d = 0$ versus $H_1: \mu_d \neq 0$ where $\mu_d =$ true mean for HDL cholesterol while on the walnut diet $-$ true mean for HDL cholesterol while on the reference diet. The test statistic is

$$t = \frac{\bar{d}}{sd/\sqrt{n}} = \frac{\bar{d}}{se} = \frac{-2.3}{0.758}$$

$$= -3.03 \sim t_{17} \text{ under } H_0$$

Therefore, the p-value $= 2 \times Pr(t_{17} < -3.03)$. From Table 5 (Appendix, text), we see that $t_{17,.995} = 2.898$, $t_{17,.9995} = 3.965$. Since $2.898 < |t| = 3.03 < 3.965$, it

follows that $1 - .9995 < p/2 < 1 - .995$ or $.0005 < p/2 < .005$ or $.001 < p < .01$. Thus, there is a significant difference between the HDL-cholesterol level while on the walnut diet versus HDL cholesterol while on the reference diet, with significantly lower levels under the walnut diet.

14.88 **(a)** Since the two-sided 95% CI excludes zero, it follows that the p-value must be $< .05$. Thus, (a) is *true*.

(b) The 90% CI must be narrower than the 95% CI. Thus, (b) must be *false* since the interval $(-4.1, -0.5)$ is wider than the 95% CI $= (-3.9, -0.7)$.

(c) The 98% CI must be wider than the 95% CI. Since the interval $(-4.2, -0.4)$ is wider than $(-3.9, -0.7)$, this is *not inconsistent* with the data in the table.

14.89 We first form the following 2×2 table:

		Chest X ray		
		Abnormal	**Normal**	**Total**
Diagnosis	Normal	0	17	17
	Bronchitis	7	11	18
		7	28	35

The smallest expected value $= 7(17)/35 = 3.40 < 5$. Thus, we must use Fisher's exact test to analyze these data.

14.90 We wish to test the hypothesis $H_0: p_1 = p_2$ versus $H_1: p_1 < p_2$, where

$p_1 =$ proportion of abnormal chest X-ray findings among normal women

$p_2 =$ proportion of abnormal chest X-ray findings among bronchitic women

Thus, the p-value $= Pr(0)$ where $Pr(0)$ is the exact probability of the 0 table (i.e., the observed table) under H_0 given that the margins are fixed. This is given by

$$Pr(0) = \frac{7! \ 28! \ 17! \ 18!}{35! \ 0! \ 17! \ 7! \ 11!} = \frac{28! \ 18!}{35! \ 11!}$$

$$= \frac{18 \times 17 \times 16 \times 15 \times 14 \times 13 \times 12}{35 \times 34 \times 33 \times 32 \times 31 \times 30 \times 29}$$

$$= \frac{1.6039 \times 10^8}{3.3892 \times 10^{10}} = 0.005$$

Thus, there is a significant association, with bronchitic women having significantly more abnormal chest X-ray findings than normal women.

14.91 This is an example of an independent-samples design since the women in the two groups are *not matched* on an individual basis.

14.92 We use the fixed-effects one-way ANOVA. We have the test statistic

$$F = \frac{\text{Between MS}}{\text{Within MS}} \sim F_{k-1,n-k} \text{ under } H_0$$

We have

Between MS = Between SS/2

$$= \frac{(205.6)^2 23 + (182.7)^2 15 + (199.8)^2 12}{2}$$

$$- \frac{[(205.6)23 + (182.7)15 + (199.8)12]^2}{50(2)}$$

$$= \frac{1,951,971.11 - 1,947,114.31}{2}$$

$$= \frac{4856.80}{2} = 2428.40$$

Within MS $= s^2$

$$= \frac{[22(25.3)^2 + 14(21.9)^2 + 11(30.3)^2]}{47}$$

$$= 657.35$$

Thus,

$$F = \frac{\text{Between MS}}{\text{Within MS}} = \frac{2428.40}{657.35}$$

$$= 3.69 \sim F_{2,47} \text{ under } H_0$$

Since $F_{2,47,.95} < F_{2,40,.95} = 3.23 < F$, it follows that $p < .05$, and there is an overall significant difference among groups.

14.93 We compute the test statistic, critical value, and level of significance using the Bonferroni approach as follows:

Thus, there is a significant difference between the mean cholesterol levels for nondrinkers versus light drinkers, but no significant difference between the mean cholesterol levels for either nondrinkers versus heavy drinkers or light drinkers versus heavy drinkers.

14.94 We use the sample-size formula given in (**7.28**) (text, p. 221) as follows:

$$n = \frac{\sigma^2 (z_{1-\beta} + z_{1-\alpha})^2}{(\mu_1 - \mu_0)^2}$$

In this case, $\mu_0 = 120$ oz, $\mu_1 = 120 + 8 = 128$ oz, $\sigma = 20$ oz, $\alpha = .05$, $\beta = 1 - .8 = .2$. Thus, we have

$$n = \frac{20^2 (z_{.8} + z_{.95})^2}{(120 - 128)^2}$$

$$= \frac{400(0.84 + 1.645)^2}{64} = 38.6$$

Thus, we would need to administer the drug to 39 pregnant women to have an 80% chance of finding a significant difference using a one-sided test with $\alpha = .05$.

14.95 We use the same formula as in Problem 14.94, but with $\beta = 1 - .9 = .1$. We have

$$n = \frac{20^2 (z_{.9} + z_{.95})^2}{(120 - 128)^2}$$

$$= \frac{400(1.28 + 1.645)^2}{64} = 53.5$$

Thus, we need 54 women in the study to have a 90% chance of finding a significant difference using a one-sided test with $\alpha = .05$.

Group	Test statistic		Critical value	p-value
Non, light	$t = \dfrac{205.6 - 182.7}{\sqrt{657.35(1/23 + 1/15)}}$	$= \dfrac{22.9}{8.51} = 2.69$	$t_{47,.9917} = 2.48$	$p < .05$
Non, heavy	$t = \dfrac{205.6 - 199.8}{\sqrt{657.35(1/23 + 1/12)}}$	$= \dfrac{5.8}{9.13} = 0.64$	2.48	NS
Light, heavy	$t = \dfrac{182.7 - 199.8}{\sqrt{657.35(1/15 + 1/12)}}$	$= \dfrac{-17.1}{9.93} = -1.72$	2.48	NS

REFERENCES

[1] Cramer, D. W., Schiff, I., Schoenbaum, S. C., Gibson, M., Belisle, S., Albrecht, B., Stillman, R. J., Berger, M. J., Wilson, W., Stadel, B. V., & Seibel, M. (1985). Tubal infertility and the intrauterine device. *New England Journal of Medicine, 312*(15), 941–947.

[2] Applegate, W. B., Hughes, J. P., & Vander Zwaag, R. (1991). Case–control study of coronary heart disease risk factors in the elderly. *Journal of Clinical Epidemiology, 44*(4), 409–415.

[3] Kuller, L. H., Hulley, S. B., Laporte, R. E., Neaton, J., & Dai, W. S. (1983). Environmental determinants, liver function, and high density lipoprotein cholesterol levels. *American Journal of Epidemiology, 117*(4), 406–418.

[4] Bain, C., Colditz, G. A., Willett, W. C., Stampfer, M. J., Green, A., Bronstein, B. R., Mihm, M. C., Rosner, B., Hennekens, C. H., & Speizer, F. E. (1988). Self-reports of mole counts and cutaneous malignant melanoma in women: Methodological issues and risk of disease. *American Journal of Epidemiology, 127*(4), 703–712.

[5] McDonough, J. R., Garrison, G. E., & Hames, C. G. (1967). Blood pressure and hypertensive disease among negroes and whites in Evans County, Georgia. In J. Stamler, R. Stamler, & T. N. Pullman (eds.), *The epidemiology of hypertension,* New York: Grune & Stratton.

[6] La Vecchia, C., Franceschi, S., Decarli, A., Fasoli, M., Gentile, A., & Tognoni, G. (1986). Cigarette smoking and the risk of cervical neoplasia. *American Journal of Epidemiology, 123*(1), 22–29.

[7] Boivin, J. F., & Hutchison, G. B. (1981). Leukemia and other cancers after radiotherapy and chemotherapy for Hodgkin's disease. *Journal of the National Cancer Institute, 67*(4), 751–760.

[8] Rinsky, R., Smith, A. B., Hornung, R., Filloon, T. G., Young, R. J., Okun, A. H., & Landrigan, P. J. (1987). Benzene and leukemia: An epidemiologic risk assessment. *New England Journal of Medicine, 316*(17), 1044–1050.

[9] Sabate, J., Fraser, G. E., Burke, K., Knutsen, S. F., Bennett, H., & Lindsted, K. D. (1993). Effects of walnuts on serum lipid levels and blood pressure in normal men. *New England Journal of Medicine, 328*(9), 603–607.